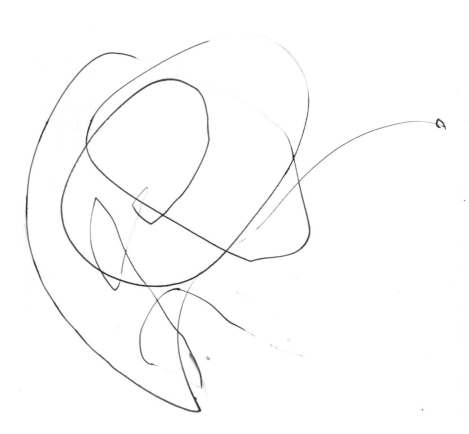

# ELECTRICAL CONTROL FOR MACHINES

## Sixth Edition

## KENNETH B. REXFORD
## PETER R. GIULIANI

**THOMSON**

**DELMAR LEARNING**　Australia　Canada　Mexico　Singapore　Spain　United Kingdom　United States

**THOMSON**

**DELMAR LEARNING**

Electrical Control for Machines
Sixth Edition
Kenneth B. Rexford and Peter R. Giuliani

**Executive Director:**
Alar Elken

**Executive Editor:**
Sandy Clark

**Senior Acquisitions Editor:**
Gregory L. Clayton

**Development:**
Dawn Daugherty

**Executive Marketing Manager:**
Maura Theriault

**Channel Manager:**
Fair Huntoon

**Marketing Coordinator:**
Brian McGrath

**Executive Production Manager:**
Mary Ellen Black

**Production Manager:**
Andrew Crouth

**Production Editor:**
Stacy Masucci

Library of Congress Cataloging-in-Publication Data:

Rexford, Kenneth B.
    Electrical control for machines /
Kenneth B. Rexford, Peter R.
Giuliani.—6th ed.
    p. cm.
Includes index.
ISBN 0-7668-6198-8
    1. Electrical controllers.
        I. Giuliani, Peter R. II. Title.

TK2851 .R47 2002
629.8′043—dc21
                            2002027107

**NOTICE TO THE READER**

*We dedicate this book to our families. Without their love and support this book would not have been possible.*

# Contents

Foreword                                        ix

Preface                                             xi

Acknowledgments                            xv

About the Authors                         xvii

Introduction to Electrical Control—The Development of Circuits      xix

**CHAPTER 1**    **Transformers and Power Supplies**           **1**
                        1.1    Control Transformers                       1
                        1.2    Transformer Regulation                  4
                        1.3    Temperature Rise in a Transformer      5
                        1.4    Operating Transformers in Parallel      5
                        1.5    Constant Voltage Regulators            5
                        1.6    Uninterruptible Power Systems          8

**CHAPTER 2**    **Fuses, Disconnect Switches, and Circuit Breakers**    **12**
                        2.1    Protective Factors                         12
                        2.2    Fuse Construction and Operation        14
                        2.3    Fuse Types                               14
                        2.4    Peak Let-Thru Current ($I_p$) and Ampere
                                     Squared Seconds ($I^2t$)                      18
                        2.5    Voltage and Frequency Surges          21
                        2.6    Circuit Breaker Types                   22
                        2.7    Programmable Motor Protection         24
                        2.8    Electrical Metering and Voltage Protection    25
                        2.9    Selecting Protective Devices            26

**CHAPTER 3**    **Control Units for Switching and Communication**    **32**
                        3.1    Oil-Tight Units                           32
                        3.2    Push-Button Switches                   33

3.3    Selector Switches                                    36
3.4    Heavy-Duty Switches                                  38
3.5    Indicating Lights                                    38
3.6    General Information on Oil-Tight Units               40
3.7    Circuit Applications                                 41
3.8    Annunciators                                         45
3.9    Light-Emitting Diodes                                45
3.10   Membrane Switches                                    47
3.11   Liquid Crystal Displays                              49

**CHAPTER 4    Relays                                        56**
4.1    Control Relays and Their Uses                        56
4.2    Timing Relays                                         61
4.3    Latching Relays                                       66
4.4    Plug-In Relays                                        69
4.5    Contactors                                            70

**CHAPTER 5    Solenoids                                     75**
5.1    Solenoid Action                                       75
5.2    Solenoid Force and Voltage                            78
5.3    Low Voltage                                           78
5.4    Overvoltage                                           79
5.5    ac Solenoids on dc                                    79
5.6    dc Solenoids on ac                                    80
5.7    50- and 60-Cycle Solenoids                            81
5.8    Solenoid Temperature Rise                             82
5.9    Circuit Applications                                  87
5.10   Variable Solenoids                                    88
5.11   Proportional Valves                                   88
5.12   Servo Valves                                          90

**CHAPTER 6    Types of Control                              94**
6.1    Open-Loop Control                                     94
6.2    Closed-Loop Control                                   95
6.3    Proportional Control                                  97
6.4    Proportional-Integral                                100
6.5    Proportional-Integral-Derivative                     101

**CHAPTER 7    Motion Control Devices                       103**
7.1    Importance of Position Indication and Control        104
7.2    Limit Switches—Mechanical                            104
7.3    Limit Switch Symbols                                 107
7.4    Circuit Applications                                 109
7.5    Proximity Limit Switches                             113
7.6    LED Indicators                                       118

|       |                                                        |     |
|-------|--------------------------------------------------------|-----|
| 7.7   | Solid-State Outputs                                    | 118 |
| 7.8   | Detection Range                                        | 118 |
| 7.9   | Hysteresis                                             | 118 |
| 7.10  | Attenuation Range                                      | 119 |
| 7.11  | Speed                                                  | 120 |
| 7.12  | Magnet-Operated Limit Switch                           | 120 |
| 7.13  | Vane Switches                                          | 122 |
| 7.14  | Linear Position Displacement Transducers               | 123 |
| 7.15  | Angular Position Displacement Transducers              | 124 |
| 7.16  | Use of ac Synchronous and dc Stepping Motors           | 128 |
| 7.17  | Servo Positioning Control                              | 137 |
| 7.18  | Sensing Theory                                         | 140 |
| 7.19  | Flow Sensors                                           | 147 |

**CHAPTER 8    Pressure Control                                              153**

| 8.1   | Importance of Pressure Indication and Control          | 153 |
| 8.2   | Types of Pressure Switches                             | 155 |
| 8.3   | Circuit Applications                                   | 160 |

**CHAPTER 9    Temperature Control                                           167**

| 9.1   | Importance of Temperature Indication and Control       | 167 |
| 9.2   | Selection of Temperature Controllers                   | 168 |
| 9.3   | Electronic Temperature Controller (Pyrometer)          | 169 |
| 9.4   | Controller Outputs                                     | 181 |
| 9.5   | Additional Terms                                       | 181 |
| 9.6   | Temperature Switches (Thermostats)                     | 182 |
| 9.7   | Temperature Sensors                                    | 185 |
| 9.8   | Circuit Applications                                   | 187 |

**CHAPTER 10   Time Control                                                  193**

| 10.1  | Selected Operations                                    | 193 |
| 10.2  | Types of Timers                                        | 195 |
| 10.3  | Synchronous Motor-Driven Timers                        | 196 |
| 10.4  | Solid-State Timers                                     | 202 |
| 10.5  | Circuit Applications                                   | 204 |

**CHAPTER 11   Count Control                                                 209**

| 11.1  | Preset Electrical Impulses                             | 209 |
| 11.2  | Circuit Applications                                   | 210 |
| 11.3  | Solid-State Counters                                   | 212 |

**CHAPTER 12   Control Circuits                                              216**

| 12.1  | Placement of Components in a Control Circuit           | 216 |
| 12.2  | Control Circuit Examples                               | 219 |

**CHAPTER 13   Motors**                                                        **243**
    13.1   ac Motors—Theory of Operation                     243
    13.2   Polyphase Squirrel-Cage Induction Motors           245
    13.3   Single-Phase Motors                                248
    13.4   Resistance Split-Phase Motors                      249
    13.5   Capacitor Start Motors                             250
    13.6   Permanent Split-Capacitor Motors                   252
    13.7   Shaded-Pole Motors                                 254
    13.8   dc Motors                                          257
    13.9   Brushless dc Motors                                263

**CHAPTER 14   Motor Starters**                                                **275**
    14.1   Contacts and Overload Relays                       275
    14.2   Across-the-Line (Full-Voltage) Starters            278
    14.3   Reversing Motor Starters                           280
    14.4   Multispeed Motor Starters                          282
    14.5   Additional Across-the-Line Starter Circuits        283
    14.6   Reduced-Voltage Motor Starters                     293
    14.7   Solid-State Motor Starters                         304
    14.8   Starting Sequence                                  306

**CHAPTER 15   Introduction to Programmable Controllers**                      **310**
    15.1   Primary Concepts in Solid-State Control            311
    15.2   Introduction to Programmable Logic Controllers     312
    15.3   Programmable Logic Controller Concepts             312
    15.4   Input/Output (I/O)                                 314
    15.5   Processor                                          317
    15.6   Memory                                             318
    15.7   Power Supplies                                     320
    15.8   Programming                                        321
    15.9   Examine On/Examine Off                             326
    15.10  Peripheral and Support Devices                     327
    15.11  Data Communication Highway                         330
    15.12  Converting from Relay Logic to PLC                 332
    15.13  PLC Application in Industry                        340

**CHAPTER 16   Industrial Data Communications**                                **347**
    16.1   Overview                                           347
    16.2   Industrial Information Technology
           Architecture                                       348
    16.3   Data Communications Network Concepts               350
    16.4   Data Transmission                                  351
    16.5   Industrial Data Highway                            352

16.6   Network Topologies                                   353
16.7   Industrial Networks                                  355
16.8   Ethernet and the Information Highway                 357
16.9   Transmission Media                                   357
16.10  Data Transfer Rate                                   358
16.11  Interference                                         358
16.12  Open Systems Versus Proprietary Systems              358
16.13  Network Layers                                       359
16.14  Typical Network Systems                              359

**CHAPTER 17   Quality Control                              363**
17.1   Defining Quality and Quality Control                 363
17.2   Electrical and Electronic Circuits Used
       in Quality Control                                   364
17.3   Quality Achieved Through Machine and
       Process Monitoring                                   364
17.4   Process Tolerance (Standards)                        367
17.5   Information Systems                                   369
17.6   Maintaining Quality                                  370

**CHAPTER 18   Safety                                       372**
18.1   Worker Safety                                        372
18.2   Machine Safety                                       374
18.3   Diagnostic Systems                                   375
18.4   Machine Safety Circuit                               375
18.5   Programmable Controllers in Safety                   377
18.6   Other Safety Conditions                              378

**CHAPTER 19   Troubleshooting                              380**
19.1   Safety First                                         380
19.2   Analyzing the Problem                                381
19.3   Major Trouble Spots                                  382
19.4   Equipment for Troubleshooting                        391
19.5   Motors                                               401
19.6   Troubleshooting a Complete Control Circuit           401
19.7   Troubleshooting the Programmable Logic Circuit       407
19.8   Electronic Troubleshooting Hints                     408

**CHAPTER 20   Designing Control Systems for Easy Maintenance   412**
20.1   Design Considerations                                412
20.2   Diagrams and Layouts                                 414
20.3   Locating, Assembling, and Installing
       Components                                           422

**APPENDIX A**   **Summary of Electrical Symbols**   **437**

**APPENDIX B**   **Units of Measurement**   **443**

**APPENDIX C**   **Rules of Thumb**   **447**
C.1   Horsepower Versus Amperes   447
C.2   Horsepower Revolutions per Minute—Torque   447
C.3   Shaft Size—Horsepower—Revolutions
per Minute   448

**APPENDIX D**   **Electrical Formulas**   **449**

**APPENDIX E**   **Use of Electrical Codes and Standards**   **451**
E.1   Major Goals   451
E.2   Recent Development of the NFPA 79 Standard   452
E.3   Tables   452

**APPENDIX F**   **Application of Electric Heat**   **469**
F.1   Calculating Heat Requirements   469
F.2   Selection and Application of Heating Elements   470
F.3   Heating Element Control Circuits   472
F.4   Heating Element Connection Diagrams   475

**APPENDIX G**   **Power Factor Correction**   **483**
G.1   Apparent Power and Actual Power   483
G.2   Magnetizing Current and Power Current   485
G.3   Determining the Amount of Correction Required   487
G.4   Typical Capacitor Design Features   491

**APPENDIX H**   **Concepts Used in Programmable and Solid-State
Controllers**   **495**
H.1   Number System   495
H.2   Logic Gate Symbols and Circuits   501
H.3   Symbols (IEEE STD. 91-1973—ANSI Y32,
14-1973; Distinctive Shape Used)   502
H.4   Circuit Applications   509

**APPENDIX I**   **Selecting a Transformer**   **517**

**GLOSSARY**   **519**

**INDEX**   **537**

# Foreword

The sixth edition of *Electrical Controls for Machines* continues the tradition of this textbook to present a balance of new and current technologies used in manufacturing and process industries. The primary change from the fifth edition is the addition of Chapter 16, Industrial Data Communications. Today, the role of a network in industry is to monitor process variables, respond quickly to process changes, and integrate production data with business data. Networks have become a critical resource for companies to be competitive in the marketplace. Therefore, a technical professional must understand networking concepts.

With each revision of this text, the abundance of products for maintaining quality control, ensuring human safety, increasing productivity, and providing management information has expanded in complexity and cost. Today's typical industrial environment can contain mature technologies (such as electrical motors) combined with expanding technologies (such as networked programmable controllers) and new technologies (such as smart sensors). To keep on top of changing technology, this text now lists recommended Web sites that provide in-depth technical and application information about the concepts presented in the chapters. Readers are encouraged to view these sites and extract the functional operation of the equipment and systems, the recommended application, and practical questions from technicians who are using the technology.

This text is designed for the student, maintenance technician, process engineer, and sales representative who need a clear and simple understanding of all elements of a complex manufacturing system. Therefore, this text can best be described as providing the reader with a practical understanding of electrical control principles. A prerequisite for the reader is knowledge of the basic theories of electricity and electrical circuits. In order to meet this prerequisite, a student should have taken a course in basic electricity. A practitioner (technician or engineer) should review a book on the principles of electrical theory and circuits.

The text is designed to be functional in either an educational or industrial environment. In education, no matter the level (vocational high school, two-year technical school, or four-year college), the entire contents should be examined and understood throughout the course. If the book is used as a reference, then appropriate chapters can be read as needed.

# Preface

We need to teach the fundamentals of process and equipment control. Those mainly affected are the thousands of small builders and users and, to some extent, the large manufacturer. Most selling organizations, both large and small, include many thousands of manufacturers' agents, distributors, sales engineers, and maintenance personnel. These are the people charged with the responsibility for selling machines and keeping them operating. The success or failure of installations may depend on the ability of these people to maintain and troubleshoot equipment properly. The personnel involved need to understand electrical components and their symbols. With this knowledge, they are in a better position to read and understand elementary circuit diagrams.

Six specific areas in which education is needed are (1) electrical and electronic components, (2) control techniques and circuits, (3) troubleshooting, (4) maintenance, (5) electrical standards, and (6) keeping current with changing technology.

1. *Components* are the building blocks of all systems. The core knowledge of the principles and application of each component in a system sets the groundwork for a complete understanding of all facets of a control system.
2. *Techniques and circuits* is the process of building the system from the components. Like building a house, without a concept and plan the structure will fail over time. Techniques and circuits are the lead to a successful plan.
3. *Troubleshooting* machine control circuits involves locating and properly identifying the nature and magnitude of a fault or error. This fault may be in the circuit design, physical wiring, or components and equipment used. The time required and the technique or system used to locate and identify the error are important. Of like importance are the time and expense involved to put the machine back into normal operating condition.
4. *Preventive maintenance* would eliminate the need for most troubleshooting. Many machines are allowed to operate until they literally fall apart.
5. A reasonable set of *standards* should be followed. If the intended result is the improvement of design and application to reduce downtime and promote

safety, electrical standards can be extremely helpful. Where should the education start? The answer is at the beginning, and keep it simple. Even a basic concept, such as the relation between a component and its symbol, can be of benefit to the user.

6. *Keeping current with changing technology* means implementing a strategy for life-long learning. Today the Internet has become a medium for presenting information and expanding knowledge. To expand your knowledge, this text notes important Web sites that should be reviewed occasionally to keep up with latest technological changes.

In addition to the Web sites for specific technologies listed at the end of each chapter, the following Web sites are broad-based sites devoted to technical and training issues.

### Standards Organizations

Electrical Inspectors Information: www.joetedesco.com

Institute of Electrical and Electronic Engineers (IEEE): www.ieee.org

International Brotherhood of Electrical Workers (IBEW): www.ibew.org

National Fire Protection Association (NFPA): www.nfpa.org

National Electrical Safety Foundation (NESF): www.nesf.org

National Joint Apprenticeship Training Committee (NJATC): www.njatc.org

National Electrical Contractors Association (NECA): www.necanet.org

National Electrical Manufacturers Association (NEMA): www.nema.org

Underwriters Laboratories Inc. (UL): www.ul.com

Council for the Harmonization of Electrotechnical Standardization of Nations of the Americas (CANENA): www.canena.org

Canadian Standards Association (CSA): www.csa.ca

### Indexes of Technical Products

Process index: www.processindex.com

Norm's Industrial Electronics: www.compusmart.ab.ca/ndyrvik/

Process Mart: www.iprocessmart.com

Graybar Electric: www.graybar.com

Electronic Engineer's Master (EEM): www.eem.com

Omega Engineering: www.omega.com

### On-Line Technical Publications

Control Engineering: www.controleng.com/

Motion: www.motion.org

Fluid Power Society—*Fluid Power Journal*: www.fluidpowerjournal.com

Allen Bradley—*View* Magazine: www.ab.com/viewanyware/the_view

*Control* Magazine: www.controlmagazine.com/ct/

Allen Bradley—*AB Journal*: www.ab.com/abjournal

ControNews and LogixNews: www.ab.com/controlnews/

Current Issues and News about Manufacturing: www.manufacturing.net

National Electrical code: www.nfpa.org/NEC/NEChome.org

**Independent Web Sites Devoted to Automation Issues**

Nerds in Control: www.control.com/control_com/index_html

PLC Tutor: www.plcs.net

NEC Information and Training: www.mikeholt.com

In addition to these links to Internet-based resources, the authors are making available to instructors and students a comprehensive Web site dedicated to enhancing and expanding the principles and teaching of electrical controls. Its Internet address is www.electricalcontrolsformachines.com. We encourage you to visit this site often and participate in its development.

# Brief Overview of the Chapters

Chapter 1 ("Transformers and Power Supplies") and Chapter 2 ("Fuses, Disconnect Switches, and Circuit Breakers") begin the text with a review of the need for adequate, reliable, protected power systems for electrically operated machines. Chapter 1 is supported by Appendix I ("Selecting a Transformer").

Chapter 3 ("Control Units for Switching and Communication") is devoted to the control circuit elements that provide both operator and process switching signals in a control system. This chapter also emphasizes the methods used to communicate status messages and problems to the operator.

Chapter 4 ("Relays") describes the construction and operation of electrically operated relays. The relay is a functional device that can be used in many different control system operations, such as logical sequencing and power control.

Chapter 5 ("Solenoids") is an overview of the operation and application of ac, dc, and proportional solenoids.

Chapter 6 ("Types of Control") covers the different types of control theories applied to controlling a process actuator. Each control method has unique characteristics that provide a reference for determining the expected controllability, which ultimately will affect the quality and productivity of the system.

Chapter 7 ("Motion Control Devices") is a comprehensive review of the control devices such as limit switches, proximity switches, and photoelectric transducers used in the control of moving actuators.

Chapter 8 ("Pressure Control") is devoted to achieving an understanding of systems that require the precise control of pressures exerted by an actuator onto a process.

Chapter 9 ("Temperature Control") analyses circuits and controllers used in industrial systems in which precise heat must be maintained within the process. Appendix F ("Application of Electric Heat") complements this chapter.

Chapter 10 ("Time Control") describes the operation and application of timers to timed, sequential controlled events.

Chapter 11 ("Count Control") covers counters (similar to timers) and their applications in control sequences that depend on counted events.

Chapter 12 ("Control Circuits") provides an array of practical applications of electrical control circuits using ladder logic diagrams.

Chapter 13 ("Motors") provides insights into the theory and operation of ac and dc motors.

Chapter 14 ("Motor Starters") explains how motor starters are an important protection and control interface between the power source and the motor. Full-voltage and reduced-voltage magnetic types as well as solid-state types are covered.

Chapter 15 ("Introduction to Programmable Control") is an introduction to the concepts and programming of PLCs.

Chapter 16 ("Industrial Data Communications") is a new chapter that explores the terminology, configuration, and issues of data communication within an industrial environment.

Chapter 17 ("Quality Control") is a review of the devices and control concepts used to monitor and control product quality in a production process.

Chapter 18 ("Safety") presents the issues and technology that affect worker and equipment safety.

Chapter 19 ("Troubleshooting") provides the principles and techniques needed to isolate a problem associated with a control circuit.

Chapter 20 ("Designing Control Systems for Easy Maintenance") is devoted to the general requirements to be considered when designing and maintaining control circuits.

In this edition, "Recommended Web Links" sections have been added to the ends of the Chapters 1 through 18. Since the Internet has become a new communication channel for companies to present the conceptual and technical information about their products, the manufacturer's Web site is a new learning environment. Students are encouraged to view the Web sites as supplements to the concepts presented in the chapters.

# Acknowledgments

The authors and Delmar Learning gratefully acknowledge the review panel for their suggestions in the development of this edition. Our thanks to:

Brett McCandless
Vincennes University
Vincennes, IN

Russ Davis
Delta College
University Center, MI

Denny Owen
Kennebec Valley Technical College
Fairfield, ME

Technical guidance and illustrations were provided by the following companies. Appreciation is expressed to them for their cooperation and assistance.

Acme Electric—Acme Transformer Division, Lumberton, NC 28358 (www.acme-electric.com)

Allen-Bradley, a Rockwell International Co., Milwaukee, WI 53204 (www.ab.com)

Allen-Bradley, a Rockwell International Co., Highland Heights, OH 44143 (www.ab.com)

Automatic Timing and Controls Co., Inc., King of Prussia, PA 19406 (www.automatictiming.com)

Banner Engineering Corporation, Minneapolis, MN 55441

Barber-Colman Co., Loves Park, IL 61132-2940 (www.barber-colman.com)

Barksdale, Division of Crane Company, Los Angeles, CA 90058-0843 (www.barksdale.com)

Chromalox, Pittsburg, PA 15238 (www.chromalox.com)

Columbus Controls, Columbus, OH 43081 (www.columbuscontrols.com)

Detroit Coil Co., Ferndale, MI 48270

Divelbiss Corporation, Fredericktown, OH 43019 (www.divelbiss.com)

Eagle Signal Controls, Austin, TX

Fenwal Inc., Ashland, MA 01721-2150 (www.fenwalcontrols.com)

Ferraz Shawmut, Newbury, MA 01950 (www.gouldshawmut.com)

Fluke Corporation, Everett, WA 98206 (www.fluke.com)

General Electric Co., Motor Sales Division, Fort Wayne, IN 46801 (www.geindustrial.com)

General Electric Co., GE Electrical Distributor of Controls, Plainville, CT 06062 (www.geindustrial.com)

Hoffman, Division of Pentair Company, Anoka, MN 55303 (www.hoffmanonline.com)

HPM Division of Taylor Industrial Services, Mount Gilead, OH 43338 (www.hpmcorp.com)

ifm efector inc., Exton, PA 19341 (www.ifmefector.com)

Liebert Corporation, Columbus, OH 43229 (www.liebert.com)

Logex, Inc., Columbus, OH 43085

McNaughton-McKay Electric Company (www.mc-mc.com)

Mercury Displacement Industries, Inc., Ewardsburg, MI 49112 (www.mdius.com)

MTS Sensors Division, Cary, NC 27513 (www.mtssensors.com)

National Fire Protection Association, Quincy, MA 02269-9101 (www.nfpa.org)

POWERTEC Industrial Motors, Rock Hill, SC 29732 (www.powertecmotors.com)

Ronan Engineering Co., Woodland Hills, CA 91367 (www.ronan.com)

Siemens Energy and Automation Inc., Programmable Controls Division, Peabody, MA 01960 (www.sea.siemens.com)

Solid Controls, Inc., Hopkins, MN 55343 (www.solidcontrols.com)

Square D/Schneider Electric, Milwaukee, WI 53201 (www.squared.com)

Standish Industries, Lake Mills, WI 53551 (www.hitekelec.com)

Superior Electric Co., Bristol, CT 06010 (www.superiorelectric.com)

Temposonics Inc., Division of MTS Systems Corporation, Plainsville, NY 11803 (www.temposonics.com)

Vickers Inc., Division of Eaton Aeroquip, Troy, MI 48084 (www.eatonhydraulics.com)

TECO-Westinghouse Motor Company, Pittsburgh, PA 15222 (www.teco-wmc.com)

XYMOX Technologies, Inc., Milwaukee, WI 53201 (www.xymoxtech.com

Yellow Springs Instrument Co., Inc., Industrial Division, Yellow Springs, OH 45387 (www.ysi.com)

# About the Authors

Kenneth B. Rexford received his professional degree from the College of Engineering at the University of Cincinnati. He is registered as a Professional Electrical Engineer in the state of Ohio.

Mr. Rexford was an electrical test engineer for the Dayton Power and Light Company and was with HPM Corporation in Mount Gilead, Ohio, for thirty-four years as Chief Electrical Engineer and Chief of Electrical Development. He has served on several national electrical standards committees.

Mr. Rexford, who is retired from industry, now devotes his time to consulting, writing, and teaching. He is a frequent guest speaker at schools and technical society meetings.

Peter R. Giuliani, Ed.D., is Professor of Computer and Information Science at Franklin University, Columbus, Ohio. Dr. Giuliani also participates in the deployment of instructional technology.

Dr. Giuliani has held engineering and engineering management positions at several system integrator companies in the central Ohio area. Recently he has provided technical training consulting services to equipment manufacturers and to electrohydraulic and process industries.

Dr. Giuliani is an avid cyclist. He can be reached through the Internet at: giuliani@franklin.edu

# Introduction to Electrical Control— The Development of Circuits

To understand electrical control circuits, it is necessary to examine three basic steps in developing a circuit.

The FIRST step is to know what work or function is to be performed. For example, a simple problem may be to light a lamp. The solution can be achieved by completing a path for electrical energy from a source such as a battery to a lamp. For convenience, a switch is used to open or close the path. When the switch is open, electrical energy is removed from the lamp, which is said to be *deenergized*. When the path of electrical energy is closed, the lamp is said to be *energized* and performs a function of illumination. See Figures 1 and 2, in which a battery is used as the source of electrical energy. These drawings are known as *pictorial* drawings because they show a picture of the actual components—battery, switch, and lamp.

In industrial electrical control circuits, symbols are used to represent the components. Figures 3 and 4 show the use of symbols for the components (battery, switch, and lamp). These diagrams are referred to as *schematic* or *elementary circuit diagrams*.

Figures 3 and 4 can be redrawn in a slightly different form as shown in Figures 5 and 6. The circuit performs exactly the same since the lamp is energized

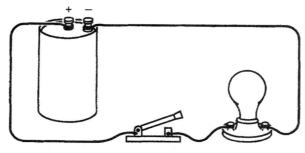

*Figure 1*   With the switch open, the path is open and the lamp is deenergized.

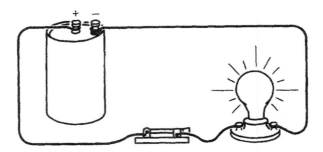

**Figure 2**   With the switch closed, the path is closed and the lamp is energized.

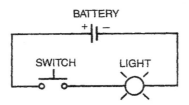

**Figure 3**   Schematic showing the switch and path open and the light deenergized.

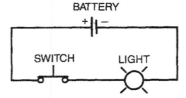

**Figure 4**   Schematic showing the switch and path closed and the light energized.

when the switch is closed. This type of drawing is called a *ladder* diagram. (The ladder-type diagram is used throughout the rest of the book.) In these drawings, the source of electrical energy is always depicted by the two vertical lines (or sides of the ladder).

Notice in Figures 5 and 6 that an important symbol has been introduced. The symbol is "conductors connected" and is shown separately in Figure 7. A similar symbol, which is not used here but appears many times in later diagrams, is "conductors not connected" and is shown in Figure 8.

The SECOND step is to know the operating conditions under which the starting, stopping, and controlling of the process is to take place. Practically all conditions fall into one or more general groups, as affected by:

- Position
- Time
- Pressure
- Temperature

In the chapters on components that follow, each of these conditions can generally be associated with certain components. In many cases, while the actual initiating of a cycle may be through the manually operated push-button switch, certain conditions must be met before the circuit can be closed. This cycle initiation could be one or any combination of the conditions previously listed.

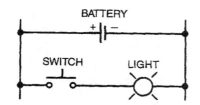

**Figure 5**  Ladder diagram showing the switch and path open and the light deenergized.

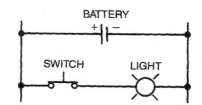

**Figure 6**  Ladder diagram showing the switch and path closed and the light energized.

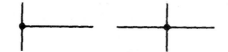

**Figure 7**  Symbols for conductors connected.

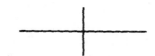

**Figure 8**  Symbols for conductors not connected.

The THIRD step in the development of a circuit is selecting the desired control conditions. There are many times that a circuit must be capable of operating under certain sets of conditions to produce the desired results. For example, a circuit may be required to operate a machine under manual, semiautomatic (single-cycle) or fully automatic (continuous-cycle) operation.

After a decision is made on which of these types of operation is to be used, a selection is made. For reasons of safety to both the machine and operating personnel, the machine must operate in the selected manner.

Review the three basic steps in developing the elementary diagram.

1. Know what work or function is to be performed.
2. Know the conditions for starting, controlling, and stopping the process.
3. Arrange for selecting the desired control conditions: manual, semiautomatic, or automatic.

As the student progresses through the study of components, the symbol for each component will be prominently displayed. It is very important that the symbol becomes closely associated with the component. In appendix A, all of the symbols used will be shown for review.

In understanding electrical control circuit diagrams, there are fundamental problems that should be recognized and overcome if progress is to be made. Some of these problems are:

- Starting with a circuit that is too large or too complicated.
- Failing to carry through a mental picture of the component into the electrical circuit.

- Failing to relate physical, mechanical, or environmental actions into devices that convert these actions into electrical signals.
- Failing to understand that an electrical circuit must perform the correct functions and not perform those actions that will result in damaged components, danger to the operator or machine, or a faulty product.

In addition, one of the biggest problems in reading circuits is gaining a clear understanding of a switch or contact condition. The condition must be properly presented in the elementary circuit diagram, and the user must properly interpret its use in a process or on a machine. Elementary or schematic circuit diagrams will be shown in each of the component sections. As a new component is introduced, the symbol will be shown in the circuit. These circuits will show methods of obtaining specific actions through the use of electrical components. Ultimately all circuits designed for specific actions will need to be assembled into a complete circuit for the overall operation of a machine.

The important point here is to become acquainted with components and their use in the small circuits and work with only one or a few components at a time.

# Chapter 1

# Transformers and Power Supplies

## Objectives

After studying this chapter, you should be able to:

- Give two reasons for energizing machine control systems at 120 volts.
- Explain how to obtain 120 volts from a higher line voltage through the use of a transformer.
- Define *turns ratio* in a transformer.
- Identify the symbol for a dual-primary, single-secondary control transformer.
- Explain what causes temperature rise in a transformer.
- Define *regulation* in a transformer.
- Draw a connection diagram for a dual-primary, single-secondary control transformer to a higher voltage line and to a 120-volt control circuit.
- Calculate the size of a transformer for a given load.
- List the uses of uninterruptible power systems during undesirable power disturbances.
- Explain the basic function of the uninterruptible power system.
- Explain the method for calculating regulation in a transformer.

## 1.1 Control Transformers

In the electrical control circuits shown in "Introduction to Electrical Control—The Development of Circuits" (Figures 1 through 6), a battery is the source of electrical energy. A battery supplies a form of electrical energy known as *direct current* (dc). Most industrial electrical controls use a form of electrical energy called *alternating current* (ac), which is supplied to industry as three phase, 240 volts (V) or 480 V, 60 hertz (Hz). The use of 480 V is most prominent. On circuit diagrams, the three-phase power source is normally shown as three parallel horizontal lines marked L1, L2, L3 (Figure 1-1).

Most electrical control systems on machines are energized at 120 V ac. There are at least two good reasons for this: safety and the use of standard-design components.

Electrical energy at this lower voltage can be obtained by using a *control transformer*. The control transformer is single phase, requiring a connection to any two of the three power lines. The transformer used in industrial machine control consists of at least two separate coils wound on a laminated steel core. The line voltage is connected to one coil, called the *primary*. The control load is connected to the other coil, called the *secondary*. Where the voltage is reduced from the primary to the secondary, the transformer is called a *step-down transformer*. In a *step-up transformer*, the voltage is increased from the primary to the secondary.

The simplest arrangement uses only two coils. One coil is used for the primary, the other for the secondary. However, multiple coils are often used, especially on the primary side, to make it easier to connect to different power voltages. Special primary windings with multiple taps for 200-208-240-480-575-V and a 120-V secondary are available.

Control transformers are available in sizes from 250 volt-amperes (VA) to 10 kilovolt-amperes (kVA) (10 kVA = 10,000 VA). The most widely used control transformers have a dual-voltage primary of 240/480 V. They have an isolated secondary winding to provide 120 V for the load.

The ratio of voltage reduction from 240 V to 120 V is 2 to 1 (2:1). From 480 V to 120 V, the ratio is 4:1. The voltage is directly proportional to the number of turns in each coil. Therefore, this ratio is often referred to as the *turns ratio*. Thus, the control voltage will always be a direct ratio of the line voltage. For example, it is possible for the line voltage in a plant to vary from approximately 460 V to 500 V or from 230 V to 250 V. If the line voltage in a given plant were to drop

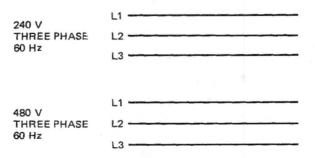

240 V
THREE PHASE
60 Hz

L1
L2
L3

480 V
THREE PHASE
60 Hz

L1
L2
L3

*Figure 1-1*  **Three-phase power line.**

from its normal 480 V to 460 V, a transformer using a 4:1 ratio would deliver 115 V to the secondary. Similarly, if a plant's power source voltage were to drop from its normal 240 V to 230 V, a transformer using a 2:1 ratio would deliver 115 V to the secondary.

The symbol for the control transformer is shown in Figures 1-2 and 1-3. Two primary coils and one secondary coil are used. When the control transformer primary coils are connected to a 240-V power source, the two coils are connected in *parallel.* When the primary coils are connected to a 480-V power source, the coils are connected in *series.*

When two coils with the same number of turns (same voltage rating) are connected in parallel, the effective number of turns for determining the turns ratio remains the same as if only one coil were used. When two coils with the same number of turns are connected in series, the numbers of turns on each coil are added together.

When two separate coils are used on the primary side, the arrangement is called a *dual primary.* When two separate coils are used on the secondary side, it is called a *dual secondary.*

The transformer is available either in an open type or in its own enclosure. At one time the general practice was to use the open type and panel mount it in the control cabinet. With panel mounted transformers, however, high allowable temperature rise may create unwanted high temperatures in the control cabinet. With this condition, the transformer in its own enclosure should be used, mounted on the outside of the cabinet.

The transformer either has screw-type terminals on the coil, or the leads are brought out to a terminal block. In some cases, fuses or circuit breakers are offered as an integral part of the transformer installation.

A typical transformer is shown in Figure 1-4. A diagram of a three-phase power supply and control transformer is shown in Figure 1-5. In that figure, the primary side of the transformer is connected to one phase of a 480-V, three-

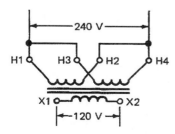

**Figure 1-2**  Primary coils connected in parallel.

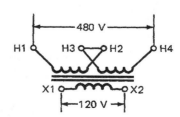

**Figure 1-3**  Primary coils connected in series.

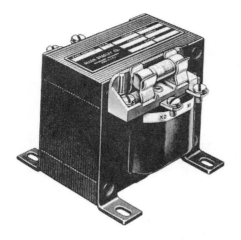

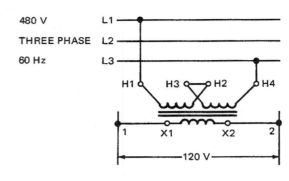

**Figure 1-4**   **Control-circuit transformer with built-in fuse block.** *(Courtesy of Allen-Bradley, a Rockwell International Company.)*

**Figure 1-5**   **Transformer connected to three-phase power line.**

phase, 60-Hz power source. The secondary is supplying 120 V ac (to the vertical sides of the ladder-type control circuit diagram).

All machine tool electrical control circuits shown in this text through Chapter 13 use the basic ladder-type diagram. The voltage source between the two vertical sides in these diagrams is 120 V ac. Therefore, the complete three-phase power circuit and transformer symbols will not always be shown, though they will generally be shown on industrial schematics.

## 1.2  Transformer Regulation

Voltage *regulation* in transformers is the difference between the no-load voltage and the full-load voltage. This value is usually expressed in terms of a percent. It can be calculated as follows:

$$\frac{\text{No-load voltage} - \text{Full-load voltage}}{\text{Full-load voltage}} \times 100$$

Using an example in which the transformer delivers 100 V at no load and the voltage drops to 95 V at full load, the regulation would be calculated as follows:

$$\frac{100 - 95}{95} = \frac{5}{95} \times 100 = 5.26\%$$

It is generally desirable to have the regulation for a control transformer in the range of 2–3%.

## 1.3 Temperature Rise in a Transformer

*Temperature rise* in a transformer is the amount by which the temperature of the windings and insulation exceeds the existing ambient or surrounding temperature. Temperature rise of a transformer is generally given on the transformer nameplate and should not be exceeded. Overloading of a transformer results in excessive temperature. This excessive temperature causes overheating, which will result in rapid deterioration of the insulation and cause complete failure of the transformer coils.

Some of the lower kVA-rated transformers (below 1 kVA) that are rated at 60 Hz can be operated satisfactorily at 50 Hz. However, at higher kVA ratings 60-Hz rated transformers operating at 50 Hz will produce a greater heat rise. Therefore, it is not advisable to use higher rated kVA transformers in 50-Hz power circuits.

## 1.4 Operating Transformers in Parallel

Single-phase transformers can be used in parallel only when their voltages are equal and impedances are approximately equal. If unequal voltages are used, a circulating current exists in the closed network between the two transformers that will cause excessive heating and result in a shorter life of the transformer. Impedance values of each transformer must be within 7.5% of one another. For example, Transformer A has an impedance of 4%. Transformer B, which is to be connected parallel to A, must then have an impedance between 3.7% and 4.3%.

## 1.5 Constant Voltage Regulators*

The constant voltage regulator (CVR) consists of a leakage reactance, ferroresonant transformer with an additional pair of magnetic shunts, and a filtering wind-

---

*Information courtesy of Acme Electric—Acme Transformer Division.

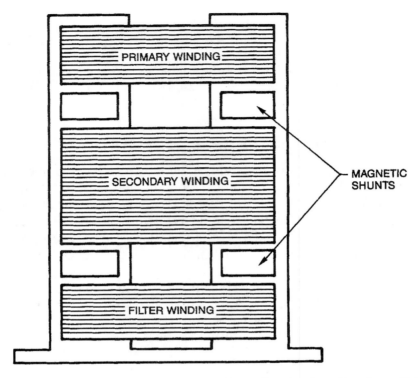

**Figure 1-6**   **Constant voltage regulator.** *(Courtesy of Acme Electric—Acme Transformer Division.)*

ing. Together they develop a regulated, low-distortion sinusoidal output. The circuit is designed so that the segment of the core under the secondary winding (Figure 1-6) will saturate and ferroresonate with the ac capacitor once each half-cycle, limiting the output voltage to a fixed value. The primary-to-secondary leakage reactance and ac capacitor are tuned to achieve ferroresonant regulation of the output over a broad range of input voltage.

The second pair of magnetic shunts and filter winding is incorporated to soften the secondary core saturation effect, cancelling the harmonic voltages that are present in conventional CVRs. The filtering winding is connected in series with the ac capacitor (Figure 1-7). This forms an LC (inductance/capacitance) trap to filter out the low order of harmonics generated by ferroresonant action. A cutaway view of the Acme transformer coil is shown in Figure 1-8.

Ferroresonance (transformer) is a phenomenon usually characterized by overvoltages and very irregular wave shapes. It is associated with the excitation of one or more saturable inductors through capacitance in series with the inductor.

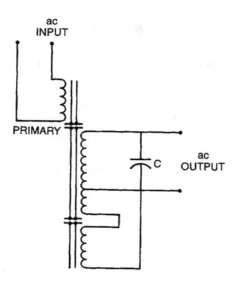

**Figure 1-7**　**Circuit for constant voltage regulator.**　*(Courtesy of Acme Electric—Acme Transformer Division.)*

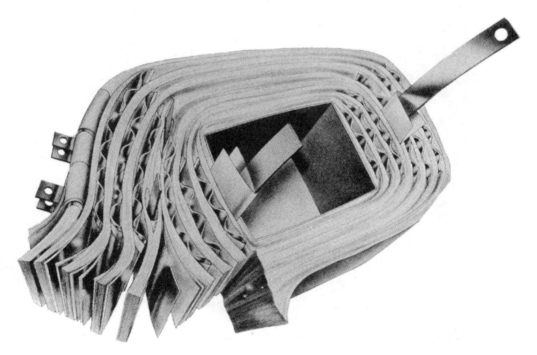

**Figure 1-8**　**Cutaway view of a CVR transformer coil.**　*(Courtesy of Acme Electric—Acme Transformer Division.)*

## 1.6 Uninterruptible Power Systems*

Certain types of electrical equipment, such as programmable logic controllers and computers, are very sensitive when it comes to the quality of their power supply. Small voltage fluctuations or variations in frequency can cause serious malfunctions, and a total power outage can result in the loss of data stored in memory.

When electrical power is generated it is both clean and stable, but during transmission and distribution it is subjected to a variety of detrimental influences. Electrical storms, noisy and largely varying loads, and accidents all lead to a less than perfect supply emerging from the utility power supply.

These supply problems can be overcome by connecting an uninterruptible power system (UPS) between the utility supply and sensitive load equipment. It will not only clean up any supply abberations, but will also maintain the critical load during a complete outage. The Liebert Corporation Uninterruptible Power System provides both power conditioning and supply backup (Figure 1-9). It takes the raw utility power and, using state-of-the-art solid-state power elec-

*Figure 1-9*   **Uninterruptible power system.** *(Courtesy of Liebert Corporation.)*

*Information courtesy of Liebert Corp.

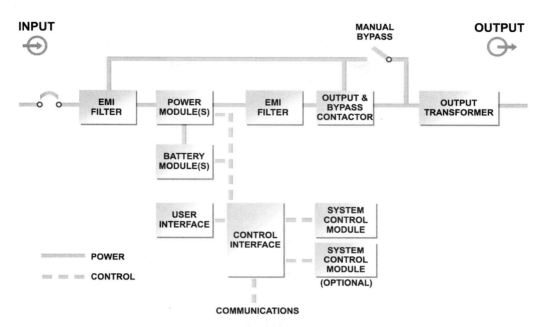

**Figure 1-10**  Block diagram of various elements of the UPS.  *(Courtesy of Liebert Corporation.)*

tronic technology, converts it into a dc form. A microprocessor-controlled inverter then reconverts the dc power into controlled ac that can be used to supply equipment. Due to this double conversion technique, from ac to dc to ac, the output power is essentially isolated from the utility supply so that the equipment will be oblivious to any utility supply variations.

Backup is provided by an internal battery that is automatically connected to the dc portion of the UPS when the input power fails. The standby battery can maintain the unit's fully rated load for 10 minutes, but this period will be longer if lighter loads are used. If the supply break exceeds the battery backup time, the UPS will shut down once the battery charge has been exhausted. However, the unit will sound an alarm to warn that this is about to occur to give adequate time to shut down the load in an orderly fashion.

This UPS will automatically restart when the utility supply returns, and the battery will quickly recharge to prepare for further use. A block diagram shows the arrangement of the various elements within this UPS (Figure 1-10).

## Recommended Web Links

Students are encouraged to view the following Web sites as a supplement to the concepts presented in this textbook. Review and analyze the array of products that are available for electrical control applications. Many of these sites offer technical information that can help in converting the principles to practical applications. To view catalogs, your PC may require Adobe Acrobat Reader software.

**Transformers**

> The Osborne Transformer Co.
> www.osbornetrans.com
> Review: Products and Applications
>
> Sola/Hevi-Duty
> www.sola-hevi-duty.com
> Review: Products, Technical Support and Power Supplies
>
> Acme Electric Corp.
> www.acmepowerdist.com
> Review: "How to Select a Transformer" and Transformer FAQ

**Power Supplies and UPS**

> Majorpower.com
> www.majorpower.com
> Review: AC and DC Solutions
>
> APC
> www.apc.com/index.cfm
> Review: Power Information
>
> Liebert Corp.
> www.liebert.com
> Review: Products—Surge Suppressors, Power Conditioners, and UPS

## Achievement Review

1. What type of electrical energy is normally used to supply industry?

2. What are the two important reasons for using 120 V in machine control systems?

3. There is 480 V ac available in a given plant. To obtain 120 V, what turns ratio is required between the primary and secondary of the control transformer?

4. You find that under unusually heavy loads, the voltage in your plant drops to 456 V. What will be the resulting secondary control voltage if you use a control transformer with a 4:1 primary-to-secondary turns ratio?

5. Draw the symbol for a dual-primary, single-secondary control transformer. Show all lead designations.

6. What parts of a transformer generally contribute to temperature rise in the transformer?

   a. The enclosure
   b. Copper winding
   c. Iron core

7. What may happen to components that are energized in a control system if, due to poor transformer regulation, the voltage drops to an unusually low level?

8. Draw a complete circuit showing the primary of a dual-primary control transformer connected to a three-phase, 480-V power line, and the single secondary connected to a 120-V control system.

9. In a given transformer the voltage is reduced from the primary to the secondary. What is this transformer called?

   a. Current transformer
   b. Step-up transformer
   c. Step-down transformer

10. Draw a block diagram showing all the standard components that make up a UPS.

11. When using an uninterruptible power system, what happens if the power is lost?

12. What factors will generally lead to a less than perfect supply emerging from the utility power system?

13. A transformer has a no-load voltage of 120 and a full-load voltage of 110. What is the percent regulation for this transformer?

14. What will generally result if you operate a transformer (1 kVA or larger) designed for 60 Hz on a 50-Hz supply?

15. What problems can exist within a utility power supply that will make the use of an uninterruptible power supply advisable?

# Chapter 2 Fuses, Disconnect Switches, and Circuit Breakers

## Objectives

After studying this chapter, you should be able to:

- Describe basic fuse construction.
- List three different types of fuses and some of their uses.
- Identify four different types of circuit breakers and uses for each.
- Describe the steps to take when first setting up electrical control and power circuits.
- Explain why time-delay fuses are used with motor starter circuits.
- List the voltage and current ratings available for fuses and circuit breakers.
- Discuss the important factors to consider when selecting protective devices.
- Draw the symbols for important protective and disconnecting devices.
- Know what is meant by *interrupting capacity*.
- Understand the use of rejection-type fuses.
- Explain the two measures of the degree of current limitations provided by a fuse.
- Explain voltage and frequency surges caused by lightning or switching.

## 2.1 Protective Factors

Once the appropriate electrical power is determined for the control circuits, methods to protect the circuit components from current and temperature surges must be considered. There are two factors to be considered when providing control circuit protection:

1. A means of disconnecting electrical energy from the circuits.
2. Protection against sustained overloads and short circuits.

The power circuit can be disconnected by using a disconnect switch or nonautomatic circuit breaker (circuit interrupter). Protection is provided by adding adequate fusing to the disconnect switch and thermal and/or magnetic trip units to the circuit interrupter.

The control circuit is normally protected by a single fuse or circuit breaker. In some cases, two or more fuses may be used. This arrangement is covered in more detail in Chapter 19, "Troubleshooting."

Figure 2-1 shows an example of a commercially available three-pole, fusible disconnect switch ganged with a handle.

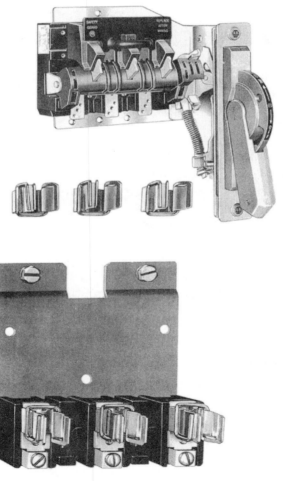

*Figure 2-1*    **Fusible disconnect switch.** *(Courtesy of Allen-Bradley, a Rockwell International Company.)*

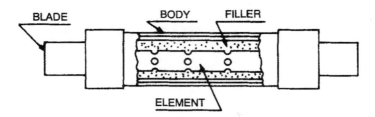

***Figure 2-2***   **Cross section through a typical fuse.**

## 2.2  Fuse Construction and Operation*

The typical fuse consists of an element surrounded by a filler and enclosed by the fuse body. The element is welded or soldered to the fuse contacts, blades, or ferrules (Figure 2-2).

The element is a calibrated conductor. Its configuration, its mass, and the materials employed are varied to achieve the desired electrical and thermal characteristics. The element provides the current path through the fuse. It generates heat at a rate that depends on its resistance and the load current.

The heat generated by the element is absorbed by the filler and passed through the fuse body to the surrounding air. A filler such as quartz sand provides effective heat transfer and allows for the small-element cross section typical in modern fuses. The effective heat transfer allows the fuse to carry harmless overloads. The small-element cross section melts quickly under short-circuit conditions. The filler also aids fuse performance by absorbing arc energy when the fuse clears an overload or short circuit.

When a sustained overload occurs, the element generates heat at a faster rate than the heat can be passed to the filler. If the overload persists, the element will reach its melting point and open. Increasing the applied current heats the element faster and causes the fuse to open sooner. Thus, fuses have an inverse time-current characteristic; i.e., the greater the overcurrent, the less time required for the fuse to open the circuit.

## 2.3  Fuse Types*

Fuses are available in numerous types; here are three common ones:

1. Standard one-time fuse (Figure 2-3)
2. Time-delay fuse (Figure 2-4)
3. Current-limiting non-time-delay fuse (Figure 2-5)

*Information courtesy of Ferraz Shawmut.

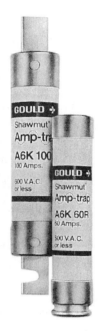

***Figure 2-3*** **One-time fuse—class K-5.** *(Courtesy of Ferraz Shawmut.)*

***Figure 2-4*** **Time-delay fuse—class RK-5.** *(Courtesy of Ferraz Shawmut.)*

***Figure 2-5*** **Current-limiting fuse—class RK-1.** *(Courtesy of Ferraz Shawmut.)*

Standard voltage ratings for fuses are 125 V, 250 V, 300 V, 480 V and 600 V. Higher-voltage fuses are available. Current ratings range from a fraction of an ampere (A) to 6000 A, in all voltage ratings.

The ability of a protective device (fuses or circuit breakers) to interrupt excessive current in an electrical circuit is important. All protective devices have a published *interrupting capacity*, which is defined as the highest current at rated voltage that a device can safely interrupt.

It is therefore important when installing or replacing a protective device that at least three items be considered:

1. Voltage rating
2. Current rating
3. Interrupting capacity

Correct fuses must be used as replacements for continued safety in the protection of equipment. To help in this area, manufacturers provide a "rejection"-type fuse. It is called Class R (R for rejection). The ferrule sizes (0–60 A) have an annular groove in one ferrule. The blade sizes (61–600 A) have a slot in one blade. Replacement of this fuse with a fuse of lower voltage or lower interrupting rating is not possible provided that this fuse is used with rejection fuse blocks. The rejection fuse block is similar to the standard fuse block except physical changes are made in the block to accommodate the annular ring in the ferrule-type fuse and the slot in blade type.

The physical configuration of the rejection-type fuse is shown in Figure 2-4 and 2-5. The rejection-type fuse has a 200,000-A interrupting rating as contrasted to class K-5 with the standard fuse configuration, ferrule or blade, shown in Figure 2-3, with an interrupting capacity of 50,000 A.

To further understand the operation of fuses, a melting time–current data curve is shown in Figure 2-6. This curve is for a typical 100-A, 250-V, time-delay fuse. It shows an inverse time relationship between current and melting time; that is, the higher the current, the faster the melting time. For example, referring to this curve, it can be seen that at 1300 A, the melting time is 0.2 seconds. At 200 A, the melting time is 300 seconds. This characteristic is desirable because it parallels the characteristic of conductors, motors, transformers, and other electrical apparatus. This equipment can carry low-level overloads for relatively long times without damage. However, under high current conditions due to short circuits, damage can occur quickly. Because of the inverse time characteristic, a properly applied fuse can provide effective protection over a broad current range from overloads to short circuits.

The standard one-time fuse (class K-5) link consists of a low-temperature-melting metal strip with several reduced area sections. On overloads that exceed the rating of the fuse, the narrow center section will melt, thus opening the circuit. The heat to melt the strip comes from excessive current passed through the fuse. In case of short circuits all the reduced cross-sectional areas will melt and vaporize simultaneously. The tube holding the strip (fuse link) must be of sufficient strength to withstand the internal pressure developed during this time.

The class K-5 fuses are suitable for the protection of mains, feeders, and branch circuits serving lighting, heating, and other nonmotor loads, provided that the interrupt capacity is adequate.

The time-delay fuses (class RK-5 and RK-1) were developed for use where a heavy overloading might exist for a short time and the fuse is not expected to open unless the overload persists. An example of a short overload is normal motor starting. These time-delay delay fuses are also suitable for general-purpose protection of transformers, service entrance equipment, feeder circuits, and branch circuits. The time-delay characteristic of an efficient fuse allows it to withstand normal surge conditions without compromising short-circuit protection.

Current-limiting time-delay fuses are used today in almost all fuse applications because in most electrical power systems, a fault can produce currents so great that much damage results. The current-limiting fuse will limit both the magnitude and duration of current flow under short-circuit conditions. This type fuse will clear available fault currents in less than one half cycle, thus

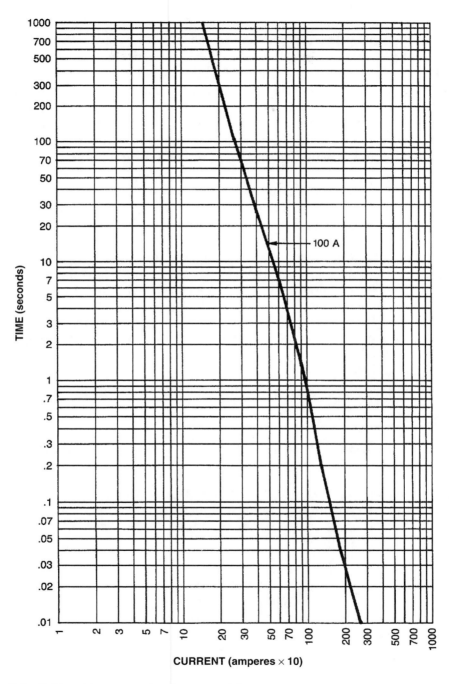

*Figure 2-6*  Melting time–current data for a typical time-delay fuse—100 amperes.

limiting the actual magnitude of current flow. It is generally supplied as a rejection-type fuse. The replacement of these fuses with a fuse of another UL class is not possible.

The current-limiting fuse (class RK-1) is available as a non-time-delay fuse or a time-delay fuse with interrupting capacity of 200,000 A and higher. The non-time-delay fuse is suitable for the protection of capacitors, circuit breakers, load centers, panel boards, switch boards, and bus ducts in which high available short-circuit currents may exist. They are also available in time delay versions for motor loads with a high degree of current limitation for minimizing short-circuit current damage.

With the increased use of solid-state power devices, fuse designs have changed to match solid-state protection demands.

Solid-state devices operate at high current densities. Cooling is a prime consideration. Cycling conditions must be considered. The ability of solid-state devices to switch high currents at high speed subjects fuses to thermal and mechanical stresses. Solid-state devices have relatively short thermal time constants. A short-circuit current that may not harm an electromechanical device can cause catastrophic failure of a solid-state device in a very short time.

Most programmable controllers (covered in Chapter 15) use a semiconductor fuse with a blown-fuse indicator. There also may be add-on switches that can be used to energize an indicator light. This light can be used for remote indication. The trigger actuator may be a part of the fuse or a field-mounted blown-fuse indicator that would be wired in parallel with the fuse being monitored.

## 2.4  Peak Let-Thru Current ($I_p$) and Ampere Squared Seconds ($I^2t$)

Current limitation is one of the important benefits provided by modern fuses. Current-limiting fuses are capable of isolating a faulted circuit before the fault current has sufficient time to accelerate to its maximum value. This current-limiting action provides several benefits:

- It limits thermal and mechanical stresses created by the fault currents.
- The magnitude and duration of the voltage drop caused by the fault currents is reduced, improving overall power quality.
- Current-limiting fuses can be precisely coordinated to minimize unnecessary service interruption.

Peak let-thru current ($I_p$) and ampere squared seconds ($I^2t$) are two measures for the degree of current limitation provided by a fuse. Maximum allowable $I_p$ and $I^2t$ values are specified in UL standards for all UL-listed current-limiting fuses and by the fuse manufacturer for semiconductor and special purpose fuses.

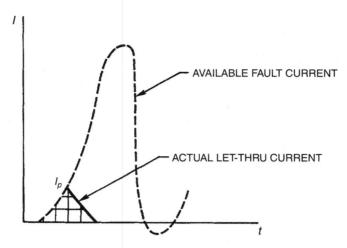

**Figure 2-7**  **Let-thru current is expressed as a peak instantaneous value ($I_p$).**  *(Courtesy of Ferraz Shawmut.)*

Let-thru current is that current passed by a fuse while the fuse is interrupting a fault within the fuse's current-limiting range (Figure 2-7). It is expressed as a peak instantaneous value ($I_p$) in amperes. $I^2t$ is a measure of the heat generated with current flow over time.

The $I_p$ data is generally presented in the form of a graph. Key information is provided in the peak let-thru graph in Figure 2-8. As shown on the figure, the important components are:

1. The $x$-axis, labeled "Available Fault Current" in rms symmetrical amperes.
2. The $y$-axis, labeled "Instantaneous Peak Let-Thru Current" in amperes.
3. The line labeled "Maximum Peak Current Circuit Can Produce," which gives the worst-case peak current possible with no fuse in the circuit.
4. The "Fuse Characteristic Line," which is a plot of the peak let-thru currents that are passed by a given fuse at various available fault currents.

Figure 2-9 illustrates the use of the peak let-thru current graph. Assume that a 200 A, class J fuse is to be applied where the available fault current is 35,000 A rms. The graph shows that with 35,000 A rms available, the peak available current is 80,500 A (35,000 × 2.3) and that the fuse will limit the peak let-thru current to 12,000 A.

You may wonder why the peak available current is 2.3 times greater than the rms available current. In theory the peak available fault current can be anywhere from 1.414 × (rms available) to 2.828 × (rms available) in a circuit in which the impedance is all reactance with no resistance. In reality all circuits include some resistance, and the multiplier has been chosen as a practical unit.

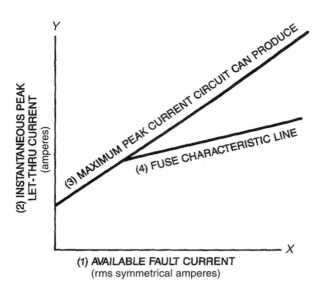

**Figure 2-8**   $I_p$ **data—peak let-thru graph.** *(Courtesy of Ferraz Shawmut.)*

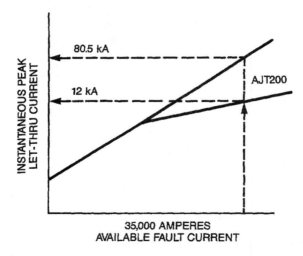

**Figure 2-9**   **Use of a peak let-thru graph.** *(Courtesy of Ferraz Shawmut.)*

The $I_p$ value is relatively useful in determining the magnitude of the peak current available downstream of a fuse for general purpose circuits. If the fuse limits the current to a value less than the available fault current, then all the equipment downstream of the fuse may be selected to withstand this lower value, rather than the full available fault current. The level of protection for equipment provided by the current-limiting fuse is not complete without considering the $I_p$

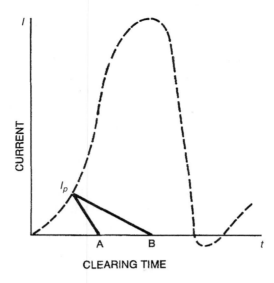

**Figure 2-10**   **Clearing time versus current.** *(Courtesy of Ferraz Shawmut.)*

and $I^2t$. $I_p$ provides the peak current value and $I^2t$ provides a measurement of the amount of heat energy that the fuse passes during opening.

Two fuses can have the same $I_p$ but different total clearing times (Figure 2 10). The fuse that clears by time A will provide greater component protection than will the fuse that clears in time B.

Fuse $I^2t$ depends on both $I_p$ and total clearing time. The $I^2t$ passed by a given fuse depends on the characteristics of the fuse and on the applied voltage. The $I^2t$ passed by a given fuse will decrease as the application voltage decreases. Unless otherwise stated, published $I^2t$ values are based on ac testing. The $I^2t$ passed by a fuse in a dc application may be higher or lower than in an ac application. The voltage available fault current and time constant of the dc circuit are the determining factors.

Fuse $I^2t$ and $I_p$ values can be used to determine the degree of protection provided for all circuit components under fault current conditions.

Manufacturers of diodes, thyristors, triacs, and cable publish $I^2t$ withstand ratings for their products. The fuse chosen to protect these products should have a clearing $I^2t$ that is lower than the withstand $I^2t$ of the device being protected.

## 2.5 Voltage and Frequency Surges

Protecting electrical equipment against voltage surges caused by switching or lightning is very important. Even low surges that exceed the equipment insulation rating can cause the insulation to eventually weaken to the point of failure. For example, the insulation in motor windings can be stressed to the point that failure will result between windings and frame or between stator coils.

In case of a surge caused by lightning, the most severe damage occurs if there is a direct hit. Surge voltages can also be caused by induction on the line due to the lightning. The voltage from a direct hit will be twenty to thirty times that caused by induction. In these cases it can be considered a high-voltage and high-frequency pulse.

A lightning arrester will limit the crest of the surge by breaking down and conducting to ground. After grounding, the arrester then returns to its initial condition of nonconducting.

Switching voltage surges are not as significant as lightning voltage surges. In most cases they will not exceed three times normal voltage. The highest overvoltage will be present when there is a ground fault in the system.

For switching voltage surges, the current-limiting fuse is used. The operating characteristic of this type of fuse is such as to interrupt in the first quarter of the cycle.

## 2.6  Circuit Breaker Types

The circuit breaker is available in four types:

1. Nonautomatic (circuit interrupter)
2. Thermal
3. Magnetic
4. Thermal magnetic

The *nonautomatic circuit breaker* (circuit interrupter) is used for load switching and isolation. Adding a *thermal trip unit* to the nonautomatic circuit breaker, by using a bimetallic element in each pole of the breaker, provides automatic tripping. This unit then carries the load current. When the conductors carry a current in excess of the normal load, the breaker thermal element increases in temperature as it carries the same current. This temperature increase deflects the thermal element and trips the breaker. Since the tripping action depends on temperature rise, a time lag is present. Therefore, the tripping action is not affected by momentary overloads.

To clear a circuit in case of a short circuit, a more rapid opening system is required. To achieve such a system, a *magnetic trip unit* is added either to the nonautomatic breaker or to the breaker with the thermal trip unit. The magnetic trip unit operates through a magnet that trips the breaker instantaneously on short-circuit current.

A combination of thermal and magnetic trip units is desirable. The thermal element provides inverse time tripping on overloads. The magnetic trip provides instantaneous trip on short circuits. *Molded-case* circuit breakers are available from 100 A to 2500 A. The voltage ratings are 240 V, 480 V, and 600 V with the interrupting capacity in some circuit breakers to 100,000 A. Figure 2-11 is a cutaway view of a molded-case circuit breaker.

The fused disconnect switch and the circuit breaker are available in many types of enclosures. They can also be obtained as open types for panel mounting. Remote operators can be mounted on the door of the control cabinet and interlock mechanically with the door. The operator may also be mounted in a "dead" or stationary portion of the cabinet. The door is mechanically interlocked with the breaker trip switch. This allows the breaker to be locked open or closed with the cabinet door open.

Figure 2-12 shows a typical molded-case circuit breaker design. It incorporates, in frame ratings 150 A to 1600 A, interrupting capacities as high as 100 kA at 480 V ac (200 kA at 240 V ac) in physical sizes normally associated with standard interrupting capacity breakers.

The L-frame circuit breaker provides higher interrupting capacities and improved current-limiting capabilities compared to previous standard line circuit breakers.

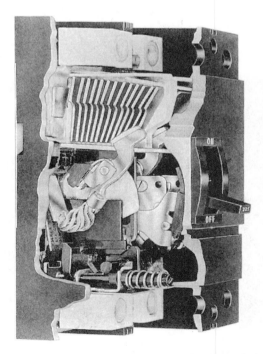

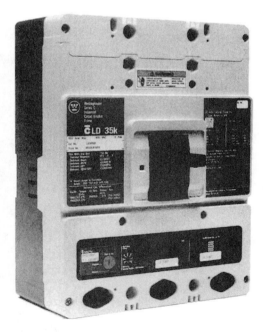

*Figure 2-11*   Cutaway view of a molded-case **circuit breaker.** *(Courtesy of Square D/Schneider Electric.)*

*Figure 2-12*   Molded-case circuit breaker. *(Courtesy of Westinghouse Electric Corporation.)*

Thermal magnetic and electronic trip unit designs are available. This circuit breaker is available with a fixed, thermal-adjustable magnetic trip unit that provides inverse time and instantaneous tripping. Also available is the electronic trip unit, including current-sensing circuits that provide an inverse time-delay tripping action for overload condition and either short delay or instantaneous tripping for protection against short-circuit conditions.

A push-to-trip button located on each trip unit provides a local means of manually exercising the trip mechanism.

High-strength glass-polyester bases and covers have excellent dielectric qualities and are inherently fungus proof. Cover design reduces the possibility of accidental contact with live terminals.

## 2.7 Programmable Motor Protection

The unit shown in Figure 2-13 is a multifunction, motor-protective relay that monitors three-phase ac current and makes separate trip and alarm decisions based on preprogrammed motor current and temperature conditions. This unit's motor protection algorithm is based on proven positive and negative (unbalance) sequence current sampling and true rms calculations. (An algorithm is a special set of rules or instructions that perform a specific operation.)

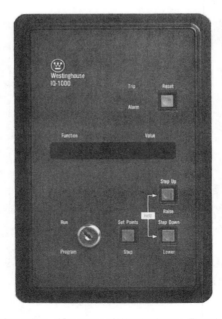

**Figure 2-13** **Motor-protective relay.** *(Courtesy of Westinghouse Electric Corporation.)*

By programming this unit with the motor's electrical characteristics (such as full-load current and locked-rotor current), the unit's algorithm will automatically tailor the optimal protection curve to the motor being monitored. No guesswork or approximation is needed in selecting a given protection curve because this unit matches the protection from an infinite family of curves to each specific motor.

Application-related motor-load problems are further addressed through the use of such functions as jam, underload, and ground fault protection.

A few of the many protective features of this unit are:

- Instantaneous overcurrent trip level and start delay
- Instantaneous overcurrent disable setting
- Locked-rotor current
- Maximum allowable stall time
- Ultimate trip current level
- $I^2t$ alarm level
- Zero-sequence ground fault trip level with start and run delays

## 2.8 Electrical Metering and Voltage Protection

The unit shown in Figure 2-14 is a microprocessor-based monitoring and protective device. It provides complete electrical metering and affords system voltage

***Figure 2-14*** **Electric metering and voltage protection.** *(Courtesy of Westinghouse Electric Corporation.)*

protection. In one compact, standard package this unit provides an alternative to individually mounted and wired ammeters, voltmeters, ammeter and voltmeter switches, wattmeters, watthour meters, and more. Protection features include:

- *Phase loss.* Voltage phase loss occurs if less than 50% of the nominal line voltage is detected. Current phase loss occurs if the smallest phase current is less than 1/16 of the largest phase current. (It updates itself twice per second and all other protection functions once per second.)
- *Phase unbalance.* Occurs if the maximum deviation between any two phases exceeds the amount of unbalance as a percent of nominal line voltage preset by DIP switches. Range: 5–40% (5% increments).
- *Phase reversal.* Occurs if any two phases become reversed for more than 1 second.
- *Overvoltage.* Dual in-line package (DIP) switch setting as a percent of nominal line voltage. Range: 105–140% (5% increments).
- *Undervoltage.* DIP switch setting as a percent of nominal line voltage. Range: 95–60% (5% increments).
- *Delay.* Allows existence of overvoltage, undervoltage, or voltage unbalance before an alarm or trip occurs. Range: 0–8 seconds (1 second increments).

## 2.9  Selecting Protective Devices

Important factors to consider when selecting protective devices are:

- Size (current rating)
- Whether a time lag is required
- Interrupting capacity
- Ambient temperature where the device is to be located
- Voltage rating
- Number of poles
- Mounting requirements
- Type of operator
- Enclosure, if required

If the line voltage and load in a circuit are known, the size and voltage rating of the protective device can be determined. The conductor size to feed the load must be properly determined. Tables are available to supply this information (refer to Appendix E).

The actual sizing of the protective device should come from sources such as the National Fire Protection Association (NFPA), the National Electrical Code*

---

*National Electrical Code*® is a registered trademark of the National Fire Protection Association, Inc., Quincy, MA 02269.

(NEC), and the NFPA-79-94 Electrical Standard for Industrial Machinery-1994. Local electrical code requirements should also be consulted. The manufacturers of protective devices can also be helpful in obtaining this information.

The problem in providing a time-lag element depends on the nature of the load. If inductive devices such as motors, motor starters, solenoids, and contactors are involved, some time lag is generally needed. Here again, the manufacturers of protective devices can help.

In selecting a protective device, you should know the available short-circuit current in your plant. The first information needed is the impedance of the transformer supplying power. This information can be read from the unit's nameplate or obtained from the manufacturer of the unit. *Impedance* is the current-limiting characteristic of a transformer and is expressed as a percent.

The impedance is used to determine the interrupting capacity of a circuit breaker or fuse employed to protect the primary of a transformer. For example, you need to determine the minimum circuit breaker trip rating and interrupting capacity for a 10-kVA, single-phase transformer with 4% impedance. It is to be operated from a 480-V, 60-Hz source. The calculation is as follows:

$$\text{Normal full-load current} = \frac{\text{Nameplate volt-amperes}}{\text{Line voltage}} = \frac{10,000}{480} = 20.8 \text{ A}$$

$$\text{Maximum short-circuit amperes} = \frac{\text{Full-load amperes}}{4\%} = \frac{20.8}{0.04} = 520 \text{ A}$$

The breaker or fuse would have a minimum interrupting rating of 520 A at 480 V.

For a second example, you need to determine the interrupting capacity in amperes required of a circuit breaker or fuse for a 75-kVA, three-phase transformer with a primary of 480 V delta and secondary of 208Y/120 V. The transformer impedance ($Z$) is 5%. If the secondary circuit is shorted, the following capacities are required:

$$\text{Normal full-load current} = \frac{\text{Volt-amperes}}{3 \times \text{Line voltage}} = \frac{75,000 \text{ VA}}{3 \times 480} = 52 \text{ A}$$

$$\text{Maximum short-circuit line current} = \frac{\text{Full-load amperes}}{5\%} = \frac{52}{0.05} = 1040 \text{ A}$$

The breaker or fuse would have a minimum interrupting rating of 1040 A at 480 V.

Ambient temperature in the location where the protective device is to be located is important. Overload protection devices can be thermally operated. Therefore, an increase in the ambient temperature can affect the trip setting of the device. In some cases, ambient temperature compensation is provided by the manufacturer in the design of the device.

Standard pole arrangements for most disconnecting devices and protective devices are two poles for single-phase lines and three poles for three-phase lines. One- and four-pole devices are also available. The multiple-pole (multipole) devices are generally ganged together with a single common operator.

Symbols for various protective and disconnecting devices are shown in Figure 2-15. By using symbols, these components can be added to power source and control source diagrams.

Figure 2-16 illustrates a complete diagram: a three-phase power source, disconnect and protection, and control transformer with protection added in the secondary circuit.

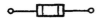

A. SINGLE-FUSE ELEMENT

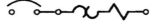

D. ONE-POLE, THERMAL-MAGNETIC CIRCUIT BREAKER

B. THREE-POLE DISCONNECT SWITCH, GANGED WITH HANDLE

E. THREE-POLE CIRCUIT INTERRUPTER, GANGED WITH HANDLE

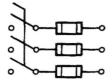

C. THREE-POLE FUSED DISCONNECT SWITCH, GANGED WITH HANDLE

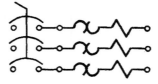

F. THREE-POLE THERMAL-MAGNETIC CIRCUIT BREAKER, GANGED WITH HANDLE

**Figure 2-15**  Symbols for protective devices.

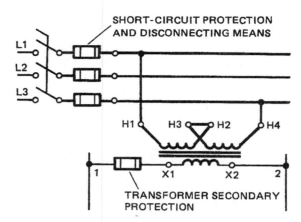

**Figure 2-16** Power and control protection. Conductors carrying load current at line voltage are denoted by heavy lines.

## Recommended Web Links

Students are encouraged to view the following Web sites as a supplement to the concepts presented in this textbook. Review and analyze the array of products that are available for electrical control applications. Many of these sites offer technical information that will help in converting the principles to practical applications. To view catalogs, your PC may require Adobe Acrobat Reader software.

### Fuses

Littelfuse
www.littelfuse.com
Review: Education, Documents, Fuseology

Ferraz Shawmut
www.gouldshawmut.com/home.html
Review: Products, General Purpose Fuses

### Disconnect Switches

Hubbell Ltd.
www.hubbell.co.uk/fuse.htm
Review: Fused and Nonfused Disconnect Switches

Advance Controls, Inc.
www.acicontrols.com/switch.html
Review: The entire Web site

**Circuit Breakers**

E-T-A
www.etacbe.com/n_america/e-t-a/etacbeframeset.html
Review: Products, Technical Info.

IDEC Corp.
www.idec.com/use/html/CircuitBreakers.html
Review: Circuit Breakers

# Achievement Review

1. What two factors must be provided for in any electrical power or control system?

2. Under what conditions would you use the time-delay fuse? Give an example.

3. What are the four types of circuit breakers?

4. Under what conditions would you want to use the magnetic trip feature in a circuit breaker?

5. List at least five important factors to consider when selecting a protective device.

6. Are more than two poles available on disconnecting devices? If so, how many?

7. Draw the symbol for each of the following:

   a. Single-fuse element
   b. Three-pole fused disconnect switch, ganged with handle
   c. Three-pole circuit interrupter, ganged with handle
   d. Three-pole, thermal-magnetic circuit breaker, ganged with handle

8. What is meant by the interrupting capacity of a fuse?

9. What does quartz sand filler in a fuse provide?

   a. Increased voltage rating
   b. Effective heat transfer
   c. An inexpensive filler

10. Why is the inverse time characteristic of a fuse important?

    a. The fuse cost is reduced.
    b. It increases the interrupting capacity.
    c. It parallels the characteristics of conductors, motors, transformers, and other electrical apparatus.

11. What are two measures of the degree of current limitations provided by a fuse?

12. How are $I^2t$ values of a fuse derived?

13. Under what condition will the highest overvoltage occur from a surge caused by switching?

14. Explain how a lightning arrester protects electrical equipment for a surge current caused by lightning.

15. Determine the interrupting capacity in amperes required of a circuit breaker or fuse for a 15-kVA, single-phase transformer with 4% impedance, to be operated from a 480-V, 60-Hz source.

# Control Units for Switching and Communication

## Objectives

After studying this chapter, you should be able to:

- Describe two methods of mounting push-button switch units.
- List four types of operators for the push-button switch.
- Explain why different colors are used for push-button switch operators.
- Draw the symbols for push-button switch units with (1) flush or extended head, (2) mushroom head, and (3) maintained contact attachment.
- List several arrangements available for selector switches.
- Draw the symbol for the selector switch.
- Draw the symbol for the foot switch.
- Discuss the advantages of push-to-test pilot lights.
- Draw the symbol and detailed circuit for the push-to-test pilot light.
- Explain the meaning of the letter in the pilot light.
- Draw simple basic circuits using selector switches.
- Discuss the use of annunciators to obtain process information.
- Explain the use of the LED.
- Discuss how the environment may affect the design of a membrane switch.
- Discuss three types of liquid crystal displays.

## 3.1  Oil-Tight Units

Almost all push-button switches, selector switches, and pilot lights offered to industry today are oil-tight units. Since this text is concerned with industrial electrical control, only oil-tight types of units are discussed.

## 3.2 Push-Button Switches

*Switches* generally consist of two parts: the contact unit and the operator. This composition allows for many combinations that cover almost every application required.

Contact units can be obtained in blocks that contain one normally open (NO) and one normally closed (NC) contact. (*Normal* can be defined as not being acted upon by an external force.) Multiples of these blocks can be assembled to obtain up to four NO and four NC contacts. With some switches that are now available, more units can be added.

The contact rating for a typical heavy-duty, oil-tight unit is:

Ac Volts: 110-125

Amperes Normal: 6.0

Amperes Inrush: 60.0

Figure 3-1A shows a cutaway section of a typical contact block. Figure 3-1B shows a cutaway section of a typical operator. Figure 3-1C also shows typical contact blocks.

In some cases, the contact block is *base-mounted* in the push-button enclosure. Thus, the units can be prewired before the cover with the operators is put in place. This method also eliminates cabling conductors to the cover.

The alternate method is *panel mounting*. The base of the operator, with the contact block attached, is mounted through an opening in the panel. It is then secured in place by a threaded ring installed from the front of the panel. The ring is part of the operator assembly. Later in the text it will be shown that this arrangement has the advantage of providing a space for a terminal block installation in the base of the enclosure. Thus, all connecting circuits can be terminated at an easily accessible checkpoint.

**Figure 3-1A** **Standard double-circuit contact block.** *(Courtesy of Eaton Corporation.)*

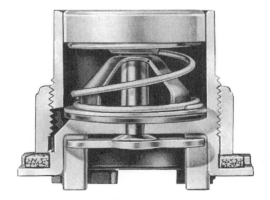

**Figure 3-1B** **Standard-duty, flush button operator.** *(Courtesy of Eaton Corporation.)*

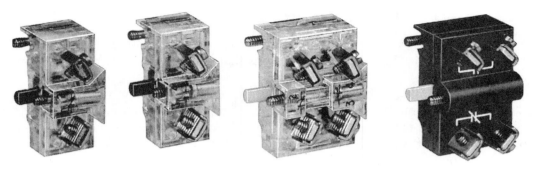

***Figure 3-1C*** **Contact blocks.** *(Courtesy of Square D/Schneider Electric.)*

There are slight differences in the way manufacturers machine mounting holes. Therefore, modifications may be required when substituting one unit for another.

Many types of operators are available to suit almost any application. They include:

- Recessed button (Figure 3-2A)
- Mushroom head (Figure 3-2B)
- Time delay (Figure 3-2C)
- Illuminated push-pull (Figure 3-2D)
- Keylock (Figure 3-2E)

*Color designation* is an important factor in push-button switch operators. Color designation not only provides an attractive panel but, more important, it lends itself to safety. Quick identification is important. Certain functions become

***Figure 3-2A*** **Recessed button push-button unit.** *(Courtesy of General Electric Company, General Purpose Control, Bloomington, IL.)*

***Figure 3-2B*** **Mushroom head push-button unit.** *(Courtesy of General Electric Company, General Purpose Control, Bloomington, IL.)*

**Figure 3-2C** Time-delay push-button unit. *(Courtesy of General Electric Company, General Purpose Control, Bloomington, IL.)*

**Figure 3-2D** Illuminated push-pull unit. *(Courtesy of General Electric Company, General Purpose Control, Bloomington, IL.)*

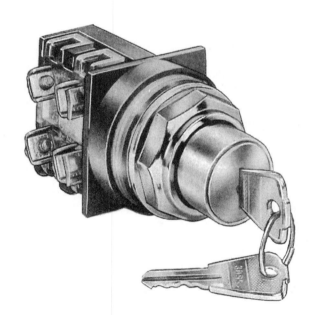

**Figure 3-2E** Cylinder lock push-button unit. *(Courtesy of General Electric Company, General Purpose Control, Bloomington, IL.)*

associated with a specific color. Standards have been developed that specify certain colors for particular functions. For example, in the machine tool industry, the colors red, yellow, and black are assigned the following functions:

Red: Stop, Emergency Stop
Yellow: Return, Emergency Return
Black: Start Motors, Cycle

A. FLUSH OR EXTENDED HEAD     B. MUSHROOM HEAD     C. TWO CIRCUIT MAINTAINED

**Figure 3-3**   Symbols for three push-button units.

Another special class of operator is the *maintained contact attachment*. Operation of one unit operator will operate the attached contacts. The contacts remain in an operated condition until the second operator is operated.

Symbols for three push-button units are illustrated in Figure 3-3.

## 3.3 Selector Switches

Selector switches can be obtained with up to four positions. They can be the maintained contact type, or the three-position switch can be arranged for spring return from the right, from the left, or from both right and left. Up to eight contacts are available per device.

The operators are available in standard knob, knob lever, or wing lever types. A cylinder lock can be used and the switch locked in any one position or all positions. The arrangement for opening or closing contacts in any one position or more depends on a cam in the operator.

Figure 3-4A shows a selector switch with a lever operator. Figure 3-4B shows a selector switch keylock operator.

In the more simple form, a selector switch with two contacts arranged for two positions may be called a *double-pole, double-throw selector switch*. Similarly, a selector switch with two contacts arranged for three positions may be called a *double-pole, double-throw with neutral selector switch*. As the number of positions increases to four and the number of contacts (poles) increases to eight, the manufacturer's reference to a specific operator is generally coded through a symbol chart or function table. Such charts or tables display which contacts are closed or open in the different positions of the selector switch operator.

Two of the simpler arrangements in selector switches are shown in Figures 3-4C and 3-4D. The symbol is shown at the left. An expansion of the symbol is provided to acquaint the reader with the actual conditions of the contacts in the positions shown in the symbol. In Figure 3-4D, note the use of the X under the position number. This indicates that the contact in line with the X is closed in that position.

Figure 3-4E shows a four-contact (pole), four-position selector switch. Both the symbol and an expansion of the symbol are included.

**Figure 3-4A**   Lever-operated selector switch.
*(Courtesy of General Electric Company, General Purpose Control, Bloomington, IL.)*

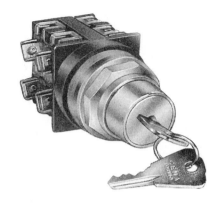

**Figure 3-4B**   Cylinder lock selector switch.
*(Courtesy of General Electric Company, General Purpose Control, Bloomington, IL.)*

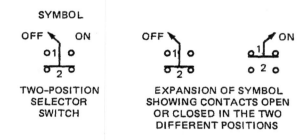

SYMBOL

TWO-POSITION
SELECTOR
SWITCH

EXPANSION OF SYMBOL
SHOWING CONTACTS OPEN
OR CLOSED IN THE TWO
DIFFERENT POSITIONS

**Figure 3-4C**   Symbols for two-position selector switch.

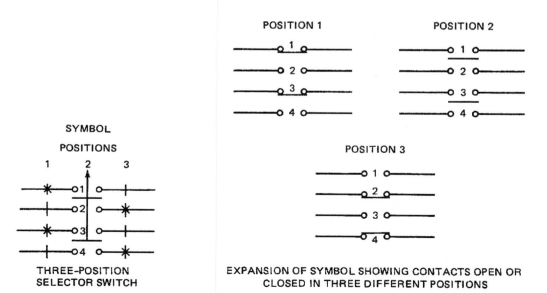

SYMBOL

POSITIONS

THREE-POSITION
SELECTOR SWITCH

EXPANSION OF SYMBOL SHOWING CONTACTS OPEN OR
CLOSED IN THREE DIFFERENT POSITIONS

**Figure 3-4D**   Symbols for three-position selector switch.

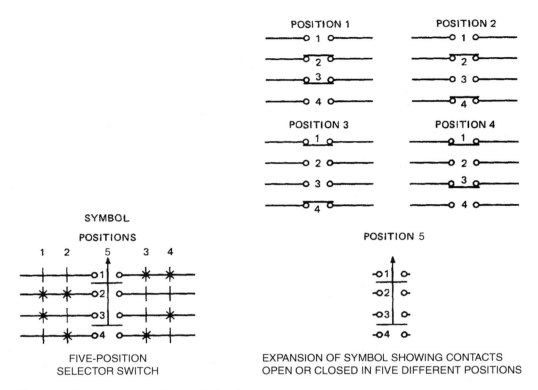

*Figure 3-4E*  **Symbols for five-position selector switch.**

## 3.4 Heavy-Duty Switches

Heavy-duty switches and special application switches under heavy-duty, special-service, oil-tight units are those in which the switch is made an integral part of the operator and enclosures. A typical example of such a switch is the foot-operated unit shown in Figure 3-5A. The symbols are shown in Figure 3-5B.

## 3.5 Indicating Lights

The indicating or pilot light is available in three basic types:

1. Full voltage
2. Resistor (low voltage)
3. Transformer (low voltage)

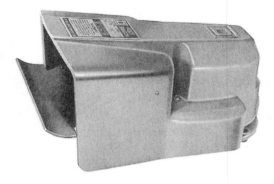

**Figure 3-5A**   Foot-operated switch.
*(Courtesy of Square D/Schneider Electric.)*

OPENS BY FOOT
PRESSURE

CLOSES BY FOOT
PRESSURE

**Figure 3-5B**   Symbols for foot-operated switch.

Due to the vibration normally present in machines, the low-voltage bulb is preferred. The low-voltage bulb operates at 6–8 V obtained through a resistance or transformer unit.

The lens is available in either plastic or glass and in a variety of colors. As for push-button operators, colors can be used for selection purposes to increase safety of operation; for example:

Red: Danger, Abnormal Conditions

Amber: Attention

Green: Safe Condition

White or Clear: Normal Condition

The push-to-test indicating light provides an additional feature. Consider, for example, an indicating lamp that will not illuminate. It may be that the bulb is not energized, or the bulb may be burned out. Depressing the lens unit will connect the bulb directly across the control voltage source. This provides a check on the condition of the bulb.

A detail circuit of the transformer-type, push-to-test indicating light is illustrated in Figure 3-6A. Figure 3-6B shows the symbol for the push-to-test indicating light. Figure 3-6C shows one type of standard transformer-type, push-to-test indicating light.

Note that the push-to-test pilot light consists of a 120:6-V transformer, a 6-V bulb covered by a lens, and a standard contact block. When the pilot light lens is depressed (operated), it mechanically operates the contact block. The normally closed contacts open and the normally open contacts close. The normally closed contacts are connected to the control circuit. The normally open contacts are connected directly to a source of 120 V. Thus, when the lens is depressed

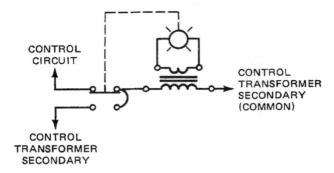

**Figure 3-6A**   Circuit for push-to-test pilot light.

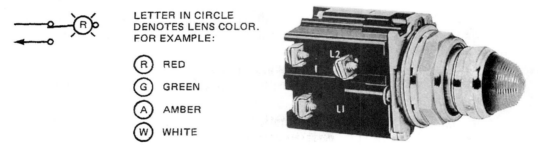

**Figure 3-6B**   Symbol for push-to-test pilot light.

**Figure 3-6C**   Standard transformer-type push-to-test indicating light. *(Courtesy of Eaton Corporation.)*

(operated), the pilot light transformer primary is connected directly across 120 V. Six volts is now applied to the bulb.

**3.5.1 Miniature Oil-Tight Units**   In addition to the standard line of indicating lights, a line of miniature oil-tight push buttons, selector switches, and pilot lights are also available. Miniature units can often be used to an advantage where space is at a premium, such as where a great deal of indicating is required. They also have the same feature selection as the standard line. Figure 3-7 shows a group of miniature oil-tight push buttons.

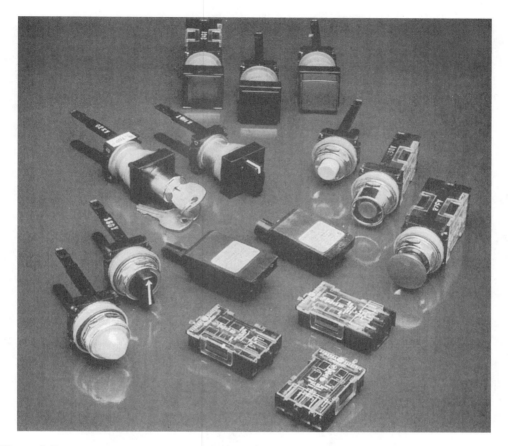

***Figure 3-7*** **New CR204 miniature oil-tight push buttons.** *(Courtesy of General Electric Company, General Purpose Control, Bloomington, IL.)*

## 3.6  General Information on Oil-Tight Units

All oil-tight operators mount on the panel of an enclosure with some type of sealing ring to give them an oil-tight feature.

A nameplate is provided to properly identify the unit. Nameplates are available in different sizes. The size generally depends on the amount of information required on the nameplate to properly describe the function of the unit. Typical nameplates for oil-tight units are shown in Figure 3-8A. A group of several different types of oil-tight units is shown in Figure 3-8B.

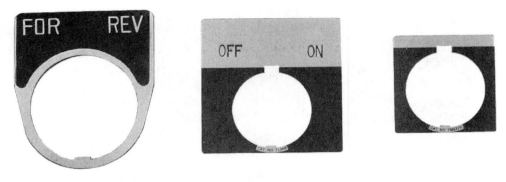

**Figure 3-8A** Nameplates for oil-tight units. *(Courtesy of Eaton Corporation.)*

**Figure 3-8B** Oil-tight units. *(Courtesy of Allen-Bradley, a Rockwell International Company.)*

## 3.7 Circuit Applications

We can now show a complete circuit, using the symbols illustrated previously. The electrical power source, electrical control source disconnecting device, protection, and an output or load that can be controlled are shown in Figures 3-9A, 3-9B, 3-9C, and 3-9D. The complete circuit is shown in Figure 3-9A. After this diagram, the primary power source and control power source are not always shown in circuit diagrams, as the control voltage available between the two vertical lines (sides of the ladder) is generally 120 V ac.

All control circuits are assigned numbers for convenience in checking the circuit. The number changes each time the circuit passes through a component.

This practice is discussed in more detail in Chapter 20, "Designing Control Systems for Easy Maintenance."

In Figure 3-9A, the selector switch is placed in the ON position, closing the contact between 1 and X1. This puts 120 V ac on the two vertical lines 1 and 2. The red pilot light can now be energized by operating the red light push-button switch. The red light remains energized as long as the push-button switch is operated.

Figure 3-9B shows the use of a maintained contact push-button switch. Operation of the green light push-button switch energizes the green pilot light. The green light push-button switch will remain in the operating position until the RESET push button is operated.

In the circuit in Figure 3-9C, a three-position, four-contact selector switch is used. The symbol shows that contacts 1 and 3 are closed in position 1. All contacts are open in position 2. Contacts 2 and 4 are closed in position 3.

The pilot light load is connected so that the red and amber lights are energized in position 1 of the selector switch, and the green and white lights are energized in position 3. None of the pilot lights is energized in position 2, as all of the selector switch contacts are open.

Figure 3-9D shows additional control added to the circuit in Figure 3-9C. Momentary contact push-button switches are added to each pilot light circuit. Now, in addition to selecting either position 1 or position 3 of the selector switch, a push-button switch must be operated to energize a pilot light.

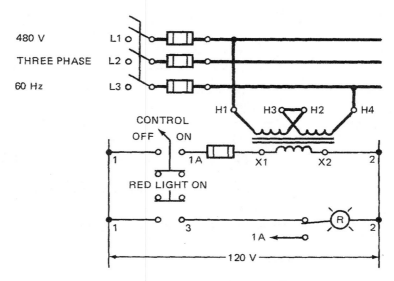

**Figure 3-9A**   Control circuit showing an electrical power source, electrical control source disconnecting device, protection, and a controllable output or load. Conductors carrying load current at line voltage are denoted by heavy lines.

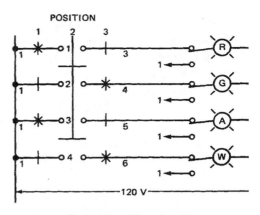

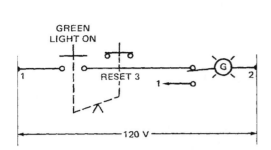

Contacts 1 and 3 are closed in
position 1. All contacts are
open in position 2. Contacts
2 and 4 are closed in position 3.

***Figure 3-9B***  **Control circuit with a**
**maintained contact push-button switch.**

***Figure 3-9C***  **Control circuit showing a three-**
**position, four-contact selector switch.**

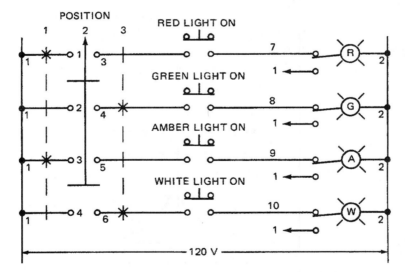

To energize the red light: Operate selector switch to position 1. Operate red
light ON push-button switch.

To energize the green light: Operate selector switch to position 3. Operate
green light ON push-button switch.

To energize amber light: Operate selector switch to position 1. Operate amber
light ON push-button switch.

To energize white light: Operate selector switch to position 3. Operate white
light ON push-button switch.

***Figure 3-9D***  **Circuit showing additional control added to Figure 3-9C.**

## 3.8  Annunciators

The visual annunciator system as produced by Ronan displays the plant or process status by lighting individual windows identifying process functions. Input to the annunciator is derived from dry or live field contacts or analog signals indicating the process conditions or status. Integral alarm logic modules operate the window lights as directed by any Instrument Society Association (ISA) or custom sequence.

The visual annunciator system architecture, configured so that the electronic modules are remote from the window display, is ideally suited to large systems in power plants and process plant control rooms. The modular logic is housed in a relay rack-mount or surface-mount chassis, connected via multiconductor wiring to the lamp display. Connection to the display and logic chassis may be made by means of individual wires to terminals or via prefabricated cable assemblies with multipin connectors. The split architecture provides ease of maintenance with access to the electronics without interfering with the control room operation.

The lamp displays are available in a large variety of construction types, overall dimensions, and window sizes. The lamp cabinets can be configured any number of windows high or wide, with window sizes ranging from 0.9 inches high by 1.6 inches wide to 3 inches high by 5 inches wide.

The semigraphic series displays the custom process or plant layout with light indicators providing operating status (or alarm) of pumps, valves, hoppers, tank levels, etc. The unique construction of the Ronan graphic allows infinite resolution in positioning the status/alarm light on the back plane. The graphic layout, contained between a clear front plexiglas and a rear diffuser, can easily be modified in case of a process change or addition of equipment. The indicator lights can be driven by either solid-state or relay-type remote logic alarm system. See Figures 3-10 and 3-11 for typical annunciators.

## 3.9  Light-Emitting Diodes

In many digital systems the output unit is in the form of a readout display, which may call for a readout of several digits. Multidigit displays are available using light-emitting diodes (LEDs). They are also used as indicators of both outgoing and incoming signals. The symbol for the LED is shown in Figure 3-12, which shows a resistor ($R$) that prevents the current from exceeding the maximum current rating of the diode. The resistor has a voltage on the left side of $V_B$ and a voltage on the right side of $V_C$. The voltage across the resistor is $V_B - V_D$. Therefore, from Ohm's law:

$$\text{Series current} = \frac{V_B - V_D}{R}$$

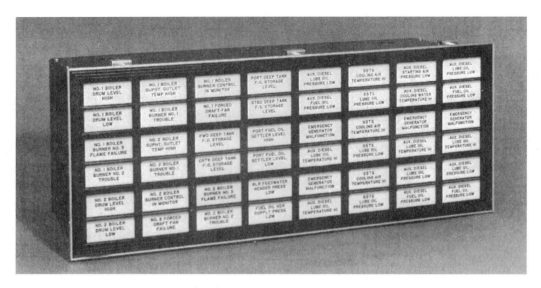

**Figure 3-10**    **6 × 8 annunciator-amp cabinet.**    *(Courtesy of Ronan Engineering Company.)*

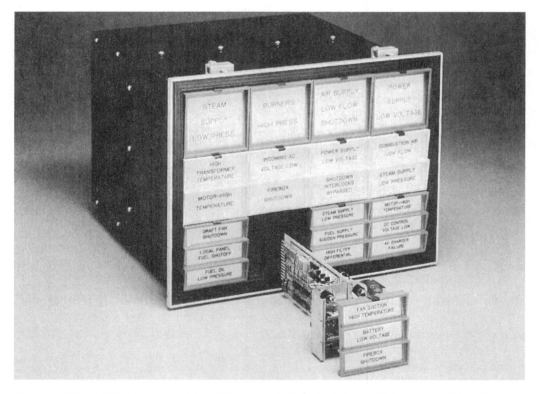

**Figure 3-11**    **Solid state annunciator with on board alarm logic for multiple alarm points.**    *(Courtesy of Ronan Engineering Company.)*

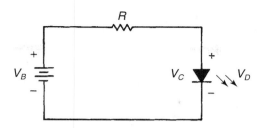

*Figure 3-12* **Circuit for an LED.**

The series current will range from 10 mA to 50 mA with a voltage drop of 1.5 V to 2.5 V.

## 3.10 Membrane Switches

The membrane switch as produced by XYMOX Technologies is a normally open, momentary contact, push-to-operate-type device. Layers of insulating materials, conductive coatings, decorative graphics, and adhesives are combined to form a completely sealed switch ideally suited to switching low-power logic signals. Multiple contacts and interconnections can be employed to provide complex switching networks that are attractive, reliable, and cost-effective.

There are two types of switches available: flexible and rigid. Outwardly, most flexible membrane switches appear to be the same. However, their construction must differ depending on the environment in which the switch must function. The construction must be chosen with attention to all factors in its operating environment, including maximum temperature, quality of atmosphere, and method of interconnection.

To satisfy these conditions three types of construction are offered. The first type consists of three layers laminated together to form the switch. These three layers are an upper circuitry layer called the *flex* and a lower circuitry layer called the *stable*. These layers are separated by a *spacer* layer. All three layers are composed of polyester substrates that are held together by pressure-sensitive acrylic adhesives. This switch is internally vented; that is, the switch cavities are vented together to create a passage for air travel between switches. In this way, no external venting is necessary. The switch package is totally sealed from outside air and, therefore, contamination. In addition, since the switch is sealed, all the circuitry can be placed on one layer, making termination on a single tail a standard. However, this construction has one drawback—an upper temperature limit of 65°C. Operation above this level, especially thermal cycling, will cause the polyester to "breathe." When brought back to ambient temperature, a pressure differential will exist, causing sensitive or, in the worst cases, collapsed switches. This failure mode is permanent and can only be corrected by externally venting the switch.

A second type of construction is used when the operation requires temperatures above 65°C. The switch is externally vented to eliminate the collapsed switch failure mode. Any time a switch is vented, certain precautions must be taken to protect the electrical integrity of the switch. One of the greatest concerns in a vented switch is the risk of contaminants in the switch cavity. Among these is condensed moisture, which can cause silver migration. To minimize this occurrence, several construction techniques are used. First, the circuitry is separated by polarity so that circuits of different potential are on different substrates. Second, exposed silver is encapsulated in a carbon or graphite overcoat that inhibits the migration process. To accomplish these design parameters the part must have two tails, one exiting from each circuit layer. These can overlap to form a single tail to minimize connection costs.

The third type of construction is less costly because it employs a single layer that is folded back on itself to form the switch. It is capable of being supplied in four different levels of completion. Level 1 is simply a single layer of printed and die-cut circuitry. Level 2 is printed and die cut into a switch configuration but is shipped in the unfolded format. Level 3 is folded and sealed using a printed spacer and an adhesive seal. Level 4 is a folded construction with an ultrasonic bond to ensure the seal. These switches in the upper levels of completion are limited environmentally to a noncondensing environment. They are externally vented but do not have protection parameters built into them like those in level 2. They are designed specifically to perform in applications in which a low-cost, quality switch will be used in a controlled environment.

The rigid type of construction consists of a graphic overlay laminated to a spacer and printed circuit board. Conductive pads are deposited on the underside of the overlay and make electrical contact with conductive tracks on the printed circuit board when the key areas are depressed. The printed circuit board provides rigidity. Terminations available include stake pins, header strips, connectors, and solder pads. See Figures 3-13A and 3-13B for a group of typical membrane switches.

Transparent LED areas are provided in switch strip and keypad component units. They can be used with custom face plates having LED windows in the keys.

Various forms of feedback are compatible with XYMOX stock membrane switches. A device that operates upon the depression of each switch will provide audible feedback and is readily available from a number of sources. Visual feedback can be provided in the form of backlighted areas.

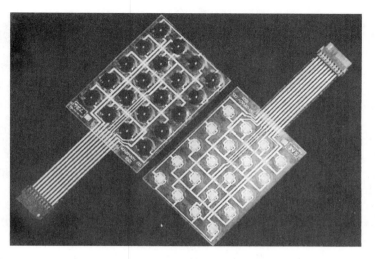

**Figure 3-13A**    **Membrane switch circuit board.** *(Courtesy of XYMOX Technologies Inc.)*

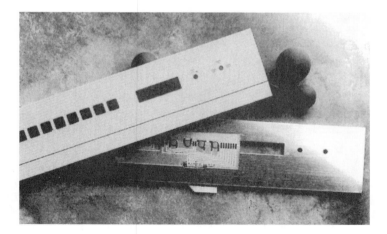

**Figure 3-13B**    **Membrane switch assembly.** *(Courtesy of XYMOX Technologies Inc.)*

## 3.11  Liquid Crystal Displays*

There are three types of liquid crystal displays (LCDs) available:

1. Basic twisted nematic
2. Supertwist
3. Active matrix

*Information courtesy of Mary Tilton, Standish Industries.

**3.11.1 Basic Twisted Nematic LCD**    A basic twisted nematic LCD consists of a layer of liquid crystal material supported by two glass plates. The liquid crystal material is a mixture of long, cylindrically shaped molecules with different electrical and optical properties depending on direction.

On the inner surfaces of the glass plates are transparent electrodes, which are patterned to form the desired visual image. The inner surfaces are coated with a polymer, which is rubbed so that the liquid crystal material at one surface lies perpendicular to the other. Across the film of liquid crystal, the molecules form a 90-degree twist.

On the outer surface of the glass plates, polarizers are placed parallel to the liquid crystal orientation and perpendicular to each other.

In the OFF state, light entering the first polarizer is guided by the liquid crystal layer twist to the second polarizer, through which it is transmitted. See Figure 3-14.

When the cell is energized, the liquid crystal material is aligned with the electric field; light transmitted through the first polarizer is blocked by the second polarizer, forming a dark image. The effect may be reversed if the polarizers are placed parallel to each other and a light image on a dark background is formed.

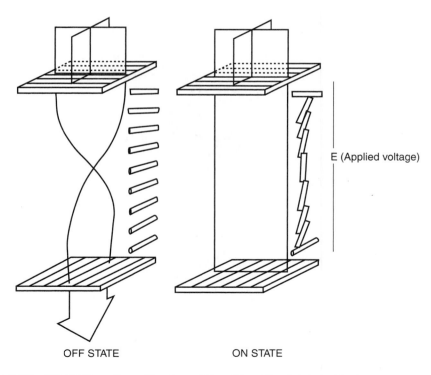

OFF STATE                    ON STATE

***Figure 3-14***    **TNLCD Operation.** *(Courtesy of Mary Tilton, Standish Industries.)*

Twisted nematic LCDs have a preferred viewing direction because the liquid crystal molecules are not truly parallel to the applied field in the energized state. Increasing the drive voltage alleviates the situation somewhat.

Improvements have been made in the viewing angle through the development of materials that allow the construction of first minimum LCDs. Figure 3-15 helps to illustrate this principle. It shows transmission versus cell gap times birefringence of a normally dark twisted nematic LCD for a fixed wavelength. The cell gap is the thickness of the liquid crystal layer, measured in microns. Birefringence is a measurement of the light scattering property of the liquid crystal material.

To make a display with the highest contrast ratio possible, the cell must be designed so that the product of the cell gap and birefringence falls in a minimum, as shown in Figure 3-15. The first three minima occur at about 0.5, 1.1, and 1.7. Typical liquid crystal mixtures have birefringence values of 0.10 to 0.18. Therefore, cell gaps from 6 to 11 microns fit the second minimum of Figure 3-15. To achieve a display in the first minimum requires cell gaps from 3 to 5 microns, which are difficult to achieve in production.

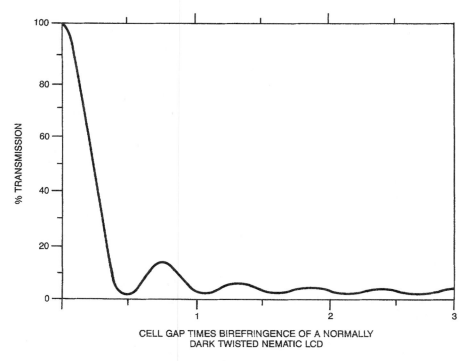

**Figure 3-15**   **Gooch-Tarry curve for wavelength equal to 550 nanometers.**   *(Courtesy of Mary Tilton, Standish Industries.)*

New materials with lower birefringence values (0.06 to 0.08) allow cells to be built in the first minimum with conventional cell spacings. The effect of these materials on the viewing angle is quite dramatic. With proper drive voltages, the cells may be viewed at all angles with good contrast. This is useful for instrument displays because viewing positions are not always predictable.

**3.11.2 Supertwist LCD**   The twisted nematic LCDs may be driven in a time-multiplexed fashion to increase the amount of information displayed. There is, however, a limit to the number of lines multiplexed because the change in transmission with voltage is not immediate. The steeper the slope of this curve, the more highly multiplexible the material.

Still, there is a limit for twisted nematic LCDs, which shows up in reduced contrast and limited viewing angle.

To achieve more highly direct multiplexed displays, supertwist technology is employed. In this type of display, the liquid crystal material undergoes more than a 90-degree twist from plate to plate; typical values range from 180 degrees to 270 degrees.

The polarizers in this case are not mounted parallel to the liquid crystal at the surface but rather at some angle. The cell, therefore, does not work on a "light-guiding" principle, as in the twisted nematic LCDs, but on a *birefringence* principle. Birefringence is a measurement of the light-scattering property of the liquid crystal material.

The position of the polarizers, the cell thickness, and the birefringence of the liquid crystal are carefully chosen to result in a particular color in the OFF state, usually a yellow-green, which gives the highest contrast ratio. To achieve a colorless appearance, a film of stretched polymer is inserted between the glass and polarizer.

**3.11.3 Active Matrix LCD**   For displays of high information content, such as a dot matrix, the viewing angle and contrast problems of directly multiplexed displays would disappear if each element were directly addressable. Because of geometry and space limitations, direct address is not possible through electrode design. The active matrix approach deals with this problem by incorporating a switch at each picture element (pixel).

An active matrix display is constructed by building a grid of switching elements on the inside surface of the glass substrate. The switching elements may be two terminal devices, such as diodes, or three terminal devices, such as transistors.

Each display pixel has a switch located at the pixel. The switches and pixels are arranged in rows and columns, with the gates of the switches of one row tied together for connection to an external voltage source. The sources of each column of switches are also tied together for external addressing.

When voltage is applied to a row of a matrix, the gates of that row allow data to be entered at the sources. The voltage at the source either turns the elements on or not. The voltage is then removed from that row, closing those switches, and the next row is addressed.

This cycle is applied to each row, then repeated, starting with the first row. Each pixel, therefore, is individually addressed.

The advantage to this method is that maximum contrast is achieved for the liquid crystal material without increasing viewing-angle dependence. Also, voltages that only partially turn on the pixels may be used to achieve gray-scale effects.

A full color display may be constructed by using color filters. An array of red, green, and blue elements is fabricated on the plate opposite the active matrix plate. Each element is red, green, or blue, and the combination of addressed elements provides a full color spectrum.


## Recommended Web Links

Students are encouraged to view the following Web sites as a supplement to the concepts presented in this textbook. Review and analyze the array of products that are available for electrical control applications. Many of these sites offer technical information that will help in converting the principles to practical applications. To view catalogs, your PC may require Adobe Acrobat Reader software.

### Control Devices

Allen-Bradley
www.ab.com/industrialcontrols/products/push_buttons
Review: Buttons and Lights

General Electric
www.geindustrial.com/cwc/products?famid=28
Review: Buttons and Lights

Signaworks
www.signaworks.com/pushbuttons.html
Review: Buttons and Lights

Square D
ecatalog.squared.com
Review: Product Catagories, Pushbuttons and Operator Interface

## Annunciators

Ronan Engineering
www.ronan.com
Review: Alarm Display Products, Annunicators

Rochester Instruments
www.rochester.com/Products/RIS/annuncia.htm
Review: The assortment of annunciators and alarm displays

## LED and LCD Displays

Lampex
www.lampex.com
Review: Products

Purdy Electronics
www.purdyelectronics.com
Review: Products, LCD, LED

# Achievement Review

1. Describe how you would panel mount an oil-tight unit.

2. Why is the low-voltage bulb in the pilot light preferred?

   a. It is much smaller in size.
   b. It will withstand vibration.
   c. It will not consume as much electrical energy.

3. Draw the symbol for a mushroom head push-button switch (one contact block, two contacts; one normally open, one normally closed).

4. Are lenses for pilot lights available in more than one color? If so, why?

5. What is the advantage of panel mounting a contact block?

   a. The mounting provides space for terminal block installation.
   b. The mounting allows for more space on the panel.
   c. The mounting is required by electrical standards.

6. Draw the symbol for a red push-to-test light.

7. Draw the symbol for a three-position, four-contact selector switch. Contacts 1 and 3 are closed in position 1, all contacts are open in position 2, and contacts 2 and 4 are closed in position 3.

8. Explain how the push-to-test pilot light functions. Draw the complete circuit diagram.

9. Draw a circuit showing a two-position selector switch and a green push-to-test pilot light. The pilot light is to be energized when the selector switch is placed in the ON position.

10. What problem might you encounter if you use the first type of XYMOX construction in an environment with temperatures above 65°C?

11. In a rigid type of construction, what contributes to the rigidity in a membrane switch?

12. What uses are made of the LEDs in a programmable controller?

13. What are some of the uses for an annunciator?

14. What does liquid crystal material consist of?

15. What three types of liquid crystal display are available?

# Chapter 4 Relays

## Objectives

After studying this chapter, you should be able to:

- Identify two main uses for the control relay.
- Show how the control relay is constructed mechanically.
- List the three published ratings for relays.
- Explain why silver is used in relay contacts.
- Discuss several factors involved with relay operation, such as contact bounce, overlap contacts, contact wipe, and split or bifurcated contacts.
- Describe the interlock circuit and explain why it is used.
- Draw the symbols for the relay coil, relay contacts, and time-delay relay contacts.
- Explain the difference between inrush and holding current in a relay coil.
- Describe how the latching relay operates.
- List the basic uses for the contactor.

## 4.1 Control Relays and Their Uses

The *relay* is basically a communication carrier. It is used in the control of fluid power valves and in many machine sequence controls such as for drilling, boring, milling, and grinding operations. Relays are also used as power amplifiers.

The relay is an electromechanical device with both stationary and moving contacts. The major parts of the relay are shown in Figures 4-1A and 4-1B. The moving contacts are attached to the plunger. Contacts are referred to as normally open (NO) and normally closed (NC). When the coil of the relay is energized, the plunger moves through the coil, closing the normally open contacts and opening the normally closed contacts. Figure 4-1A shows the contacts in the normal or

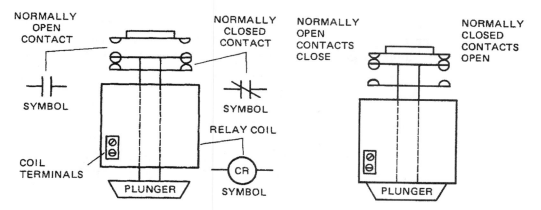

**Figure 4-1A** Deenergized condition of relay showing contact position.

**Figure 4-1B** Energized condition of relay. With coil energized, magnetic force pulls the plunger up, operating the contacts.

deenergized condition of the coil, as well as the symbols for the relay coil and contact.

When the coil of the relay is connected to a source of electrical energy, the coil is energized. When the coil is energized, the plunger moves up, as shown in Figure 4-1B, due to a force produced by a magnetic field. The distance that the plunger moves is generally short—about 1/4 inch or less.

There is a difference in the current in the relay coil from the time the coil is first energized and when the contacts are completely operated. When the coil is energized, the plunger is in an out position. Due to the open gap in the magnetic path (circuit), the initial current in the coil is high. The current level at this time is known as *inrush current*. As the plunger moves into the coil, closing the gap, the current level drops to a lower value called the *sealed current*. The inrush current is approximately six to eight times the sealed current.

Coils are available to cover most standard voltages from 24 V to 600 V. (The standard voltage used in machine control is 120 V.) Relay coils are typically made of a molded construction to reduce moisture absorption and increase mechanical strength.

The level of voltage at which the relay coil is energized to move the contacts from their normal, unoperated position to their operated position is called *pick-up voltage*. After the relay is energized, the level of voltage on the relay coil at which the contacts return to their unoperated condition is called *drop-out voltage*.

Coils on electromechanical devices such as relays, contactors, and motor starters are designed so as not to drop out (deenergize) until the voltage drops to a minimum of 85% of the rated voltage. The relay coils also will not pick up (energize) until the voltage rises to 85% of the rated voltage. This voltage level is set by the National Electrical Manufacturers Association (NEMA). As a rule, 85% is found to be a conservative figure. Most electromechanical devices will not drop out until a lower voltage level is reached. Also, most electromechanical devices will pick up at a lower rising voltage level. Generally, coils on electromechanical devices will operate continuously at 110% of the rated voltage without damage to the coil.

The two important parts in a relay are the coil and contacts. Of these, the contacts generally require greater consideration in practical circuit design.

There are some single-break contacts used in industrial relays. However, most of the relays used in machine tool control have double-break contacts. The rating of contacts can be misleading. The three ratings generally published are:

1. Inrush or "make contact" capacity
2. Normal or continuous carrying capacity
3. The opening or break capacity

For example, a typical industrial relay has the following contact ratings:

- 10-A noninductive continuous load (ac)
- 6-A inductive load at 120 V (ac)
- 60-A make and 60-A break inductive load at 120 V (ac)

A resistance is an example of a noninductive load, such as would occur in a resistance unit used as a heating element. An inductive load is basically a coil, such as a solenoid, contactor coil, or motor starter coil.

The point to remember is that in determining the contact rating, it must be clear what rating is given.

Relay contacts are usually silver or a silver-cadmium alloy. This material is used because of the excellent conductivity of silver. Silver oxide, which forms on the contacts is also a good conductor. Adding a small amount of cadmium to the silver increases the interrupting capacity of the contacts. Some manufacturers offer gold-plated contacts as optional features. Gold plating improves shelf life and provides improved contact reliability in low-energy, low-duty-cycle applications. The circuit designer should consider the following points in the use of control relays:

- *Changing contacts normally open to normally closed, or vice versa.* Most machine tool relays have some means of making this change, with either a simple flip-over contact, to replacement contacts, or spring loaded switch.
- *Universal contacts.* The term *universal* refers to relays in which the contacts may be NO or NC. Only one type may be selected in the use of each contact, but not both types.
- *Split, or bifurcated, contacts.* As the name implies, a split or bifurcated contact is divided into two parts. This division provides for double the

contact make points, improving reliability of contact and reducing contact bounce.

- *Contact bounce.* All contacts bounce (spring apart or vibrate) on closing. The problem is to reduce the bounce to a minimum. In rapid-operating relays, contact bounce can be a source of trouble. For example, an interlock in a transfer circuit can be lost by a contact opening momentarily. The use of split contacts, overlapping contacts, other type of antibounce contacts or a circuit change may help.
- *Overlap contacts.* To overlap contacts, one contact can be arranged to operate at a different time relative to another contact on the same relay. For example, the NO contact can be arranged to close before the NC contact opens. This arrangement is called "make before break."
- *Contact wipe.* This factor results from the relative motion of the two contact surfaces after they make contact, that is, one contact wipes against the other. One advantage to be gained from this motion is that it induces self-cleaning of the contacts.

Panel mounting space has been a problem for the user of relays. With increasing complexity of control applications, the number of required relays has grown and thus more panel space is required. The need for smaller relays led to the development of miniature 300-V relays and compact 600-V relays. This change along with a modular form of construction results in a compact relay assembly. Relays shown in this book are of the 600-V design. Figure 4-2 shows a typical four-pole, 600-V control relay, with plug-in contact cartridges. (The reference to 300-V and 600-V relays applies to the voltage being carried by the contacts. The standard relay coil voltage in the machine control field is 120 V.)

**Figure 4-2** Industrial control relays. *(Courtesy of Square D/Schneider Electric.)*

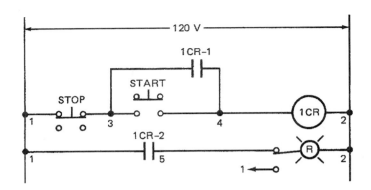

*Figure 4-3* Control circuit using a relay with two NO contacts.

These relays are generally available with up to 12 contacts in combinations of NO and NC contacts. They may be either convertible or fixed. In most cases, the relays can be combined with a pneumatic timing head or mechanical latch. However, when adding the pneumatic timing head, the relay is restricted to four instantaneous contacts.

Figure 4-3 shows a circuit diagram using a relay with two NO contacts. One contact, 1CR-1, is used as an interlock around the START push button. Thus an *interlock circuit* is a path provided for electrical energy to the load after the initial path has been opened. The second relay contact, 1CR-2, is used to energize a light. Remember that when a relay coil is energized, the NO contacts close. The circuit can be deenergized by operating the STOP push-button switch.

Figure 4-4 shows the addition of a selector switch, fuse, pilot light, and a second relay to the circuit shown in Figure 4-3. When the selector switch is operated to the ON position, electrical energy is available at the two vertical sides of the circuit. The green light is energized, indicating that the operation has been completed. One additional relay contact is added in the circuit from relay 1CR. This contact closes when the relay 1CR is energized and it, in turn, energizes a second relay coil, 2CR. The operating circuit can be deenergized by operating the STOP push-button switch.

Starting in this chapter it can be observed that a numbering system has been introduced in the circuits. This system is used in all the remaining circuits in the text. The two vertical sides of the diagram, representing the source of electrical energy (control voltage), are numbered 1 and 2. While practice may vary with the manufacturer, this text uses 1 for the hot (protected) side and 2 for the common side (generally grounded). There may be another variation of this system in the industry. See Chapter 17 and Appendix E.

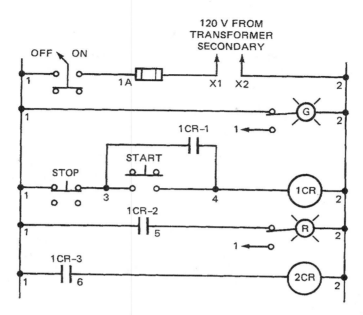

*Figure 4-4* Control circuit from Figure 4-3 expanded.

Note that starting at one side or the other on each line of the circuit, the number changes each time you move through a contact or coil. Thus, each contact (push-button switch, selector switch, limit switch, pressure switch, temperature switch, and so on) will have two numbers, one for the incoming conductor and one for the outgoing conductor. Two numbers are also used for all coils (relays, timers, contactors, motor starters, and so on).

While this practice does not seem to have any importance in the operation of the circuit, it is invaluable when the components are wired on a machine in accordance with the specific circuit design.

This practice is not only a must for electromechanical control but also for solid-state and programmable controllers.

## 4.2  Timing Relays

The pneumatic timing head can be added to the control relay. The pneumatic timing relay is also available either as an add-on unit to the 300-V or 600-V relay line or as a separate unit.

In circuit work it is an advantage to have a timing contact as well as instantaneous contacts from the same energized relay coil. The timing contact can be arranged to delay after energizing or deenergizing the coil.

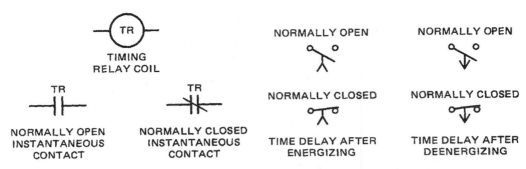

**Figure 4-5** Symbols for the timing relay coil and instantaneous contact.

**Figure 4-6** Symbols for timing contacts.

The symbols for the timing relay coil and the instantaneous contact are shown in Figure 4-5. The final schematic description comes with the symbols for the timing contact. There are four separate symbols:

1. NO time delay after energizing
2. NC time delay after energizing
   (These contacts are sometimes referred to as *ON time-delay contacts.*)
3. NO time delay after deenergizing
4. NC time delay after deenergizing
   (These contacts are sometimes referred to as *OFF time-delay contacts.*)

With the time-delay coil energized, the ON time-delay contacts remain in their unoperated state through the preset time period. At the end of this period, the contacts change to their operated state. They remain in this state until the time-delay coil is deenergized.

The OFF time-delay contacts change to their operated state immediately when the time-delay relay coil is energized. They remain in the operated state during the period the time-delay coil is energized.

The contacts return to their unoperated condition after a preset time delay following the deenergizing of the time-delay coil. See Figure 4-6 for the symbols.

Figure 4-7A shows a time-delay attachment, and Figure 4-7B shows a complete time-delay relay.

The adjustable timing range on the 300-V or 600-V relay line is on the order of 0.1 second to 60 seconds. Accuracy varies quite widely, depending on the size, type, and manufacturer. It is in the range of 10%.

Figures 4-7C, 4-7D, and 4-7E display solid-state timing relays. Timing indication is provided by an LED that flashes during timing, glows steadily after timing, and is off when the timer is deenergized. Some of the features include convertible ON–OFF delay, timing from 0.05 second to 10 hours, repeat cycle version, ±1% repeat accuracy.

**Figure 4-7A** Time-delay attachment for a pneumatic time-delay unit. *(Courtesy of Allen-Bradley, a Rockwell International Company.)*

**Figure 4-7B** Complete ac timing relay. *(Courtesy of Square D/Schneider Electric.)*

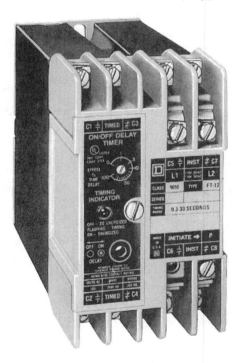

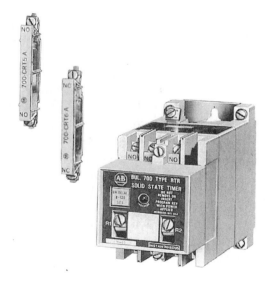

**Figure 4-7C** Solid-state timing relay with fixed timing range. *(Courtesy of Square D/Schneider Electric.)*

**Figure 4-7D** Solid-state timing relay with replacement contacts. *(Courtesy of Allen-Bradley, a Rockwell International Company.)*

***Figure 4-7E*** **Solid-state timing relay with replacement timing units.** *(Courtesy of Allen-Bradley, a Rockwell International Company.)*

Figure 4-8A shows a timing relay added to the circuit shown in Figure 4-4. The timing relay coil replaces the relay coil 2CR. An ON time-delay contact (NC) is placed in series with relay coil 1CR. The sequence of operations proceeds as follows:

1. Operate the START push-button switch.
2. Relay coil 1CR is energized.
   a. Relay 1CR contact 1CR-1 closes, interlocking around the START push-button switch.
   b. Relay contact 1CR-2 closes, energizing the green pilot light.
   c. Relay contact 1CR-3 closes, energizing the time-delay relay coil 1TR.
3. After a preset time on 1TR, the NC ON time-delay contact 1TR opens, deenergizing relay coil 1CR.
   a. Relay 1CR contact 1CR-1 opens, opening the interlock circuit around the START push-button switch.
   b. Relay 1CR contact 1CR-2 opens, deenergizing the green pilot light.
   c. Relay 1CR contact 1CR-3 opens, deenergizing the time-delay relay coil 1TR.

Figure 4-8B shows another timing relay and control relay added to the circuit shown in Figure 4-8A. The new timing relay 2TR will have one NO contact

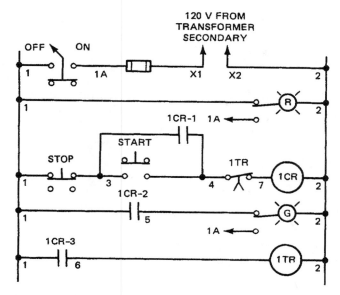

**Figure 4-8A** Control circuit from Figure 4-4 with a timing relay added.

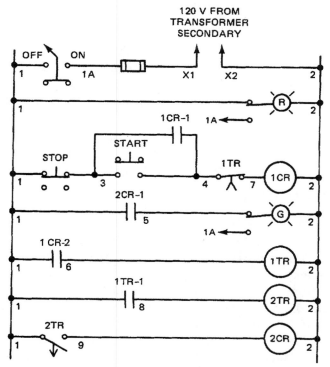

**Figure 4-8B** Control circuit from Figure 4-8A with additional timing and control relays.

arranged for time delay after deenergizing (OFF time). The second relay coil is 2CR. The sequence of operations proceeds as follows:

1. Operate the START push-button switch.
2. Relay coil 1CR is energized.
   a. Relay 1CR contact 1CR-1 closes, interlocking around the START push-button switch.
   b. Relay 1CR contact 1CR-2 closes, energizing timing relay coil 1TR.
   c. Instantaneous contact 1TR-1 on timing relay 1TR closes, energizing timing relay coil 2TR.
   d. The OFF delay contact 2TR closes, energizing relay coil 2CR.
3. The 2CR relay contact 2CR-1 closes, energizing the green pilot light.
4. Preset time on timing relay 1TR expires.
   a. Timing contact 1TR opens, deenergizing relay 1CR.
   b. The 1CR relay contact 1CR-1 opens, opening the interlock circuit around the START push-button switch.
   c. The 1CR relay contact 1CR-2 opens, deenergizing timing relay 1TR.
   d. Instantaneous contact on timing relay 1TR opens, deenergizing timing relay 2TR.
5. After a preset time delay as set on 2TR expires, OFF time-delay contact 2TR opens, deenergizing relay 2CR.
   a. Relay contact 2CR-1 opens, deenergizing the green pilot light.

Another timing relay is now available with up to two timed and two instantaneous contacts in various NO and NC combinations. Convenient program keys (8) provide a selection of four timing ranges in either ON-delay or OFF-delay mode. Timers are available with either a self-contained adjustment potentiometer or with provision for remote adjustment. Timers feature an LED (light-emitting diode) indicator light, which is off, flashing, or on to indicate timer operating status. Figure 4-7D illustrates this timing relay.

## 4.3  Latching Relays

A mechanical latching attachment can be installed on the control relay (Figure 4-9A). This setup provides an interesting variation of the control capabilities of the relay. The latching relay is electromagnetically operated. It is held by means of a mechanical latch. By energizing a coil of the relay, called the *latch coil,* the relay operates and results in the relay NO contacts closing and the NC contacts opening. The electrical energy can now be removed from the coil of the relay (deenergized), and the contacts remain in their operated condition. Now a second coil on the relay, the *unlatch coil,* must be energized in order to return the contacts to their unoperated condition. This arrangement is often referred to as a *memory relay.*

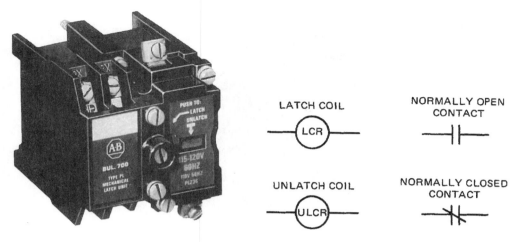

**Figure 4-9A** Latching attachment for a relay.
*(Courtesy of Allen-Bradley, a Rockwell International Company.)*

**Figure 4-9B** Symbols for the latching relay.

The latching relay has several advantages in electrical circuit design. For example, it may be necessary to open or close contacts early in a cycle. At the same time, it may also be desirable to deenergize the section of the circuit responsible for the initial energizing of the relay latch coil. Later in the cycle, the unlatch coil can be energized to return the contacts to their original or unoperated condition. The circuit is then set up for the next cycle.

The symbol for the latching relay is shown in Figure 4-9B. To further explain the symbols:

- Contacts should be shown in the unlatched condition; that is, as if the unlatch coil were the last one energized.
- The latch and unlatch coils are on the same relay and always have the same reference number.
- The contacts are always associated with the latch coil as far as reference designations are concerned.

Another use for the latching-type relay involves power failure, when it may be necessary that the contacts remain in their operated condition during the power-off period. Conditions in this case are the same after the power failure as they were before.

If quietness of operation is desired in a long cycle, this can be provided. The coil can be deenergized, thus eliminating the usual hum. Figure 4-9C illustrates a typical latching relay. The conventional two-coil circuit for the latching relay is shown in Figure 4-9D.

*Figure 4-9C* **An ac latching relay.** *(Courtesy of Square D/Schneider Electric.)*

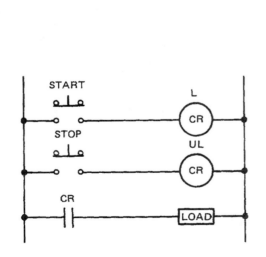

*Figure 4-9D* Control circuit for a latching relay.

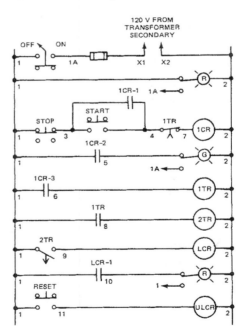

*Figure 4-9E* Control circuit.

Refer again to the circuit shown in Figure 4-8B. The latch coil of a latching relay replaces the coil of control relay 2CR.

In Figure 4-9E, the closing of timing contact 2TR energizes the latch coil LCR. NO contact LCR-1 closes, energizing a red pilot light. When the timing relay 2TR times out and the timing contact opens, the latch coil is deenergized. However, the light remains energized until the RESET push-button switch is operated, energizing the unlatch coil ULCR.

## 4.4 Plug-In Relays

Some industrial machine relays are available as plug-in types. They are designed for multiple switching applications at or below 240 V. Coil voltages span standard levels from 6 V to 120 V. They are available for ac or dc. Mounting can be obtained with a tube-type socket, square-base socket mounting, or flange mounting using slip-on connectors.

A typical plug-in relay is shown in Figure 4-10.

The plug-in relay has a distinct advantage in allowing relays to be changed without disturbing the circuit wiring. In critical operations in which the relay service is very hard and downtime is a premium, the plug-in relay may have some advantages. Assessing the actual operating conditions in specific cases is the best way to determine their need.

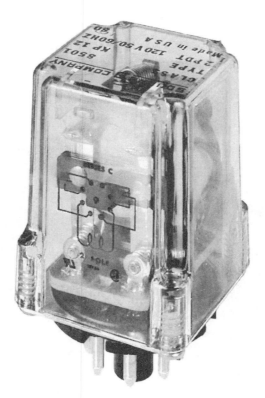

*Figure 4-10* **Plug-in relay.** *(Courtesy of Square D/Schneider Electric.)*

## 4.5 Contactors

The *contactor,* in general, is constructed similarly to the relay. Like the relay, it is an electromechanical device. The same coil conditions exist in that a high in-rush current is available when the coil of the contactor is energized. The current level drops to the holding or sealed level when the contacts are operated. Generally the contactor is supplied in two-, three-, or four-pole arrangements. The coil of the contactor is generally energized at 120 V. The major difference is in the size range available with contactors. Contactors capable of carrying current in the range of 9 A through approximately 2250 A are available. For example, a size 00 contactor is rated at 9 A (200–575 V); a size 9 contactor is rated at 2250 A (200–575 V).

One normally open auxiliary contact is generally supplied as standard on most contactors. This contact is used as a holding contact in the circuit, for example, around a normally open push-button switch. Additional normally open and normally closed auxiliary contacts can be obtained as an option. They can be supplied with the contactor from the manufacturer or ordered as a separate unit and mounted in the field.

Figures 4-11A and 4-11B show two typical electromechanical contactors. Figure 4-11A is a size 1 with a 27-A rating; Figure 4-11B is a size 9 with a 2250-A rating.

The use of the auxiliary contact, which is generally rated at 10 A, is shown in Figure 4-11C. It can be seen that the auxiliary contact is used like the relay in-

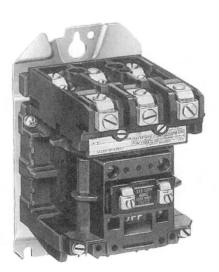

**Figure 4-11A** Size 1 contactor. *(Courtesy of Allen-Bradley, a Rockwell International Company.)*

**Figure 4-11B** Size 9 contactor. *(Courtesy of Allen-Bradley, a Rockwell International Company.)*

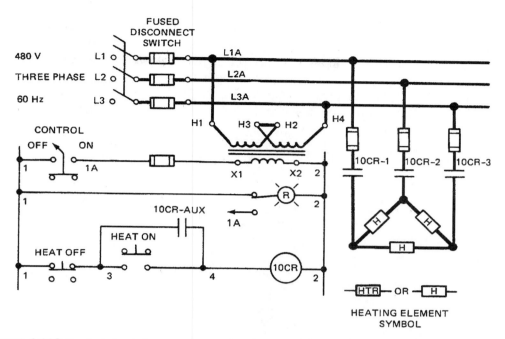

*Figure 4-11C* Control circuit for an electromechanical contactor. Conductors carrying load current at line voltage are denoted by heavy lines.

terlock or holding circuit. In the circuit in Figure 4-11C, the light is connected across the source of electrical energy (circuit lines 1 and 2). The light in the circuit in Figure 4-11C, therefore, is energized when the control ON–OFF selector switch is operated to the ON position.

The balance of the sequence is as follows:

1. Operate the HEAT ON push-button switch.
2. Coil of contactor 10CR is energized.
3. Power contacts 10CR-1, 10CR-2, 10CR-3 close, energizing the heating elements at line voltage.
4. Auxiliary contact 10CR-AUX closes, interlocking around the HEAT ON push-button switch.
5. Operate the HEAT OFF push-button switch, deenergizing the coil of 10CR contactor.
   a. Contacts 10CR-1, 10CR-2, 10CR-3 open, deenergizing the heating elements.
   b. Auxiliary contact 10CR-AUX opens, opening the interlock circuit around the HEAT ON push-button switch.

In addition to the conventional electromechanical contactor, a mercury-to-metal contactor is available. Figure 4-12 shows a cross section through such a contactor. The load terminals are isolated from each other by the glass in the hermetic seal. The plunger assembly, which includes the ceramic insulator, the magnetic sleeves, and related parts, floats on the mercury pool. When the coil is powered, creating a magnetic field, the plunger assembly is pulled down into the mercury pool, which in turn is displaced and moved up to make contact with the electrode. This action closes the circuit between the top and bottom load terminal, which is connected to the stainless steel can.

Figure 4-13 shows a group of one-, two-, and three-pole contactors. They are available up through 100-A capacity.

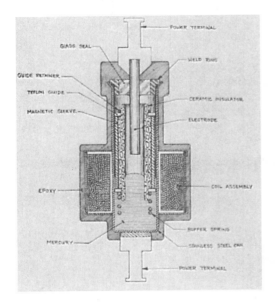

**Figure 4-12** Cross section through a mercury-to-metal contactor. *(Courtesy of Eagle Signal Controls.)*

**Figure 4-13** Mercury contactors. *(Courtesy of Eagle Signal Controls.)*

The basic use for the contactor is for switching of power in resistance heating elements, lighting, magnetic brakes, or heavy industrial solenoids. Contactors can also be used to switch motors if separate overload protection is supplied.

## Recommended Web Links

Students are encouraged to view the following Web sites as a supplement to the concepts presented in this textbook. Review and analyze the array of products that are available for electrical control applications. Many of these sites offer technical information that can help in converting the principles to practical applications. To view catalogs, your PC may require Adobe Acrobat Reader software.

### Relays

Allen-Bradley
www.ab.com/industrialcontrols/products/relays_timers/index.html
Review: Product Line

General Electric
www.geindustrial.com/cwc/products?famid=32
Review: General Purpose and NEMA Relays

Moeller Electric Ltd.
www.moeller.co.uk/easy_overview.htm
Review: Programmable Control Relays

### Contactors

Allen-Bradley
www.ab.com/manuals/ms/NCSO.htm
Review: Contactors, Manual and Magnetic Starters, and Overload Relays

Mercury Displacement Industry
http://www.mdius.com
Review: Relays

## Achievement Review

1. What is the difference between inrush current and holding current in the coil of a control relay?

2. What are the three ratings generally published concerning control relay contacts?

3. Explain contact wipe, and explain how contact bounce can affect the operation of an electrical control circuit.

4. Draw the symbols for a control relay coil, normally open relay contact, and normally closed relay contact.

5. Show the use of a control relay contact in an interlock circuit. Use a control relay, two push-button switches, and a red push-to-test pilot light.

6. What advantages can be obtained from the use of a time-delay relay as compared to a standard control relay?

7. Identify the following time-delay relay contacts as to whether they operate before or after the relay coil is energized and whether they are normally open or normally closed.

A.                     B.                     C.                     D.

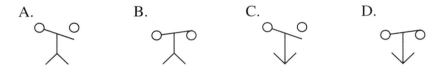

8. Follow the instructions in question 5, but substitute a time-delay relay for the standard control relay. In this case, the red pilot light is to energize at some predetermined time after the relay coil is energized.

9. Describe how a latching relay operates.

10. How many poles are generally available on contactors? What is the range in current-carrying capacity?

11. What is one advantage in using the plug-in relay?

   a. No circuit numbers are required.
   b. The best application of the plug-in relay is in power circuits.
   c. Relays can be changed without disturbing the circuit wiring.

12. Coils on relays are designed not to drop out (deenergize) until the voltage drops to what percentage of the rated voltage?

   a. 60%
   b. 85%
   c. 100%

# Chapter 5  Solenoids

## Objectives

After studying this chapter, you should be able to:

- Explain why it is necessary for the plunger in a solenoid to complete its stroke.

- Discuss the two important problems to consider in the application of a solenoid.

- Describe the application of solenoids to operating valves.

- Draw the symbol for the solenoid.

- Draw a control circuit showing the energizing of a solenoid through the closing of a relay contact, using a control relay, two push-button switches, and a solenoid.

- Know the difference between sealed current and inrush current in a solenoid.

- Know the function of the proportional solenoid.

- Know the function of the force motor in a servo valve.

## 5.1 Solenoid Action

The general principle of the solenoid action is very important in machine control. Like the relay and contactor, the *solenoid* is an electromechanical device. In this device, electrical energy is used to magnetically cause mechanical movement.

Review the explanation given in Section 4.1 for the relationship between plunger travel and coil current. A similar condition exists for the solenoid. It is important that the plunger complete its stroke when the solenoid is energized. Otherwise, the current in the coil will be high, resulting in damage to the coil. Figure 5-1 shows a typical curve of the relationship between current and stroke and shows what happens to the current when the stroke is not completed to zero.

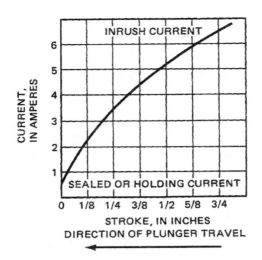

**Figure 5-1**  Graph of current versus stroke.

The current in amperes at the open position is called *inrush current.* The current in amperes at the closed position is called *sealed,* or *holding, current.* The ratio of inrush current to sealed current generally varies from approximately 5:1 in small solenoids to as much as 15:1 in large solenoids.

When the solenoid coil is energized, the plunger is in an out position. Due to the open gap in the magnetic path (circuit), the initial current in the coil is high. As the plunger moves into the coil, closing the gap, the current level drops to a lower value.

As shown in Figure 5-2, the solenoid is made up of three basic parts:

1. Frame
2. Plunger
3. Coil

The frame and plunger are made up of laminations of a high-grade silicon steel. The coil is wound with an insulated copper conductor.

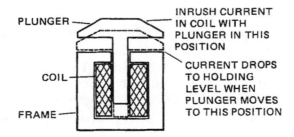

**Figure 5-2**  Cross section through air-gap solenoid.

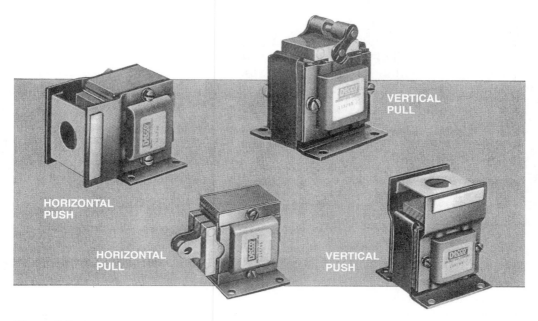

*Figure 5-3* **Industrial solenoids.** *(Courtesy of Detroit Coil Company.)*

Typical industrial solenoids are shown in Figure 5-3.

Solenoids for ac use are available as oil-immersed types. Heat dissipation and wear conditions are improved with this design. They are also available with a plug-in base. A typical example of the oil-immersed design is shown in Figure 5-4.

*Figure 5-4* **An oil-immersed solenoid.** *(Courtesy of Detroit Coil Company.)*

When the coil of a solenoid is energized, a magnetic field is created about the coil. This magnetic field produces a force that acts on the solenoid plunger. Due to this force, the plunger moves into the coil. This force on the plunger is called *pull.* The pull in solenoids varies widely. It may be as low as a fraction of an ounce or as high as nearly 100 pounds.

Connections to the coil may be supplied in the following ways:

- Pigtail leads
- Terminals on the coil
- Terminal blocks
- Plug-in connections

There are two important problems to be considered in the application of a solenoid:

1. The pull of the solenoid must at all times exceed the load. If the pull is a little less, the solenoid action will be sluggish and may not complete the stroke. There are also conditions that the user may not always be able to control, such as low voltage or increased loading through friction or pressure. Therefore, it is generally advisable to overrate the solenoid by 20–25%. **Caution:** Too much pull will result in the plunger slamming, resulting in damage to the plunger and frame.
2. The duty cycle of the work load should be known. Some applications require a duty cycle with only an occasional operation. Other cases require up to several hundred operations per minute. **Caution:** The operation of the solenoid above its maximum cycling rate will result in excessive heating and mechanical damage.

## 5.2  Solenoid Force and Voltage*

The pull-in force of a solenoid decreases rapidly as the voltage decreases below the coil nominal rating. As the voltage increases over the nominal value, the pull-in force increases, but the solenoid temperature may also rapidly increase.

## 5.3  Low Voltage

From a low-voltage standpoint, solenoid size selection should allow for adequate force at some arbitrary low voltage level. This allowance will ensure adequate solenoid force even during periods of low voltage and will prevent failure to pull-in and consequent coil burn out.

*Courtesy of Detroit Coil Company.

Design practices vary but low voltage levels are usually set at 85% or 90% of rated or nominal levels.

Forces at other reduced voltages can be closely approximated by means of the following formula:

$$F_1 = F \times \left(\frac{V_1}{V}\right)^2$$

where   $F_1$ = solenoid force at a reduced voltage $V_1$
   $F$  = solenoid force at rated voltage $V$

## 5.4  Overvoltage

The extra force resulting from overvoltage (or voltage above the coil nominal rating) will normally last for short periods of time. Under these conditions the mechanical life of a solenoid will not be seriously affected.

The coil temperature rise and the consequent ultimate temperature of the solenoid will increase. Unless the ultimate temperature exceeds the class "A," 105°C rating of the insulation, the application will be satisfactory. The solenoid temperature can be lowered by mounting on a surface, such as an aluminum plate that will conduct the heat away.

## 5.5  ac Solenoids on dc*

A solenoid designed to operate on alternating current can also be operated on direct current. There are, however, some limitations.

A solenoid operating on ac draws a high inrush current when the solenoid is open. As the solenoid closes, this current decreases to a low holding current when the solenoid is fully closed. This current characteristic of an ac current is extremely important because the high inrush current provides a high initial force that is usually desirable to overcome the load on the solenoid. When the solenoid is held closed, the current is at the low holding level. This low holding current generates very little heat, so the solenoid remains cool. In short, an ac solenoid has a built-in current valve, which provides for a high force pull-in and cool holding.

Now suppose an ac solenoid is operating on dc. On dc the current flow is constant regardless of whether the solenoid is open or closed. The inrush and holding

*Courtesy of Detroit Coil Company.

currents are the same. Because of this constant current feature of dc, a compromise between the pull-in force and holding temperature must be made. Specifically, if enough current is provided to give the same pull-in force as ac would provide, the solenoid may overheat if held energized. If the current is reduced to a level that will prevent overheating when held closed, the pull-in force will be greatly reduced. Therefore, the question of whether or not a given ac unit performs satisfactorily on dc depends on how pull-in force and overheating can be balanced.

Some applications may not require that the solenoid be held energized. In this case, an ac design unit might be operated on either ac or dc power. A dc current could be supplied to give the high pull-in force, equal to the force obtained from ac power. Overheating during the brief energized periods on dc would not be a consideration.

In applications in which the stroke is extremely short, an ac unit can usually be operated successfully on dc because at short strokes on ac power, the inrush or open current is only slightly greater than the holding current. The constant current characteristic of dc will not be different enough to cause problems of pull-in force or overheating.

In many cases, ac solenoids can be operated on dc power with the addition of a switch and resistor. The switch is arranged to be opened when the solenoid closes. When the switch opens, the resistor is in series with the solenoid coil. The addition of this resistance reduces the coil current so the solenoid can be held energized without burning out. A high current which will produce a high pull-in force is then possible.

## 5.6  dc Solenoids on ac*

Within limits, dc solenoids can be operated on ac. See Figure 5-5 for typical dc solenoids. For reasons of economy and flexibility, dc solenoids are usually made with solid iron parts. When operated on ac, eddy current and eddy current losses are introduced. These losses in solid iron parts are high, and high temperatures can be developed.

Therefore, to prevent overheating, the use of ac power on dc solenoids should be limited to applications where a low current is adequate. The use of ac power is also practical where the solenoid is on and off in so short a time that the eddy current losses cannot generate excessive heat.

Also, if a dc unit is to be used on both ac and dc power, it should be equipped with shading coils. These coils, common on ac designs, keep the solenoid from buzzing when the ac sine wave goes through zero. A dc-design solenoid with shading coils is sometimes termed an ac-dc unit.

*Courtesy of Detroit Coil Company.

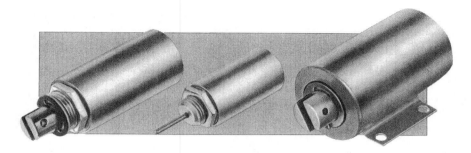

***Figure 5-5*** **Tubular dc solenoids.** *(Courtesy of Detroit Coil Company.)*

## 5.7 50- and 60-Cycle Solenoids*

With the exceptions of Canada and most South American countries, nearly all foreign countries operate on a 50-cycle power supply.

Coils for solenoids can be wound so that they may be used with either 50- or 60-cycle current. However, for best performance it is recommended that coils be wound for the specific frequency on which they will be used.

So-called *dual frequency* (50–60) coils are actually wound for 50 cycles and their use on 60 cycles is limited to applications in which their reduced force is adequate to operate the mechanism.

Sometimes one coil can serve as a dual frequency coil if the 50-cycle operating voltage is lower than the 60-cycle nominal voltage, for example, 120 V at 60 cycles, 100 V at 50 cycles. In this case, the winding for both specifications is the same so the same coil can be used successfully in both applications. Looking at the situation in another way, the twenty extra volts available at 60 cycles will provide enough extra power to offset the power loss at 50 cycles mentioned in the previous paragraph.

Tapped or three-lead dual frequency (50–60 cycle) coils are a more expensive but practical solution. These coils have one common lead, with a 60-cycle lead tapped in near the end of the coil and a 50-cycle lead at the coils end. Such coils are usually wound for 60-cycle voltage at a multiple of 115 or 120 V, with a 50-cycle tap at a multiple of 110 V. For example, a coil could be wound for 50 cycles at 110 V with a tap or third lead for 60 cycles, 115-V operation.

Ideally, a 50-cycle solenoid should be manufactured with more laminations in both plunger and field, but because of the limited demand for 50-cycle solenoids, manufacturers use a 60-cycle plunger and field, and alter the coil only. This compromise results in a 50-cycle solenoid with slightly less than normal force. Force reduction is roughly 10% at $\frac{1}{2}$-inch stroke, and 5% at $\frac{1}{4}$-inch stroke. Holding force is affected very little.

*Courtesy of Detroit Coil Company.

A 50-cycle power supply is not always available for preshipment testing of equipment before export. If this happens, 50-cycle power can be simulated on 60-cycles by adjusting your 60-cycle voltage to a level of 6/5 rated voltage.

## 5.8 Solenoid Temperature Rise*

When current flows in a solenoid coil, resistance in the copper wire causes the coil temperature to rise in both ac and dc solenoids. In an ac solenoid, eddy currents and hysteresis losses in the iron caused by the alternating magnetic field also generate heat. The resulting temperature increase of the solenoid is called *heat rise*. The ultimate temperature of a solenoid is the ambient temperature surrounding the solenoid plus the solenoid temperature rise. It is important to know this ultimate solenoid temperature for two reasons: First, it must be kept below the temperature rating of the solenoid's electrical insulation. Otherwise the solenoid will burn out. Second, the solenoid's force decreases as its temperature increases.

According to Ohm's law a change in applied voltage changes the current flow. So heat is directly related to voltage input. It is known that changes in frequency affect current flow, so heat rise also varies with frequency. Specifically, a 60-Hz coil will have a higher heat rise when run on 50 Hz, for a given voltage.

Heat rise also depends on ambient temperature. The same solenoid in a high ambient environment will have less heat rise than in a lower ambient because in a higher ambient the coil resistance is higher, restricting the current flow. Lower current flow means lower temperature rise.

Solenoid temperature rise will also be affected by the solenoid mounting. A heavy metal mounting will serve as a heat dispenser or heat sink and will conduct heat away from the solenoid, thus reducing the temperature rise.

Solenoid temperature rise can be measured in two ways: (1) by a thermocouple or (2) by the change-in-resistance method.

In method (1) we know that the highest temperatures are found at the heart of the coil. Therefore, to get an accurate reading the thermocouple or other temperature-sensing element must be placed near the center of the coil. You can see that this method is impractical as it would require a special coil, with a thermocouple wound in to get an accurate reading.

Method (2) is based on the fact that a coil's resistance is directly related to its temperature. Therefore, the temperature rise from one condition to another can be computed if the resistance at these two conditions is known. These two conditions are usually termed *hot and cold resistances*. This is the easiest and most practical method by which you can measure temperature rise. See Figure 5-6.

*Courtesy of Detroit Coil Company.

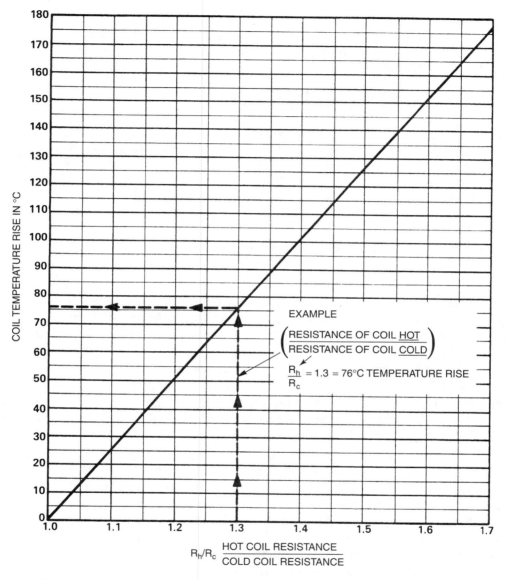

**Figure 5-6**   **Measuring temperature rise by change in resistance.** *(Courtesy of Detroit Coil Company.)*

A convenient circuit that permits fast deenergizing of the coil and resistance measurement is shown in Figure 5-7. The solenoid is connected to the center of a double-throw, double-pole switch. The applied voltage is connected to one side and the bridge to the other. The switch can be quickly thrown to disconnect the power supply and connect the resistance measurement bridge.

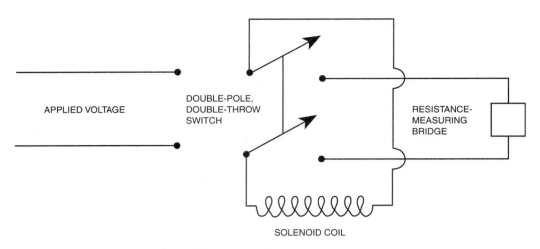

**Figure 5-7**  **Circuit for checking resistance.** *(Courtesy of Detroit Coil Company.)*

When measuring temperature rise, be sure the solenoid temperature is stabilized in the ambient so the cold resistance ($R_c$) is accurate. Similarly, energize the solenoid for about two hours so its temperature will stabilize before measuring ($R_h$).

By far the most satisfactory resistance measuring device is an accurate bridge. Volt-ohmeters are usually not accurate enough.

The solenoid as applied to valves is covered several times in this book. Therefore, a brief explanation of simple valve action is in order. The pressure referred to can be either hydraulic (fluid) or pneumatic (air). The pressure *ported* (directed) to a cylinder-piston assembly is used to explain the control of position and pressure changes on machines.

Figure 5-8A shows a single-solenoid, spring-return operating valve in its normal, deenergized condition. The symbols for the solenoid and valve are also shown. In Figure 5-8B, the solenoid valve is shown in the energized condition. Note that the spring is loaded (depressed). When the solenoid is deenergized, the spring pressure returns the valve spool to its normal, deenergized condition.

Note that when solenoid A is deenergized, pressure is free to flow to port A. Port B is open to tank. When solenoid A is energized, pressure is free to flow to port B, and port A is open to tank.

The double-solenoid valve is shown in Figure 5-8C. In the deenergized condition as shown, no pressure is available to either port A or port B, as it escapes around the land areas of the valve spool to the tank port. The centering springs keep the valve spool in this position, with both solenoids deenergized.

Figure 5-8D shows the double-solenoid operating valve with solenoid A energized. The force or pull exerted on the plunger in solenoid A moves the valve spool to the right. The spring in solenoid B is compressed (loaded). The land areas on the valve spool are now located so that pressure available at port P is free to flow into port A. Port B is open to the tank port T. As there is generally little or no pressure connected at the tank port, it follows that there will be little or no pressure at port B.

Figure 5-8E shows the double-solenoid operating valve with solenoid B energized. The force or pull exerted on the plunger in solenoid B moves the valve spool to the left. The spring in solenoid A is compressed (loaded). The land areas on the valve spool are now located so that pressure available at port P is free to flow into port B. Port A is open to the tank port T. As there is generally little or no pressure connected at the tank port, it follows that there will be little or no pressure at port A.

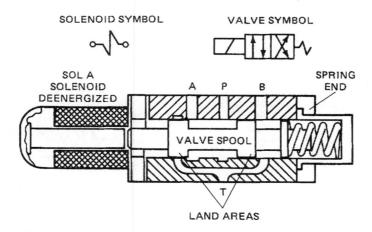

**Figure 5-8A** Cross section through solenoid in normal, deenergized condition. A and B are ports, P stands for pressure, and T is the tank.

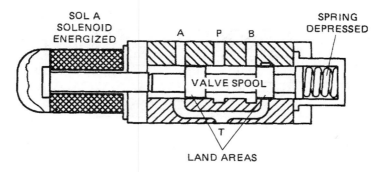

**Figure 5-8B** Cross section through solenoid in energized condition. A and B are ports, P stands for pressure, and T is the tank.

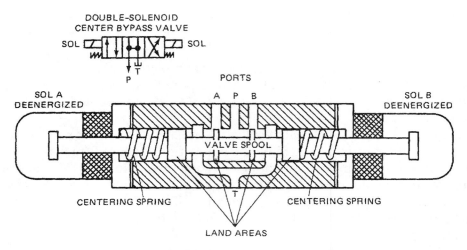

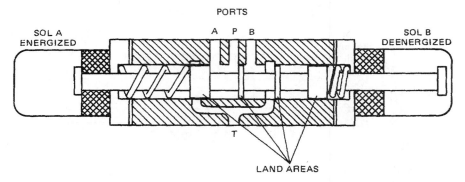

*Figure 5-8C* Cross section through double-solenoid valve with both solenoids deenergized. A and B are ports, P stands for pressure, and T is the tank.

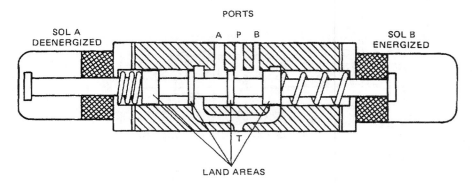

*Figure 5-8D* Cross section through double-solenoid valve with solenoid A energized and solenoid B deenergized. A and B are ports, P stands for pressure, and T is the tank.

*Figure 5-8E* Cross section through double-solenoid valve with solenoid A deenergized and solenoid B energized. A and B are ports, P stands for pressure, and T is the tank.

## 5.9 Circuit Applications

Refer to the basic relay circuit shown in Figure 4-3. The same circuit is shown in Figure 5-9A, except that a solenoid is substituted for the red pilot light.

The sequence of operations is as follows:

1. Operate the START push-button switch.
2. Coil of relay 1CR is energized.
3. Relay contact 1CR-1 closes, interlocking around the START push-button switch.
4. Relay contact 1CR-2 closes, energizing solenoid A.
5. Operate the REVERSE push-button switch.
6. Relay 1CR is deenergized.
7. Relay contact 1CR-1 opens, opening the interlock circuit around the START push-button switch.
8. Relay contact 1CR-2 opens, deenergizing solenoid A.

In Figure 5-9B, a time-delay relay is added to the circuit. The sequence for this circuit proceeds as follows:

1. Operate the START push-button switch.
2. Coil of relay 1CR is energized.
3. Relay contact 1CR-1 closes, interlocking around the START push-button switch.
4. Relay contact 1CR-2 closes, energizing solenoid A.
5. Relay contact 1CR-3 closes, energizing timing-relay coil TR.
6. After a preset time on TR expires, the timing contact TR closes, energizing solenoid B.
7. Operate the REVERSE push-button switch, deenergizing relay coil 1CR.
8. Relay contact 1CR-1 opens, opening the interlock circuit around the START push-button switch.
9. Relay contact 1CR-2 opens, deenergizing solenoid A.
10. Relay contact 1CR-3 opens, deenergizing the coil of timing relay TR.
11. Timing-relay contact TR opens, deenergizing solenoid B.

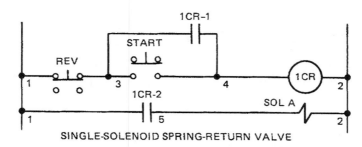

SINGLE-SOLENOID SPRING-RETURN VALVE

*Figure 5-9A*  Control circuit with solenoid.

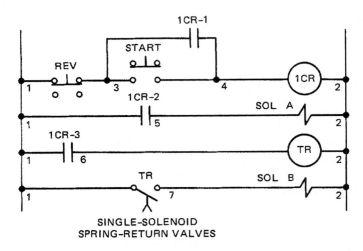

**Figure 5-9B**   Control circuit with solenoid and time-delay relay.

Note that if the REVERSE push-button switch is operated before the preset time expires on TR, solenoid B will not be energized.

## 5.10  Variable Solenoids*

To this point the solenoids that have been discussed are of the ON/OFF type. These are used in the traditional type of solenoid valve to position the spool into one of two or three discrete positions.

With proportional and servo valves it is necessary to position the spool at an infinite number of intermediate positions. In the case of directional valves, spool position will provide both directional and flow control since the amount of spool movement away from center (valve opening) determines the flow rate through the valve (assuming constant pressure drop).

## 5.11  Proportional Valves*

*Proportional valves* use a proportional solenoid to directly or indirectly position the main spool and are normally used for open-loop speed control of actuators.

*Information courtesy of Eaton Corporation.

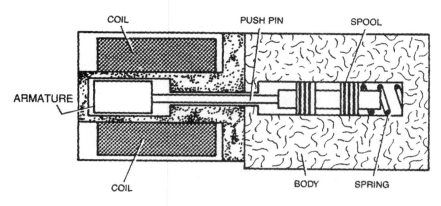

*Figure 5-10* **Proportional solenoid.** *(Courtesy of Eaton Corp.)*

Proportional solenoids operate in much the same way as ON/OFF dc solenoids, the main difference being that the solenoid current is varied to determine the solenoid force. A cross-sectional view of a proportional solenoid is shown in Figure 5-10.

A force is produced by the solenoid in relation to the current passing through the solenoid coil, which pushes the spool across in the valve body. The spool will continue to move until the spring compression force balances out the solenoid force. By varying the solenoid current, therefore, the spool can be moved to a greater or lesser amount in the valve body.

Proportional solenoids can be used where relatively large amounts of spool movement are required (up to 5 mm), but they will operate in only one direction of movement; i.e., if the spool has to be positioned either side of center, then two are required, one at each end.

A cross-sectional view of a proportional directional control valve, single-stage design, without integral feedback transducers is shown in Figure 5-11.

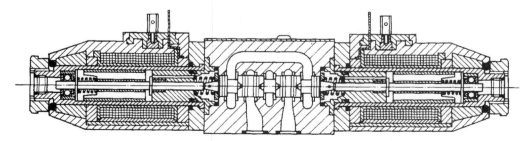

*Figure 5-11* **Proportional directional control valve, single-stage design without integral feedback transducers.** *(Courtesy of Eaton Corp.)*

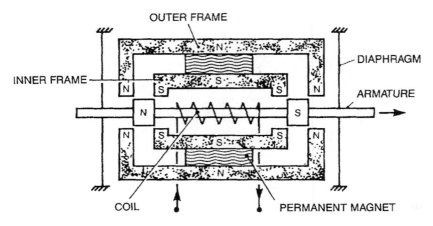

**Figure 5-12**   **Force motor.**  *(Courtesy of Eaton Corp.)*

## 5.12 Servo Valves*

Servo valves use a force motor to indirectly position the main spool and are used mainly for high-performance, closed-loop control of speed or position.

A cross-sectional view of a force motor is shown in Figure 5-12. The coil of the force motor is wound onto an armature, which is surrounded by an inner and outer nickel-iron frame. A permanent magnet magnetizes the two frames such that the inner one is, say, south and the outer one north. If a current is passed through the coil so as to make the left end of the armature north and the right end south, a force is created (by attraction and repulsion of the poles) toward the right. The armature is supported on diaphragms that provide the spring force, so the armature will again move until the magnetic force and spring force balance out.

Force motors give a more linear force/current relationship than proportional solenoids, but they are normally confined to short-stroke applications (typically ±0.5 mm). Reversing the direction of current through the coil will reverse the armature polarity and, hence, the force direction. Unlike proportional solenoids, therefore, force motors can be used for spool movement either side of center (push/pull).

## Recommended Web Links

Students are encouraged to view the following Web sites as a supplement to the concepts presented in this textbook. Review and analyze the array of products that

*Information courtesy of Eaton Corporation.

are available for electrical control applications. Many of these sites offer technical information that can help in converting the principles to practical applications. To view catalogs, your PC may require Adobe Acrobat Reader software.

## Solenoids

Cliftronics, Inc.
www.cliftronics.com/product_lines.html
Review: Solenoids

Bicron Electronics
www.solenoid.com
Review: Solenoids and Technical Guide (pdf)

Ledex
www.ledex.com
Review: Products, Application

Detroit Coil
www.detroitcoil.com
Review: Products, and "What Is a Solenoid"

## Fluid Power—Electrically Operated Valves

Moog
www.moog.com/imc/product/ind_hyd/default.asp
Review: Directional Control Valves (pdf)

## Solenoid Valves

IQS
www.solenoid-valves.net
Review: The links to more than 10 producers of solenoid valves

## Proportional Valves

Moog
www.moog.com/noq/_product_index__C489T2204
Review: Valve Construction and Electrical Characteristics

## Electrohydraulic Proportional Valve

AxioMatic
www.axiomatic.com/connector-amplifier.html
Review: The link to amplifier

**Servohydraulic Valves**

Rexroth-Bosch
www.boschfluidpower.com/products_valves_electronic.html
Review: Ten publications available in pdf format. Also see
"Downloadable documents in PDF format" for additional electronic
control servo components.

# Achievement Review

1. Draw the symbol for the solenoid coil.

2. What happens in a solenoid if the plunger is prevented from completing its stroke?

3. At what position of the solenoid plunger does inrush current appear?

4. Why must the pull of a solenoid always exceed the load?

5. What will happen in a solenoid if the operation exceeds the solenoid design?

6. In the application of a solenoid to a spring-return operating valve, assume that the solenoid has been energized. What happens to the valve spool when the solenoid is deenergized?

7. Draw a control circuit using two push-button switches, a control relay, and a solenoid. The solenoid is to be energized when a normally open push-button switch is operated. It is to be deenergized when a normally closed push-button switch is operated.

8. What is the solenoid current in amperes with the plunger in the closed position?

   a. Maximum current
   b. Zero (no current will be indicated)
   c. Sealed or holding current

9. Which of the following statements is true of the spring when a solenoid is energized in a single solenoid, spring-return valve?

   a. It remains in the same condition as it was before the solenoid was energized.
   b. It is depressed (loaded).
   c. It aids the solenoid in moving the valve piston.

10. What causes the force acting on the plunger of a solenoid when the solenoid is energized?

    a. Hydraulic pressure
    b. Gravity
    c. A magnetic field that is produced about the coil

11. What is the difference between the operation of a proportional solenoid and the ON/OFF dc solenoid?

12. In using the proportional solenoid, how can the spool be moved to a greater or lesser amount?

13. In a force motor, how can the force direction be reversed?

14. What can you say about the linear force/current relationship in the proportional solenoid as compared to the force motor?

15. What will happen to the solenoid coil temperature if it is subjected to extended periods of overvoltage?

16. What is the relationship between inrush and holding current when an ac solenoid is operated on dc?

17. When a dc solenoid is used on both ac and dc, why should it be equipped with shading coils?

18. What two methods are used to measure solenoid temperature rise? Which is the more practical?

# Chapter 6  Types of Control

## Objectives

After studying this chapter, you should be able to:

- Understand the basic control system.
- Explain the advantages of closed-loop control.
- Explain the use of a sensor.
- Describe the difference between proportional, derivative, and integral types of control.
- Explain steady-state error and what can be done to correct it.
- Explain the use of a position transducer.

## 6.1  Open-Loop Control

In Chapters 7 through 11 we will be discussing control involving motion, pressure, temperature, time, and counting. In the simplest form of open-loop control, the machine function reaches its preset position or condition from only the original or preset values. For example, in Figure 6-1 a hydraulically powered piston is to be at position A for start conditions. On the operation of a CYCLE START push-button switch, the piston is to move to position B, stop, and return to position A. A limit switch is used to control both positions. In many cases this type of open-loop control is accurate enough to satisfy a given production requirement. Figure 6-2 shows a block diagram of basic control system. However, there are many potential problems, involving things such as friction, oil temperatures in hydraulic systems, valve-shifting time, material being processed, and the like. All these conditions may lead to errors that cannot be tolerated in a given production machine. It then becomes necessary to account for the errors and adopt some method to minimize (hopefully eliminate) the errors. This may be accomplished by using closed-loop control.

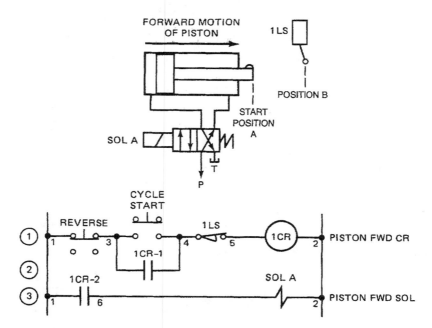

**Figure 6-1**   A basic control circuit.

**Figure 6-2**   Block diagram of basic control system.

## 6.2  Closed-Loop Control

Closed-loop control for an automatic feedback control system can be shown in block form (Figure 6-3). In this figure it can be seen that the sensor is the first item to consider as it gathers information from the process.

A *sensor* is a piece of hardware that measures system variables. Normally the sensor changes or transduces the measured quantity into another form of energy. It is the instrument or device that senses process condition in any process control system. Sensors provide the feedback signals by detecting the present condition, state, or value of a process variable. They receive and respond to a stimulus or signal. An antenna, load cell, or photoelectric cell designed to detect and respond to some force, change, or radiation for purpose of information or control are good examples of sensors.

In any automatic process the corresponding components of sensing, information processing, and action can be distinguished. A simple industrial process might be the filling of a water tank. Figure 6-4 shows such a system using two sensors.

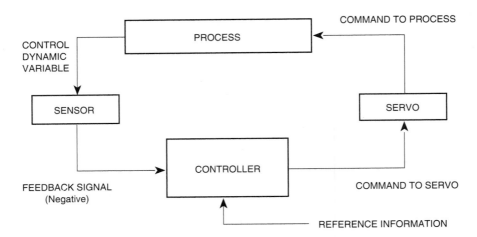

**Figure 6-3**   **Automatic feedback control system.**

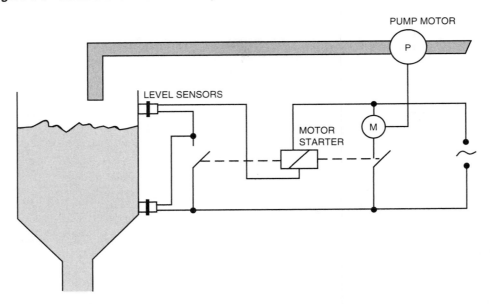

**Figure 6-4**   **Filling level control system of a water tower.** *(Courtesy of ifm efector, inc.)*

They detect the minimum and maximum permissible levels and signal the information processor. In this system, the information processor is a relay that determines from the signals when the actuator (pump) should be switched on and off.

This automated process performs its task without interruption within the reliability limitations inherent in technical systems. However, its capabilities are limited. For example, if no water is available, the "intelligence" is not sufficient to switch off the pump. If only dirty water were available, the control system would be equally incapable of interrupting the filling process. The information processor

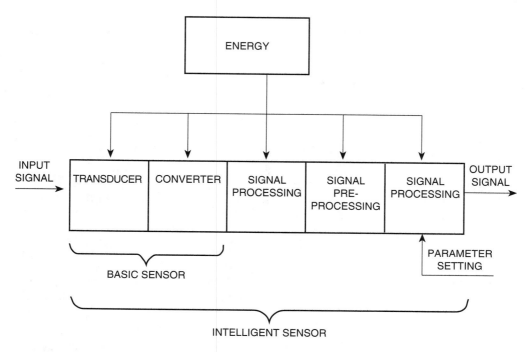

**Figure 6-5** **Structure of a sensor.** *(Courtesy of ifm efector, inc.)*

does not recognize conditions such as no water being available or dirty water in the line because no sensors are provided in the example shown in Figure 6-4.

Numerous sensor tasks are required when automating technical processes. However, all technical sensors share a common structure, as shown in Figure 6-5. Of course, not every sensor contains all the components shown, hence the large number of designs that are available.

## 6.3  Proportional Control

The control signal can be thought of as the error between the input and output. While an error exists, the servo valve will operate to move the load. When the load reaches the desired position, the error signal will be zero. The servo valve centers, and the load stops moving. This type of control is known as *proportional* control because the control signal is proportional to the difference between the input (demand) and feedback (actual) signals.

In examining the response of such a system, a step change in input signal may produce a characteristic as shown in Figure 6-6. Obviously the load current cannot move instantaneously to follow the input signal, so there will be a time delay between applying the input signal and the load reaching the desired position.

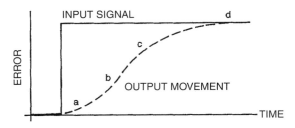

**Figure 6-6** **Graph of input signal versus output movement.** *(Courtesy of Eaton Corp.)*

During the first part of the load movement (a), the system is overcoming the inertia of the load and accelerating it to the required speed. There then may be a period of constant speed movement (b) in which the servo valve is wide open and the system is moving the load as fast as it is able. As the load approaches the required position (c), the speed reduces (curve flattens out), and the load gradually moves to the demand position at point (d). The flattening out at the end is due to the fact that the control signal to the servo valve is basically the error between input and output; as the error decreases, the control signal reduces, thus slowing down the movement of the load (Figure 6-7).

It may be possible to speed up the response of the system by increasing the gain of the amplifier so that for a given error, the amplifier produces a larger control signal. Increasing the gain, however, may produce a result like that shown in Figure 6-8. Although the output arrives at the desired position faster, it overshoots since the load is traveling so fast it cannot be stopped quickly enough when it reaches the required position. There may in fact be several overshoots and undershoots before it settles down to the final position. In many applications, an overshoot may be undesirable, so an alternative method of improving the system response is to modify the control signal such that it is proportional not only to the error, but also to the rate at which the error is changing (error-rate):

Control signal = Error signal × Gain + Rate of change of error × Gain

When the error is reducing, for example, the load is moving toward the required position; the error-rate term will then be negative and tend to reduce the

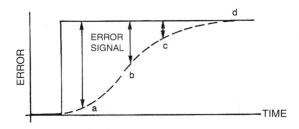

**Figure 6-7** **Change in error signal.** *(Courtesy of Eaton Corp.)*

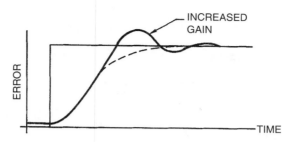

**Figure 6-8**  **Response to increased gain.** *(Courtesy of Eaton Corp.)*

control signal (Figure 6-9). Between points (a) and (b) on the output curve, the error is changing relatively slowly so the large error term outweighs the small error-rate term and provides a large control signal to accelerate the load. From (b) to (c), however, the error is reducing quickly, so the error-rate term predominates and starts to reduce the control signal, which means the load starts to decelerate earlier, hence the overshoot can be eliminated (Figure 6-10).

Mathematically, the rate of change of a quantity is known as the derivative, so this type of control is termed *proportional plus derivative* and is expressed as:

Control signal = [Proportion signal (error) × Proportional gain] +
[Derivative signal (error rate) × Derivative gain]

Consider now a situation in which the load has to be positioned in a vertical plane or in which the load is subjected to some external force (Figure 6-11). Imagine that the load has moved to its required position. When actual position and demand position are the same, the error will be zero, and the servo valve will center. However, due to the load on the piston and leakage across the servo valve spool, the piston will tend to creep down. To counteract this, therefore, the servo valve spool must be displaced slightly away from center to prevent movement of the load. To offset the valve spool, a small control signal is required. In order for there to be a control signal, an error must exist between input and output; that is, the load can never be exactly in the required position. This condition is known as the *steady-state error* (Figure 6-12).

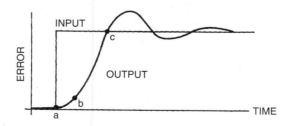

**Figure 6-9**  **Relative change in error.** *(Courtesy of Eaton Corp.)*

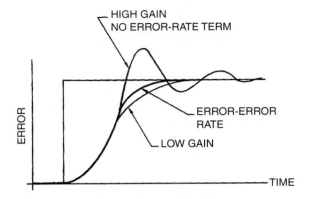

**Figure 6-10**   **Proportional plus derivative.**  *(Courtesy of Eaton Corp.)*

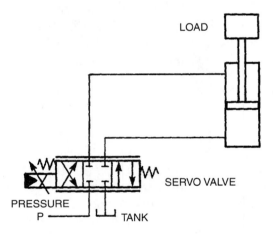

**Figure 6-11**   **Vertical load situation.**  *(Courtesy of Eaton Corp.)*

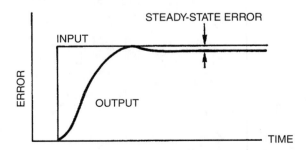

**Figure 6-12**   **Steady-state error.**  *(Courtesy of Eaton Corp.)*

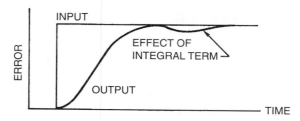

**Figure 6-13**   **Effect of integral term.** *(Courtesy of Eaton Corp.)*

## 6.4 Proportional-Integral

The steady-state error can be reduced by again modifying the control signal such that it is now proportional to the error plus an *error × time* term. The error × time component is known as the *integral term* and is effectively equal to the error multiplied by the length of time the error has existed. Thus, the longer the time, the larger the term becomes. This component can, therefore, be added to the proportional signal to virtually eliminate the steady-state error:

Control signal = [Proportional signal (error) × Proportional gain] +
(Integral signal (error time) × Integral gain]

If a steady-state error exists, the integral term will increase with time until it is large enough to generate the control signal required to correct the error (Figure 6-13). Adjustment of the proportional and integral gains allows the system to be tuned for the best results. In some applications it may be necessary to reset the integral term to zero each time a new demand signal is applied to the system.

## 6.5 Proportional-Integral-Derivative

The performance of a basic (proportional) control system can be significantly improved by adding in:

- A derivative (error-rate) term to improve the dynamic response and reduce overshoot
- An integral (error-time) term to reduce the steady-state error

In fast-acting systems, the integral term tends to make the system less stable, so it is often used in conjunction with the derivative term, which tends to improve the stability. The system would then be described as having proportional, integral, and derivative (PID) action. This configuration is sometimes referred to as three-term control.

## Recommended Web Links

Students are encouraged to view the following Web sites as a supplement to the concepts presented in this textbook. Review and analyze the array of products that are available for electrical control applications. Many of these sites offer technical information that can help in converting the principles to practical applications. To view catalogs, your PC may require Adobe Acrobat Reader software.

The following links are to articles from *Control Engineering's* (www.controleng.com) Current/Past Issues. The articles have been selected to complement the concepts presented in Chapter 6. Students are encouraged to read additional articles.

To access articles:

1. Click on "Issue Archive"
   Click on year "2000"
   Click on "February"
   Look under "Control Engineering"
   Article: "Understanding PID Control"

2. Click on "Issue Archive"
   Click on year "1998"
   Click on "October"
   Look under "Back to Basics"
   Article: "Open Loop Control Offers Some Advantages"

## Achievement Review

1. Explain how the transducer is used in a closed-loop control system.

2. Explain proportional control.

3. In a closed-loop control system, what is the effect of increasing the gain of an amplifier?

4. In a closed-loop control system, how do you compensate for leakage across the valve spool?

5. Explain the term *error* × *time*.

6. What is meant by three-term control?

# Chapter 7

# Motion Control Devices

## Objectives

After studying this chapter, you should be able to:

- List the three basic classes of limit switches.
- Explain where rotating-cam limit switches are used.
- Explain the following terms relative to limit switches:
  - Operating force
  - Release force
  - Pretravel or trip travel
  - Overtravel
  - Differential travel
- Discuss the proximity limit switch and explain how it is used.
- Describe how the vane limit switch operates.
- Draw the limit switch symbols for four different conditions.
- Design an electrical operating circuit showing how the operation of a normally closed limit switch contact can be used to deenergize a solenoid, using two push-button switches, a normally closed limit switch contact, a relay, and a solenoid.
- List several methods for achieving proximity switching.
- List some uses for the linear transducer.
- List the information required for ac synchronous motor application in angular position control.
- List some of the advantages in using a dc stepper motor.
- Discuss the four types of photoelectric transducers.
- Explain the principle used in flow monitors.

# 7.1 Importance of Position Indication and Control

In the subject of electrical control for machines, position plays an important part. The problem is to accurately and reliably supply position information by providing an adequate electrical signal. This information may be used for both indication and control. Indication becomes a powerful tool in troubleshooting a machine or process.

In many cases the relative position of a machine part or process product may not be critical. However, in some cases position information must be reliable to 0.001 inch or less. The machine designer must know the operation of the machine and/or the product process. The degree of accuracy required can then be determined.

There are many components that are used to obtain position information. Starting with the mechanical limit switch, most of the components are covered in this chapter.

# 7.2 Limit Switches—Mechanical

To describe the various limit switches available on the market today and their uses would require an entire book. Thus, we cover here only a few widely used units.

Mechanical-type switches can be subdivided into those operated from linear motion and those operated from rotary motion. Figure 7-1A is activated by a linear motion whereas Figure 7-1B is activated by a rotary motion.

There are large and small switches that operate from linear motion. The *precision limit switch* is a small switch. This switch varies from the larger size

***Figure 7-1A*** Linear activated limit switches. *(Courtesy of Allen-Bradley, a Rockwell International Company.)*

***Figure 7-1B*** Rotary activated limit switches. *(Courtesy of Allen-Bradley, a Rockwell International Company.)*

mainly in a lower operating force and shorter stroke. The operating force may be as low as 1 pound. The stroke may be only a few thousandths of an inch.

Limit switches operated by rotary motion are generally called *rotating-cam limit switches*. These are control-circuit devices used with machinery having a repetitive cycle of operation in which motion can be correlated to shaft rotation. They are used to limit and control the movement of a rotating machine and to initiate functions at various points in the repetitive cycle of the machine. A rotating-cam limit switch arrangement is shown in Figure 7-2.

The switch assembly consists of one or more snap-action switches. The cams are assembled on a shaft. The shaft, in turn, is either directly or through gearing driven by a rotary motion on the machine.

The cams are independently adjustable for operating at different locations within a complete 360-degree rotation. In some cases the number of total rotations available is limited. In other cases the rotation can continue and at speeds up to 600 revolutions per minute (rpm).

In selecting a limit switch, it is important to determine its application in the electrical circuit. The following factors must be considered:

- Contact arrangement
- Current rating of the contacts
- Slow or snap action
- Isolated or common connection
- Spring return or maintained
- Number of NO and NC contacts required

**Figure 7-2** **Rotating-cam limit switch.** *(Courtesy of Allen-Bradley, a Rockwell International Company.)*

In most cases the switch consists of double-break, snap-action, silver-tipped to solid silver contacts. The contact current rating will vary from 5 A to 10 A at 120 V ac continuous. The make-contact rating will be much higher, and the break-contact rating will be lower. Isolated NO and NC contacts are available. In some cases, multiple switches in the same enclosure operated by the same mechanical action are used.

A second important factor is the type of mechanical action available to operate the switch. Here the *operator* is the major decision. Length of travel, speed, force available, accuracy, and type of mounting possible are some of the considerations.

In discussing the action of limit switches, specific terms are used. A knowledge of these terms is helpful:

- *Operating force*—the amount of force applied to the switch to cause the "snap over" of the contacts
- *Release force*—the amount of force still applied to the switch plunger at the instant of "snap back" of the contacts to the unoperated condition
- *Pretravel or trip travel*—the distance traveled in moving the plunger from its free or unoperated position to the operated position
- *Overtravel*—the distance beyond operating position to the safe limit of travel; usually expressed as a minimum value
- *Differential travel*—the actuator travel from the point where the contacts snap over to the point where they snap back
- *Total travel*—the sum of the trip travel and the overtravel

Figure 7-3 shows pretravel, overtravel, differential travel, and total travel in diagram form.

A. Pretravel        B. Overtravel            1. Actuator — free position

C. Total travel     D. Differential travel   2. Actuator — operating position

                                             3. Overtravel — limit position

                                             4. Actuator — release position

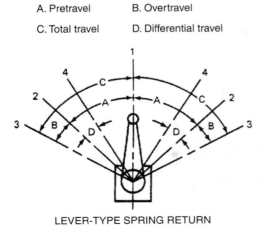

LEVER-TYPE SPRING RETURN

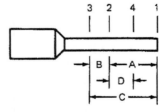

PUSH-TYPE SPRING RETURN

*Figure 7-3*  **Limit switch operating movement.**

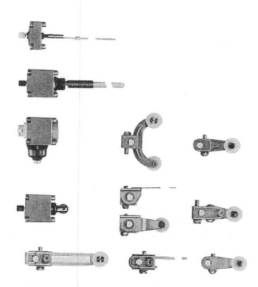

**Figure 7-4** **Operators for limit switches.** *(Courtesy of Allen-Bradley, a Rockwell International Company.)*

Most manufacturers of limit switches list force and travel information for certain switches in their specifications. Accuracy of switch operators at the point of snapover varies with different types and manufacturers. In general, it is in the range of 0.001 inch to 0.005 inch.

The operator that probably has the greatest use is the roller lever. It is available in a variety of lever lengths and roller diameters. The next most frequently used operator is the push rod. It can consist of only a rod, or it can be supplied with a roller in the end. In most cases, particularly with the oil-tight machine-tool limit switch, the head carrying the operator can be rotated to four positions, 90 degrees apart. It can also be either top- or side-mounted. Two other operators used in machine control are the fork lever and the wobble stick. Some of the various operators available are shown in Figure 7-4.

## 7.3 Limit Switch Symbols

Mechanical limit switch symbols are used in applying the limit switch to the schematic or elementary circuit diagram (Figure 7-5A). These represent the switch in four different conditions. Basically, the switch has a normally open contact and/or a normally closed contact. In some switches, there are two NO and two NC contacts.

Circuit diagrams illustrate the condition of switches when the mechanical system is at rest. Therefore, some switches may be shown in an unoperated condition and others may be shown in an operated condition

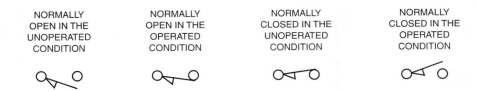

**Figure 7-5A**   Limit-switch symbols.

**Figure 7-5B**   Symbol used to show both normally open and normally closed contacts on the same limit switch.

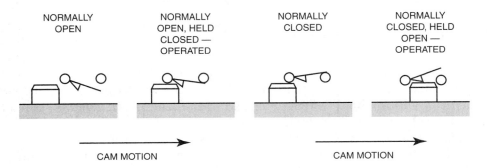

**Figure 7-5C**   Operation of the switch in relation to the symbol.

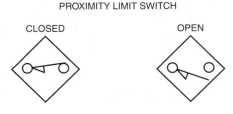

**Figure 7-5D**   Symbol for proximity limit switch.

There are times when both the NO and NC contacts on a given limit switch may be used in a circuit. Under this condition, it helps to join the two contact symbols with a broken line. This line indicates that they are contacts on the same limit switch (Figure 7-5B).

To further illustrate the operation of the switch in relation to the symbol, consider the case in Figure 7-5C. Assume the cam to be moving to the right as shown. As the cam contacts the switch, the switch operates and changes from the unoperated to the operated condition. Note that Figure 7-5C has no use other than to provide help in remembering the symbols.

There is a slight change in the symbol for the proximity limit switch as compared to the mechanical limit switch (Figure 7-5D).

## 7.4 Circuit Applications

Limit switches in machine-tool electrical control are used to gather information relative to the position of a machine part. To illustrate this, it is helpful to show a means of providing motion. In Chapter 5, the operating valve was introduced. The valve, along with a cylinder-piston assembly, is used in the limit switch application circuits.

At this point it is helpful to explain the solenoid operating valve symbol as it is used and shown in circuit problems.

Figure 7-6A shows the symbol for a single-solenoid, spring-return operating valve in the deenergized condition. Figure 7-6B shows the symbol for the same valve with solenoid A energized. Note the change in the flow of pressure through the valve to the piston-cylinder assembly.

Figure 7-6C shows the symbol for the double-solenoid operating valve with both solenoids deenergized. Figure 7-6D shows the symbol with solenoid A energized. Figure 7-6E shows the symbol with solenoid B energized. Here again, as with the single-solenoid operating valve, note the change that takes place when either solenoid is energized. Remember that when either solenoid is deenergized, centering springs in the valve return the piston to the center position.

The limit switch circuit shown in Figure 7-7 is built on the basic solenoid circuit shown in Figure 5-9A. A cylinder-piston assembly is shown as a means of moving a cam on the piston from position A to position B. If the piston is in position A, solenoid A is deenergized. If the piston is in position B, solenoid A is energized. Limit switch 1LS contact, shown in the circuit of Figure 7-7, is normally closed.

The sequence of operations proceeds as follows:

1. Operate the START push-button switch.
2. Relay coil 1CR is energized.
   a. Relay contact 1CR-1 closes, interlocking around the START push-button switch.
   b. Relay contact 1CR-2 closes, energizing solenoid A.

The valve spool shifts, permitting pressure to enter the main cylinder area (left-hand end). The pressure medium (water, oil, or air) in the right end of the cylinder (rod end) is free to return to the tank. The piston moves from its start position at A to a new position at B.

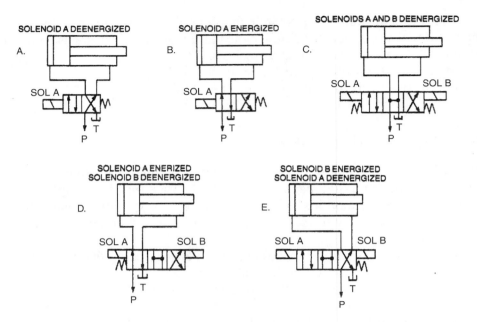

**Figure 7-6**  Solenoid valve operation with piston-cylinder assembly. P, pressure; T, tank.

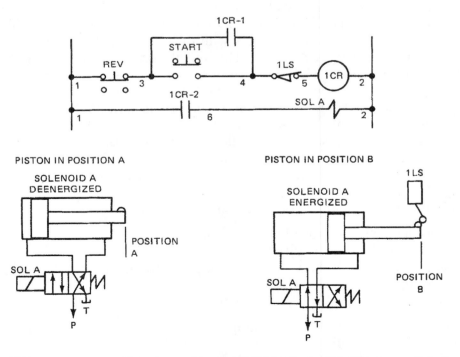

**Figure 7-7**  Control circuits. One piston with extension detection switch. P, pressure; T, tank.

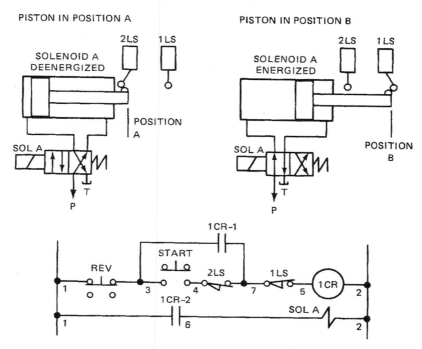

**Figure 7-8** Control circuit. P, pressure; T, tank.

3. At position B, limit switch 1LS is operated, opening its normally closed contact.
4. Relay coil 1CR is deenergized.
    a. Relay contact 1CR-1 opens, opening the interlock circuit around the START push-button switch.
    b. Relay contact 1CR-2 opens, deenergizing solenoid A.

With solenoid A deenergized, the valve spring returns the valve spool to its initial position. This permits pressure to enter the rod end (right end), returning the piston to position A.

In the circuit shown in Figure 7-7, it is assumed that the piston was at position A to start. To ensure that the piston is at position A to start a cycle, a second limit switch, 2LS, is placed at position A (Figure 7-8). A normally open limit switch contact is used. It is held operated (contact closed) by a cam on the piston. Note that 2LS limit switch contact is placed inside the START push-button interlock circuit formed by relay contact 1CR-1. Otherwise, as soon as the cam moves off limit switch 2LS, the limit switch contact opens, deenergizing the circuit.

The operation of this circuit is identical to that shown in Figure 7-7, except the piston must be in position A, operating limit switch 2LS for start conditions.

A third circuit is shown in Figure 7-9. This circuit uses a double-solenoid, spring-return-to-center valve. A limit switch performs a double duty. The NO

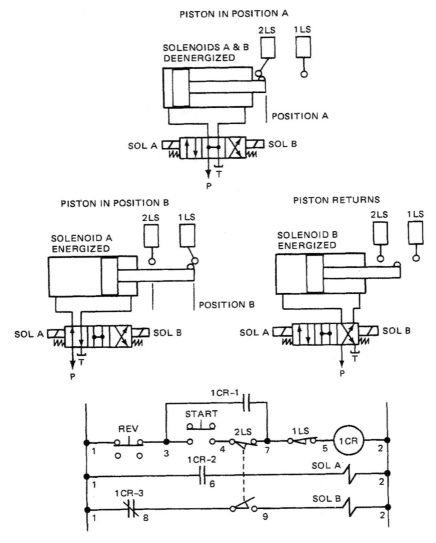

**Figure 7-9**   Control circuit. One piston with extension-retraction switches. P, pressure; T, tank.

contact on limit switch 2LS, held closed in position A (start position), ensures that the piston is at position A for start conditions. The NC contact on limit switch 2LS is held open at the start position. This condition opens the circuit to solenoid B.

The sequence of operations for the circuit shown in Figure 7-9 is as follows:

1. Operate the START push-button switch.
2. With limit switch 2LS normally open, contact held closed, relay 1CR is energized.

a. Relay contact 1CR-1 closes, interlocking around the START push-button switch.
b. Relay contact 1CR-2 closes, energizing solenoid A.
c. Relay contact 1CR-3 opens.

As the piston moves off limit switch 2LS, the normally open contacts open and the normally closed contacts close. However, since relay contact 1CR-3 is now open, solenoid B remains deenergized.

3. Piston moves to position B, operating limit switch 1LS, opening its normally closed contact.
4. Relay 1CR is deenergized.
    a. Relay contact 1CR-1 opens, opening the interlock circuit around the START push-button switch.
    b. Relay contact 1CR-2 opens, deenergizing solenoid A.
    c. Relay contact 1CR-3 closes, energizing solenoid B.

The piston now moves back to position A, operating limit switch 2LS.

5. The normally closed limit switch contact 2LS opens, deenergizing solenoid B.

## 7.5 Proximity Limit Switches*

The proximity switch is a generic term; it is a device capable of acting as an electronic switch when in the presence or close proximity of an object. The important distinction that differentiates it from a mechanical switch is that it does not require physical contact with anything else to operate. The methods of achieving this operation have been far-ranging. They have made use of the effect of objects on radio frequency (RF) fields, magnetic fields, capacitive fields, acoustic fields, and light rays. RF fields are usually altered by the presence of ferrous materials that absorb energy by any currents produced in them by the field. Magnetic fields have been utilized to simply close reed switches by bringing the magnet up close to the switch or by introducing a magnetic material between the magnet and the switch. Magnets have also been used to alter electrical fields in devices making use of the Hall effect. (The effect is produced when a magnetic field applied to a conductor carrying a current produces a voltage across the conductor. This voltage is known as the Hall voltage. The Hall voltage is proportional to the product of the current and the field. That is, a device that exhibits the Hall effect is a multiplier. With constant current, the Hall voltage will be proportional to the magnetic field. With the magnetic field constant, the Hall voltage will be proportional to the current flow.)

*Courtesy of ifm efector, inc.

Capacitive devices make use of the change in capacity that occurs when the object to be sensed acts as a plate of a capacitor for which the sensor acts as the other plate when detecting metallic objects or as an alteration to the dielectric between plates when detecting nonmetallic objects. Sonic devices utilize sound fields that are either interrupted by the object to be detected or detect the reflection of sound from objects. Photoelectric devices work in a similar manner except that light rays are detected rather than sound waves.

All of these methods have inherent strengths and weaknesses in actual application. For example, 2-wire DC proximity switches (Figure 7-10) eliminate the potential for wiring mistakes during installation because they are not sensitive to voltage polarity. The proximity switches have two black wires, exactly like a mechanical switch. To install the switch, one wire is connected to the power supply and the other connected to the load.

**7.5.1 Inductive Sensors**   The class of proximity switch using an RF field typically employs one half of a ferrite pot core whose coil is part of an oscillator circuit. When a metallic object enters this field, at some point the object will absorb enough energy from the field to cause the oscillator to stop oscillating. It is this difference between oscillating and not oscillating that is detected as the difference between an object present or not present. Obviously, there are several variables that determine the distance at which this detection will take place, including:

- The diameter of the pot core (distance varies directly with the core diameter)
- Size of object to be sensed (distance varies directly with size)
- Kind of metal (distance is greatest with iron, less for other metals)
- Circuit sensitivity (distance is established by circuit design)

Sensing distance will also be affected by voltage and temperature to some extent. However, in a well-designed switch, these effects can be minimized. In some units, the sensing distance can be adjusted; usually by means of a multi-turn trimmer potentiometer.

**7.5.2 Capacitive Sensors**   Capacitive sensors also contain oscillators. However, they usually begin oscillating in the presence of an object to be detected when that object creates enough capacitance in a critical part of the oscillator circuit to cause oscillation. In both detection methods (capacitive and inductive), the presence or absence of oscillation is detected and this information is used to do useful work by operating a load directly through a solid-state output circuit or indirectly through a relay.

Figure 7-11 shows two capacitive proximity limit switches. These two capacitive proximity limit switches are used with a separate control unit to provide an intrinsically safe sensing system, approved for Factory Mutual Class I, II, III,

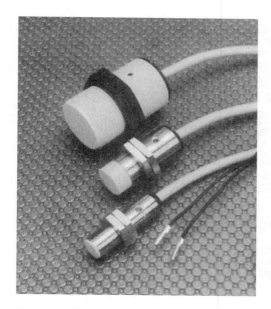

**Figure 7-10**   Two-wire dc proximity
switches. *(Courtesy of ifm efector, inc.)*

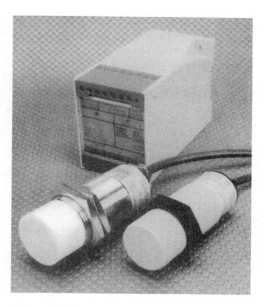

**Figure 7-11**   Capacitive proximity limit
switches. *(Courtesy of ifm efector, inc.)*

Division 1, Group A-C hazardous locations. Typical applications include high/low level detection of liquids or granular solids in tank, hoppers, silos, etc.

See Figures 7-12A through 7-12E for sketches showing typical industrial applications. Figures 7-12F and 7-12G show photos of industrial applications.

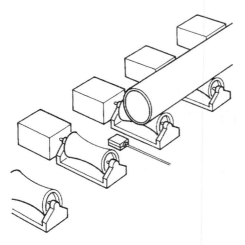

**Figure 7-12A**   Pipes on a conveyor.
*(Courtesy of ifm efector, inc.)*

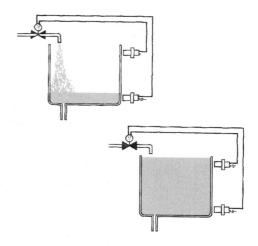

**Figure 7-12B**   Levels in a plastic tank.
*(Courtesy of ifm efector, inc.)*

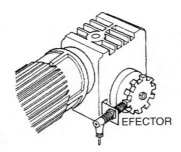

**Figure 7-12C** **Pulse pickup.** *(Courtesy of ifm efector, inc.)*

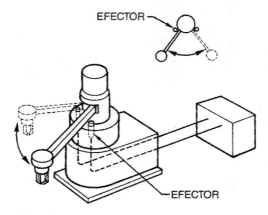

**Figure 7-12D** **Limit switch on robot arm.** *(Courtesy of ifm efector, inc.)*

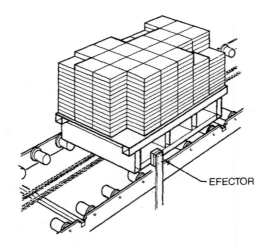

**Figure 7-12E** **Tiles on a conveyor system.** *(Courtesy of ifm efector, inc.)*

***Figure 7-12F*** **Proximity switch in a machine application.** *(Courtesy of ifm efector, inc.)*

***Figure 7-12G*** **Proximity switches in a manufacturing process.** *(Courtesy of ifm efector, inc.)*

## 7.6  LED Indicators

LED indicators are usually available on the outside of the switch to show whether a switch is closed or open, whether an object to be sensed is present or not present, and whether there is line power available. It is possible to use three LEDs to indicate all three conditions, but it is most usual to use only one LED indicating that a switch is closed or open.

## 7.7  Solid-State Outputs

Solid-state output devices respond differently to excessive currents and wrong voltage polarities than electromechanical output devices do. Electromechanical contacts are not particularly affected by excessive currents if those currents are not sustained for too long a period of time, and the contacts do not have to break an inductive circuit. These current conditions that a contact takes in stride might destroy a solid-state output. The solid-state output, however, can be designed so that excessive current conditions cannot occur even under short circuit conditions. Outputs also use reverse voltage protection circuits.

## 7.8  Detection Range

When detecting an object, the sensing range of the switch is an important specification. The nominal sensing range given in a manufacturer's literature typically refers to the distance at which a reasonably large (compared to the sensing area of the switch) piece of mild steel is used as a target. All other metals will be sensed at distances less than this reference metal. The distance at which any target will be sensed depends on its positioning in front of the sensor area. The farther the object to be sensed is from the switch, the more closely it must be centered on the sensing area of the switch. For instance, if the material is very close to the switch, it need not cover the entire sensing area in order to be sensed. However, if the target is at the maximum sensing distance of the switch, the entire sensing area must be covered for the object to be detected.

## 7.9  Hysteresis

As a test target is brought from infinity toward the sensing area, the switch will transfer at some distance from the sensing face within the attenuation range. As the target is removed to infinity, the switch will switch back at some distance from the sensing face. (See Figure 7-13.) These two distances will be different by a small amount. This difference is called *hysteresis* and is usually expressed as a

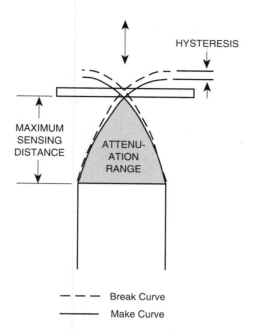

**Figure 7-13**   **Hysteresis.** *(Courtesy of ifm efector, inc.)*

percentage of the switching distance. The percentage is the difference between the two sensing distances divided by the inbound sensing distance times 100. Hysteresis is introduced to prevent indecision and rapid on/off switching of the device when the target is right on the edge of the trip point.

## 7.10  Attenuation Range

The target can also approach the sensing area laterally in a plane parallel to the sensing area. (See Figure 7-14.) A point will be reached in from the edge of the sensing area where the switch will trip. The closer the object is to the sensing surface, the closer to the edge of the surface the trip point will be. There will be an identical point on the other side of the surface as the target leaves the sensing area. The distance between the pickup and drop-out point is called the *attenuation range*. Notice that the sensor picks up on the make curve on one side and drops out on the break curve on the other side. Obviously, the attenuation range decreases as the distance from the sensing area increases until a point is reached where the attenuation range is zero.

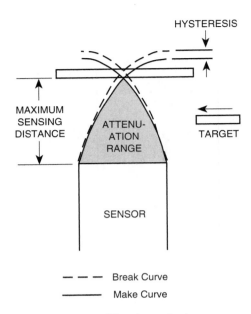

**Figure 7-14**   **Attenuation range.** *(Courtesy of ifm efector, inc.)*

## 7.11 Speed

No electronic switch produces an output immediately upon detecting an object. There is always some delay. The time that is required to detect the presence of an object and produce a full output to the load is called the *response time* of the switch.

The response time of a proximity switch is a function of the frequency of the oscillator in the switch. The higher the oscillator frequency, the shorter the response time and the higher the potential switching frequency. However, in ac switches, the limiting factor in the number of times that the output can switch depends on the frequency of the alternating voltage of the line. In general, dc units, can switch their outputs at higher rates than ac units. Small switches with short distances usually have higher switching rate capability than larger units.

## 7.12 Magnet-Operated Limit Switch

The magnet-operated proximity limit switch (Figure 7-15) operates by passing an external magnet near the face of the sensing head to actuate a small, hermetically sealed reed switch. The 120-V ac pilot-duty model also includes an epoxy-encapsulated triac output.

**Figure 7-15**   **Magnet-operated limit switch.** *(Courtesy of General Electric Company, General Purpose Control, Bloomington, IL.)*

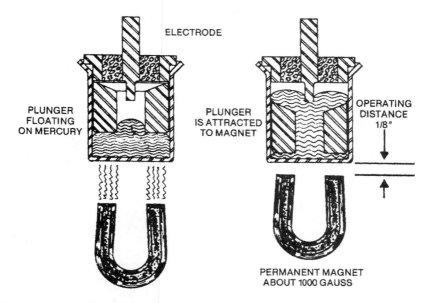

**Figure 7-16**   **Mercury proximity limit switch. This magnetic proximity switch requires a magnetic field of about 1000 gauss to attract the floating magnetic plunger down, thereby displacing mercury upward to the electrode. Rating: 1/3 hp @ 120 V ac.** *(Courtesy of Eagle Signal Controls.)*

The mercury proximity limit switch operates by passing a permanent magnet of sufficient strength past the switch. The sketch shown in Figure 7-16 displays the internal working parts and the relationship between the strength of the magnet and the operating distance of a mercury switch.

## 7.13  Vane Switches

The vane-operated limit switch is actuated by passage of a separate steel vane through a recessed slot in the switch (Figure 7-17). Either vane or switch can be attached to the moving part of the machine. As the vane passes through the slot, it changes the balance of the magnetic field, causing the contacts to operate. The switch is available with either a normally open or normally closed contact.

The switch can detect very high speeds of vane travel without detrimental effects such as arm or mechanism wear or breakage. There is no physical contact between vane and switch. Therefore, the upper limit on vane speed is governed by factors other than the switch.

The vane switch offers excellent accuracy and response time. Provided the path of the vane through the slot is constant, repeatability is constant within ±0.0025 inch or less. The time required for the switch to operate after the vane has reached the operating point (response time) is less than a millisecond.

Figure 7-18 shows an oil-tight and dust-tight vane-operated limit switch. It provides high reliability and long life, with an electrical rating capable of handling high inductive loads.

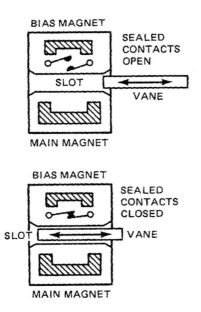

*Figure 7-17*  **Cross section of vane-operated limit switch.**

*Figure 7-18*  **Vane-operated limit switch with indicating light.** *(Courtesy of General Electric Company, General Purpose Control, Bloomington, IL.)*

# 7.14 Linear Position Displacement Transducers

In closed-loop control applications it is necessary to convert some physical property into an electrical signal in order to provide a feedback signal to the amplifier. The components that carry out this conversion are known as *transducers*.

There are many different devices available to measure linear movement. Factors such as accuracy required and environment will normally determine which one to use.

The simplest device would be a linear potentiometer in which the wiper is connected to the moving component and provides a voltage proportional to its position along the potentiometer winding. Problems of linearity, limited resolution, mechanical wear, etc. may occur in practice. The use of linear potentiometers as feedback transducers is, therefore, rather limited, though they are very inexpensive devices and may be suitable for simpler applications.

To overcome problems of mechanical wear, a noncontact device is required such as that shown in Figure 7-19, which illustrates a linear variable differential transformer (LVDT). It consists of a primary and two secondary coils surrounded by a soft iron core connected to the moving component. The primary coil is connected to a high-frequency ac supply, and voltages are induced in the two secondaries by transformer action. If the two secondary coils are connected in opposition, with the core centralized, the induced voltages in each coil will cancel out and produce zero output. As the core is moved away from the center, the voltage induced in one secondary will increase, while that in the other will reduce. This condition now produces a net output voltage, the magnitude of which is proportional to the amount of movement and the phase determined by the direction. The output can then be fed to a phase-sensitive rectifier (known as a *demodulator*), which will produce a dc signal proportional to movement and polarity, depending on direction.

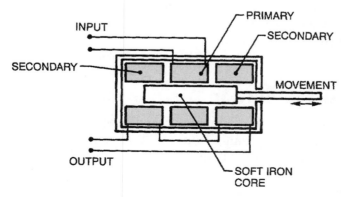

*Figure 7-19*  Linear variable differential transformer. *(Courtesy of Eaton Corp.)*

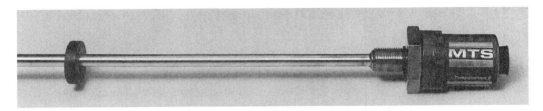

**Figure 7-20**  **Linear displacement transducer.** *(Courtesy of MTS Sensors Division.)*

The solid-state transducer shown in Figure 7-20, by precisely sensing the position of an external magnet, is able to measure linear displacements with infinite resolution. Since there is no contact between the magnet and the sensor rod, there is no wear, friction, or degradation of accuracy. The outputs of a solid-state transducer represent an absolute position, rather than an incremental indication of position change. Either digital or analog output is available.

The measuring principle by which this transducer operates can be explained as follows: when a current pulse is sent through a wire (which has been threaded through a tube and returned outside), the resulting magnetic field is concentrated in the tube, which acts as a wave guide. If this tube is then passed through a doughnut-shaped magnet, the two magnetic fields interact. A tube made of magnetostrictive material will experience a local rotary strain where these fields interact. This strain will continue for the duration of the electrical pulse. The rotary strain pulse travels along the wave guide element at ultrasonic speed and can be detected at the end of the tube.

By measuring the time from the generation of the initial electrical pulse until the ultrasonic pulse is detected, the distance of the external magnet from reference point can be determined.

This transducer is used for either position readout or closed-loop control. It may be mounted inside hydraulic cylinders or externally on machines.

The applications are very broad, covering small winding machines, machine tools, plastic forming machines, etc.

## 7.15 Angular Position Displacement Transducers

In many industrial applications, angular displacement is an important consideration or requirement. It may range from a few degrees to many revolutions of a shaft. The motion required for a specific application must be accurate in very small increments of change.

The applications for motion control are many. They include machining, positioning of tools, testing, inspection, welding, and assembly. In each application

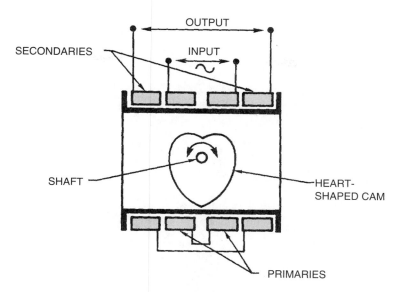

**Figure 7-21** **Rotary LVDT (RVDT).** *(Courtesy of Eaton Corp.)*

a rotary variable differential transformer or drive motor and control combination is designed to satisfy the requirements.

As with linear motion, rotary potentiometers can be used for measuring angular position, but the same problems of wear, etc. apply, so again they are confined to relatively simple applications. The rotary equivalent of an LVDT is known as an RVDT (rotary variable differential transformer) (Figure 7-21).

A specially shaped cam of magnetic material is rotated inside the primary and secondary coils by the input shaft. The profile of the cam determines the amount of magnetic coupling between primary and secondary; hence, as with the LVDT, an output signal proportional to shaft rotation is provided. The limitation in this case is the maximum angular movement that can be achieved and still produce a linear output, which in practice may be of the order of ±60 degrees.

**7.15.1 Rotary Encoder**   Another approach to position control that involves the transfer of angular position to a usable signal is the *rotary encoder,* which can be either electromechanical, electronic, optical, or a combination of all three. It can be used to monitor the rotary motion of a device. There are two types:

1. The *incremental* encoder, which transmits a specific quantity of pulses for each revolution of a device.
2. The *absolute* encoder, which provides a specific code for each angular position of the device. It may be in binary coded decimal (BCD) or Gray code (see number system in Appendix H). The *Gray code* is defined as sequential numbers by binary values in which only one value changes at a time.

**Figure 7-22** **Rotary shaft encoder.** *(Courtesy of Automatic Timing and Controls Company, Inc.)*

The rotary encoder shown in Figure 7-22 is an optical incremental rotary shaft encoder. It uses an infrared LED and precision optical and mechanical components to produce a series of pulses corresponding to shaft rotation. Output options include standard square wave, directional high/low, and quadrature. It also features reverse polarity and output short-circuit protection.

The wiring circuit for this unit is shown in Figure 7-23A. The three optional outputs are shown in Figure 7-23B. The unit shown in Figure 7-24 uses optical sensing to provide incremental conversion of rotary shaft motion into a digital output signal. An LED and phototransistor circuit with a slotted disk produces square wave output pulses. Some models with 10 pulses per revolution to 1200 pulses per revolution are available in unidirectional and bidirectional (quadrature) output types. Output pulse rate is 20,000 pulses per second maximum, and shaft speed is 4000 rpm continuous and 6000 rpm maximum.

The unit shown in Figure 7-25 offers a noncontacting optical design that allows high-speed, low-torque operation and includes an LED light source. This optical programmable controller encoder (absolute) is specifically designed to interface with programmable controllers and contains all the necessary electronics to provide a latched output on command from a programmable controller. A data-ready output signals the programmable controller when the encoder data is

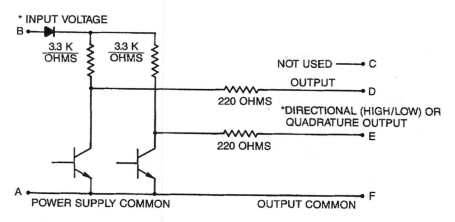

*Directional (High/Low) or quadrature are available output options

Concerning safety — ATC makes every effort to build a safe product.
We try to state specifications accurately. But every product made will eventually
fail, so we design our products into equipment so that they fail safely.

**Figure 7-23A** **Wiring circuit for encoder.** *(Courtesy of Automatic Timing and Controls Company, Inc.)*

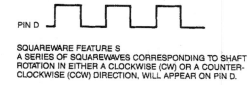

SQUAREWARE FEATURE S
A SERIES OF SQUAREWAVES CORRESPONDING TO SHAFT
ROTATION IN EITHER A CLOCKWISE (CW) OR A COUNTER-
CLOCKWISE (CCW) DIRECTION, WILL APPEAR ON PIN D.

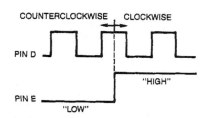

DIRECTIONAL HIGH/LOW FEATURE D
SQUAREWAVE WILL APPEAR ON PIN D REGARDLESS OF
DIRECTION OF SHAFT ROTATION. PIN E WILL BE "HIGH"
FOR CW (UP COUNT) AND "LOW" FOR CCW (DOWN COUNT).

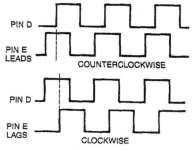

QUADRATURE FEATURE Q
QUADRATURE OUTPUT ON PIN E WILL LEAD THE PIN D
SQUAREWAVE OUTPUT FOR CCW SHAFT ROTATION AND LAG
THE PIN D SQUAREWAVE OUTPUT FOR CW SHAFT ROTATION.

**Figure 7-23B** **Three optional outputs for rotary encoder.** *(Courtesy of Automatic Timing and Controls Company, Inc.)*

**Figure 7-24**   Rotary encoder.
*(Courtesy of Eagle Signal Controls.)*

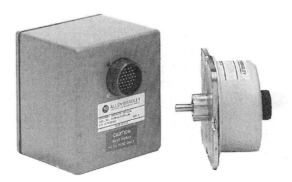

**Figure 7-25**   Optical programmable controller
**encoder.** *(Courtesy of Allen-Bradley, a Rockwell
International Company.)*

latched. The encoders are capable of being multiplexed, allowing one programmable controller with one set of input cards/modules to accept data from many encoders. The multiplex and latch circuits operate independently.

The encoder shown in Figure 7-25 is a single-turn (0–359 degrees), with accuracy of ±1 degree and resolution of 1 degree.

In summary, then, it can be said that the encoder electronically digitizes shaft motion of a rotating element; that is, it converts mechanical motion to an electronic digital format for a controller. By coupling the encoder shaft to a lead screw, rack and pinion, or rotating element, its electronic output can be used to measure distance from a known position and direction of rotation.


# 7.16  Use of ac Synchronous and dc Stepping Motors

One supplier of motion control equipment offers an ac synchronous motor and a dc stepping motor made under the trademark of SLO-SYN®*. SLO-SYN ac synchronous motors are permanent-magnet ac motors with extremely rapid starting, stopping, and reversing characteristics. They have a basic shaft speed of 72 or 200 rpm, depending on the motor selected. See Figure 7-26 for an exploded view.

Three lead types need only a single-pole, three-position switch for complete forward, reverse, and OFF control. When operating a SLO-SYN motor from a single-phase source, a phase-shifting network consisting of a resistor and capacitor is used (Figure 7-27).

*Information on SLO-SYN® motors is courtesy of Superior Electric Company.

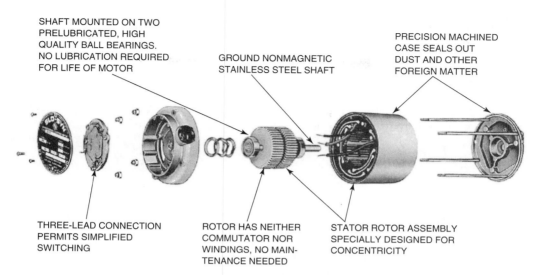

SHAFT MOUNTED ON TWO
PRELUBRICATED, HIGH
QUALITY BALL BEARINGS.
NO LUBRICATION REQUIRED
FOR LIFE OF MOTOR

GROUND NONMAGNETIC
STAINLESS STEEL SHAFT

PRECISION MACHINED
CASE SEALS OUT
DUST AND OTHER
FOREIGN MATTER

THREE-LEAD CONNECTION
PERMITS SIMPLIFIED
SWITCHING

ROTOR HAS NEITHER
COMMUTATOR NOR
WINDINGS, NO MAIN-
TENANCE NEEDED

STATOR ROTOR ASSEMBLY
SPECIALLY DESIGNED FOR
CONCENTRICITY

**Figure 7-26**  **Exploded view of a SLO-SYN® ac synchronous motor.** *(Courtesy of Superior Electric Company.)*

Standard SLO-SYN motors are available in torque ratings of 22–1500 ounce-inches. The output torque varies with changes in the input voltage.

When a SLO-SYN motor is energized, ac current flows only through the windings. Current does not flow through the rotor or through brushes, since the motor is of a brushless construction. Therefore, it is not necessary to consider high inrush currents when designing a control for a SLO-SYN motor, because starting and operating current are, for all practical purposes, identical.

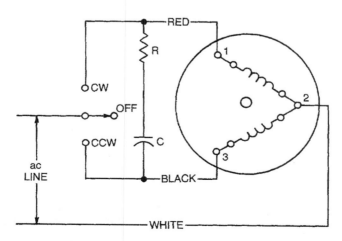

**Figure 7-27**  **Input connections for ac operation.**

The permanent-magnet construction of a SLO-SYN motor provides a small residual torque that holds the motor shaft in position when the motor is deenergized. When holding torque is required, a dc voltage can be applied to one winding when the ac voltage is removed. If necessary, dc voltage can be applied to both windings with a resulting increase in holding torque (Figure 7-28).

In determining the ac synchronous motor required for a specific application, the following information must be known:

1. *Motor shaft speed.* SLO-SYN motors are available in shaft speeds of 72 and 200 rpm at 60 Hz. Gearing can be used to obtain other speeds. It is also necessary to consider the effects of gears (if used) on torque and inertial load characteristics of the load.

2. *Load characteristics.* Examples that follow show how to determine the torque and moment of inertia characteristics of the load.

   a. *Torque* in ounce-inches = *Fr* (Figure 7-29), where *F* is force in ounces and *r* is radius in inches. Example: Using a 4-inch-diameter pulley it is found that a 2-pound pull on the scale is required to rotate the pulley:

$$F = 2 \text{ pounds} = 32 \text{ ounces}$$

$$r = \frac{4}{2} = 2 \text{ inches}$$

The torque is, then, $32 \times 2 = 64$ ounce-inches

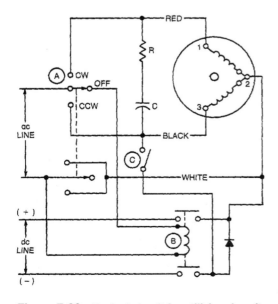

| MODE | (A) | (B) | (C) |
|---|---|---|---|
| HOLDING | "OFF" | CLOSED | CLOSED |
| START | OPEN | OPEN | CLOSED |
|  | OPEN | OPEN | OPEN |
| RUN | CW OR CCW | OPEN | OPEN |
| STOP | OPEN | OPEN | CLOSED |
|  | "OFF" | CLOSED | CLOSED |

SWITCHING SEQUENCE

*Figure 7-28*  Typical circuit for utilizing dc voltage to provide holding torque.

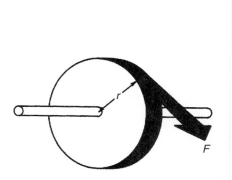

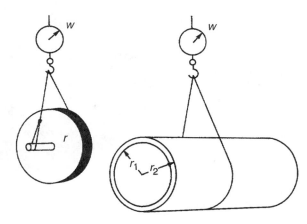

**Figure 7-29**  Torque.                    **Figure 7-30**  Moment of inertia.

b. *Moment of inertia* in lb.-in.$^2$ (Figure 7-30):

$$I\,(\text{lb.-in.}^2) \;=\; \frac{Wr^2}{2} \qquad \text{for a disc}$$

$$I(\text{lb.-in.}^2) \;=\; \frac{W}{2}\,(r_1^2 + r_2^2) \qquad \text{for a cylinder}$$

where $W$ is the weight in pounds and $r$ is the radius in inches. Example: Using a load of an 8-inch-diameter gear weighing 8 ounces:

$$W = \frac{8}{6} = 0.5 \text{ lb.}$$

$$r = \frac{8}{2} = 4 \text{ in.}$$

$$\text{Moment of inertia} = \frac{0.5 \times (4)^2}{2} = 4 \text{ lb.-in.}^2$$

With voltage and frequency known, the user should refer to the manufacturer's tables of torque and moment of inertia to select the motor that best suits the requirements.

Another motor offered by Superior Electric Company is the SLO-SYN dc stepping motor. It is a permanent-magnet motor that converts electronic signals into mechanical motion (Figure 7-31).

SLO-SYN stepping motors operate on phase-switched dc power. Each time the direction of current in the motor windings is changed, the motor output shaft rotates a specific angular distance. The motor shaft can be driven in either direction and can be operated at very high stepping rates.

**Figure 7-31**  **A dc SLO-SYN® stepping motor.** *(Courtesy of Superior Electric Company.)*

The motor shaft advances 200 steps per revolution (1.8 degrees per step) when a four-step input sequence (full-step mode) is used, and 400 steps per revolution (0.9 degrees per step) when an eight-step input sequence (half-step mode) is used. Power transistors connected to flip-flops or other logic devices are normally used for switching, as shown in the wiring diagram in Figure 7-32. The four-step input sequence is also shown here. Since current is maintained on the motor windings when the motor is not being stepped, a high holding torque results.

A SLO-SYN dc stepping motor offers many advantages as an actuator in a digitally controlled positioning system. It easily interfaces with a microcomputer or microprocessor to provide opening, closing, rotating, reversing, cycling, and highly accurate positioning in a variety of applications.

Mechanical components such as gears, clutches, brakes, and belts are not needed because stepping is accomplished electronically.

A SLO-SYN dc stepping motor applies holding or detent torque at standstill to help prevent an unwanted motion.

The motors are available in a range of frame sizes and with standard step angles of 0.72 degrees, 1.8 degrees, and 5 degrees. Step accuracies are 3% or 5%, noncumulative. They can be driven at rates to 20,000 steps per second with a minimum of power input.

In determining the motor required for a specific dc stepping motor application, the following information must be determined:

- Output speed in steps per second
- Torque in ounce-inches
- Load inertia in lb.-in.$^2$
- Required step angle
- Time to accelerate in milliseconds
- Time to decelerate in milliseconds
- Type of drive system to be used
- Size and weight considerations

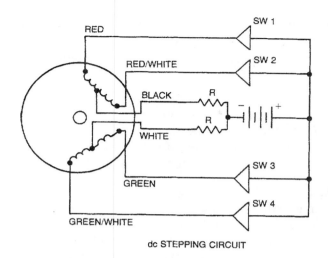

dc STEPPING CIRCUIT

FOUR-STEP INPUT SEQUENCE*
(FULL STEP SEQUENCE)

| STEP | SW1 | SW2 | SW3 | SW4 |
|------|-----|-----|-----|-----|
| 1 | ON | OFF | ON | OFF |
| 2 | ON | OFF | OFF | ON |
| 3 | OFF | ON | OFF | ON |
| 4 | OFF | ON | ON | OFF |
| 1 | ON | OFF | ON | OFF |

\* Provides CW rotation as viewed from nameplate end of motor. To reverse direction of motor rotation perform switching steps in the following order; 1,4,3,2,1.

**Figure 7-32**   A dc stepping circuit.

1. *Torque* in ounce-inches = $Fr$, where $F$ (in ounces) is force required to drive the load and $r$ (in inches) is radius.
2. *Moment of inertia*:

$$I(\text{lb.-in.}^2) = \frac{Wr^2}{2} \qquad \text{for a disc}$$

$$I(\text{lb.-in.}^2) = \frac{W}{2}(r_1^2 + r_2^2) \qquad \text{for a cylinder,}$$

where $W$ is the weight in pounds and $r$ is the radius in inches.
3. *Equivalent inertia*. A motor must be able to:

   a. Overcome any frictional load in the system.

b. Start and stop all inertia loads including that of its own rotor.
The basic rotary relationship is:

$$T = \frac{I\alpha}{24}$$

where  $T$ = torque in ounce-inches
$I$ = moment of inertia
$\alpha$ = angular acceleration in radians per second$^2$
24 = conversion factor from ounce-inch-sec$^2$ × 24 lb.-in.$^2$

*Measuring angles in radians.* A radian is the angle at the center of a circle that embraces an arc equal in length to the length of the radius. The value of the radian in degrees equals 57.296 degrees. Then, 3.1416 radians denotes an angle of 180 degrees. Note that it is often convenient to measure angles in radians when dealing with angular velocity. If $\omega$ = angular velocity per second of the revolving body, in radians; $V$ = velocity of a point on the periphery of the body, in feet per second; and $r$ = radius in feet, then:

$$\omega = \frac{V}{r}$$

For example, if the velocity of a point on the periphery is 10 feet per second and the radius is 1 foot, then:

$$\omega = \frac{10}{1} = 10 \text{ radians per second}$$

Angular acceleration ($\alpha$) is a function of the change in velocity ($\omega$) and the time ($t$) required for the change:

$$\alpha = \frac{\omega_2 - \omega_1}{t}$$

Or, if starting from 0:

$$\alpha = \frac{\omega}{t}$$

where $\omega$ = angular velocity in rad/sec and $t$ = time in seconds.
Since:

$$\omega = \frac{\text{Steps per second}}{\text{Steps per revolution}} \times 2\pi$$

angular velocity and angular acceleration can also be expressed in steps per second ($\omega'$) and steps per second$^2$ ($\alpha'$), respectively.

## SAMPLE CALCULATIONS

1. To calculate the torque required to rotationally accelerate an inertia load (Figure 7-33):

$$T = 2 \times I_o \left( \frac{\omega'}{t} \times \frac{\pi\theta}{180} \times \frac{1}{24} \right)$$

where  $T$ = torque required in ounce-inches
$I_O$ = inertial load in lb.-in.$^2$
$\pi$ = 3.1416
$\theta$ = step angle in degrees
$\omega'$ = step rate in steps per second

**Example 1:** Assume the following conditions:

- Inertia = 9.2 lb.-in.$^2$
- Step angle = 1.8 degrees
- Acceleration = 0–1000 steps per second in 0.5 seconds

Then:

$$T = 2 \times 9.2 \times \frac{1000}{0.5} \times \frac{\pi \times 1.8}{180} \times \frac{1}{24}$$

$$= 48.2 \text{ ounce-inches to accelerate inertia}$$

2. To calculate the torque required to accelerate a mass moving horizontally and driven by a rack and pinion or similar device (Figure 7-34):

   The total torque that the motor must provide includes the torque required to:

   a. Accelerate the weight, including that of the rack
   b. Accelerate the gear
   c. Accelerate the motor rotor
   d. Overcome frictional forces

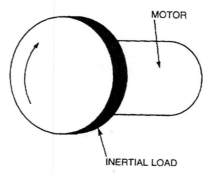

**Figure 7-33**  Torque problem 1.

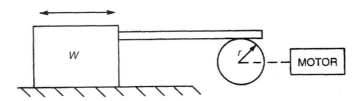

***Figure 7-34*** **Torque problem 2.**

To calculate the rotational equivalent of the weight:

$$I_{eq} = Wr^2,$$

where $W$ is the weight in pounds and $r$ is the radius in inches.

**Example 2:** Assume the following:

- Weight = 5 lb.
- Gear pitch diameter = 3 inches
- Gear radius = 1.5 inches
- Velocity = 15 feet per second
- Time to reach velocity = 0.5 second
- Pinion inertia = 4.5 lb.-in.$^2$
- Motor rotor inertia = 2.5 lb.-in.$^2$

$$
\begin{aligned}
I_{eq} \;=\; Wr^2 = 5(1.5)^2 \;&=\; 11.25 \text{ lb.-in.}^2 \\
\text{Pinion} \;&=\; 4.5 \text{ lb.-in.}^2 \\
\text{Rotor} \;&=\; 2.5 \text{ lb.-in.}^2 \\
\text{Total} \;&=\; 18.25 \text{ lb.-in.}^2
\end{aligned}
$$

Velocity is 15 feet per second, with a 3-inch pitch diameter gear. Therefore:

a.  Speed $= \dfrac{15 \times 2}{3 \times 3.1416} = 19.1$ revolutions per second

b. The motor step angle is 1.8° (200 steps per revolution)

c. Velocity in steps per second $\omega' = 19.1 \times 200 = 3820$ steps per second

d. To calculate torque to accelerate the system:

$$T = 2 \times I_o \times \frac{\omega}{t} \times \frac{\pi\theta}{180} \times \frac{1}{24}$$

$$= 2 \times 18.25 \times \frac{3820}{0.5} \times \frac{3.1416 \times 1.8}{180} \times \frac{1}{24} = 365 \text{ ounce-inches}$$

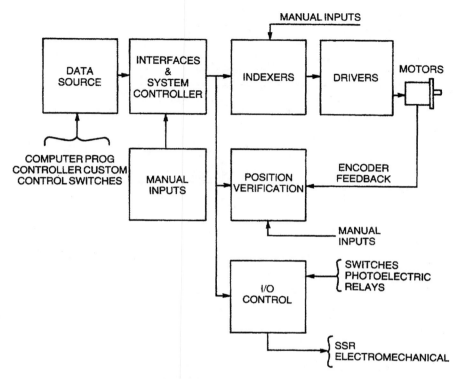

**Figure 7-35** Typical system block diagram.

To calculate torque needed to slide the weight, assume a frictional force of 6 ounces. Then $T_{\text{friction}} = 6 \times 1.5 = 9$ ounce-inches. Therefore total torque required $= 365 + 9 = 374$ ounce-inches.

There are many ways in which these motors can be controlled. In general, the design of the control will depend on the particular application.

A typical system block diagram is shown in Figure 7-35, which illustrates the versatile building block approach provided by Superior Electric's MODUL-YNX motor controls. This approach allows the user to select the combination of options and features needed for a specific application.

## 7.17 Servo Positioning Control

The servo is controlled in a closed-loop configuration similar to Figure 7-35. Typical controllers are designed to control from one to three independent servo motors. For example, Figure 7-36 illustrates the ability of a three-axis controller

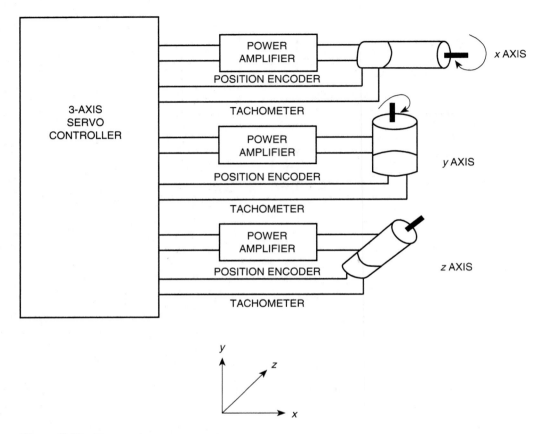

***Figure 7-36***   **Three-axis servo control.**

to simultaneously monitor the position feedback of each axis's motor, and provide appropriate current and voltage output to the axis's motor power amplifier.

The role of the controller is to translate the profile of travel desired by the operator into the exact position of the servos. Typically, controllers are applied to:

- Machining centers with multiple cutting devices
- Robots with multiple coordinates of travel

Figure 7-37 shows a three-axis robot that would need a three-axis controller to position each of its arms to move to a desired path. Figure 7-38A and 7-38B illustrate two examples of the programmed path of an object as it is extended and retracted. Note the changes in acceleration and deceleration as the object travels.

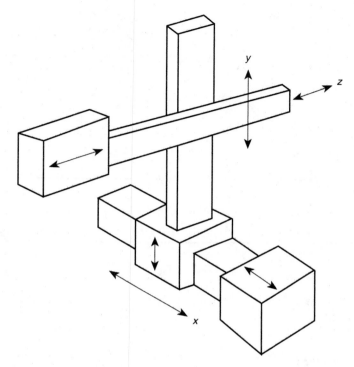

***Figure 7-37*** **Three-axis robot.**

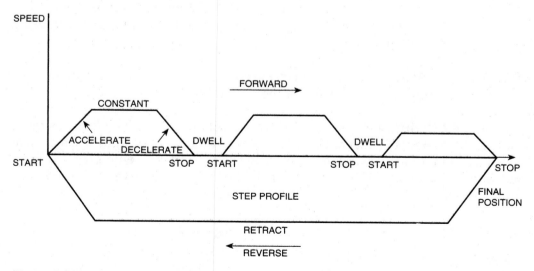

***Figure 7-38A*** **Movement profile involving a dwell time.**

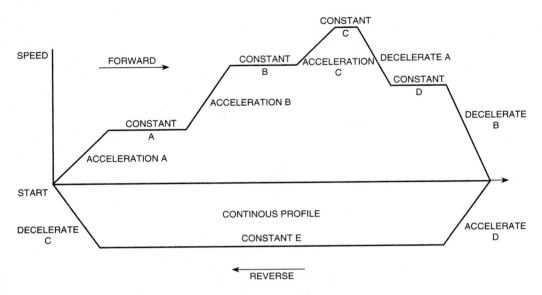

**Figure 7-38B**   Movement profile involving multiple acceleration and deceleration rates.

# 7.18  Sensing Theory*

Like nearly any other technology, photoelectric and ultrasonic sensing have their own sets of buzzwords. This section introduces some of the more important terms. Some are not universal and a few definitions have developed several names. Since it is unlikely that these terms will become standardized in the near future, it is best to be familiar with all of them and to be comfortable with using the synonyms interchangeably.

**Photoelectric Sensor**   A photoelectric sensor is an electrical device that responds to a change in the intensity of the light falling on it.

**LED (Light Emitting Diode)**   An LED is a solid-state semiconductor, similar electrically to a diode, except that it emits a small amount of light when current flows through it in a forward direction. LEDs can be built to emit green, yellow, red, or infrared light. (Infrared light is invisible to the human eye.) LEDs are usually built into sensors as a light source to change with process conditions.

*Courtesy of Banner Engineering Corp.

**Phototransistor**    The phototransistor is the most widely used optoelement in industrial photoelectric sensor design. It offers the best trade-off between light sensitivity and response speed compared to photoresistive and other photojunction devices. Photocells are used whenever greater sensitivity to visible wavelengths is required, as in some color registration and ambient light detection applications. Photodiodes are generally reserved for applications requiring either extremely fast response time or linear response over several magnitudes of light level change.

**Modulated LED Sensors**    Unlike their incandescent equivalents, LEDs can be turned "on" and "off" (or modulated) at a high rate of speed, typically at a frequency of several kilohertz. (See Figure 7-39.) This modulating of the LED means that the amplifier of the phototransister receiver can be "tuned" to the frequency of modulation and amplify only light signals pulsing at that frequency.

    This process is analogous to the transmission and reception of a radio wave of a particular frequency. A radio receiver tuned to one station ignores other radio signals that may be present in the room. The modulated LED light source of a photoelectric sensor is usually called the *transmitter* (or *emitter*), and the tuned photodevice is called the *receiver*.

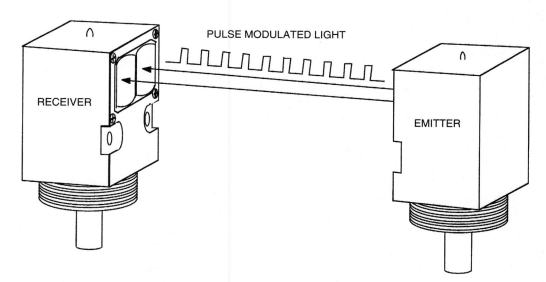

**Figure 7-39**    **A modulated (pulsed) light source.** *(Courtesy of Banner Engineering Corporation.)*

**Ambient Light Receiver**   One type of nonmodulated photoelectric device still found in frequent use is the ambient light receiver. Products like red-hot metal or glass emit large amounts of infrared light. As long as these materials emit more light than the surrounding light level, they may be reliably detected by an ambient light receiver (see Figure 7-40).

**Ultrasonic Sensors**   Ultrasonic sensors emit and receive sound energy at frequencies above the range of human hearing (above about 20kHZ). Ultrasonic sensors are categorized by transducer type: either electrostatic or piezoelectric. Electrostatic types can sense objects up to several feet away by reflection of ultrasound waves from the object's surface. Piezoelectric types are generally used for sensing at shorter ranges (see Figure 7-41).

**Remote Photoelectric Sensors**   Remote photoelectric sensors contain only the optical components of the sensing system. The circuitry for system power, amplification, and output are all at another location, typically in a control panel. Consequently, remote sensors are generally smaller and more tolerant of hostile sensing environments than are self-contained sensors.

**Self-Contained Photoelectric Sensors**   Self-contained photoelectric sensors contain the optics along with all the electronics. Their only requirement is a

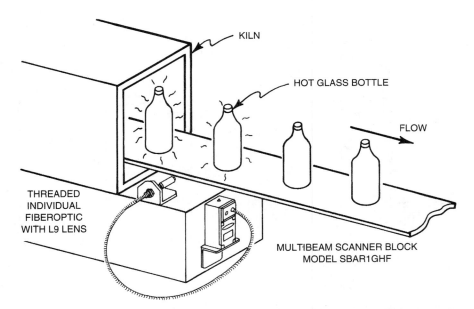

*Figure 7-40*   Counting bottles during the manufacturing process by detecting infrared (heat) radiation with an ambient light receiver. *(Courtesy of Banner Engineering Corporation.)*

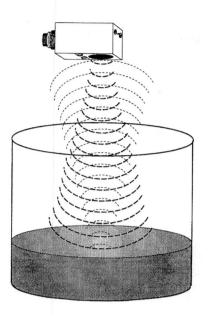

**Figure 7-41**   **Detection of ultrasonic sound energy.** *(Courtesy of Banner Engineering Corporation.)*

source of voltage for power. The sensor itself does all of the work, which includes modulation, demodulation, amplification, and output switching.

**Fiber Optics**   There are many sensing situations where space is too restricted or the environment too hostile even for remote sensors. For such applications, photoelectric sensing technology offers fiber optics as a third alternative in sensor "packaging." Fiber optics are transparent strands of glass or plastic that are used to conduct light energy into and out of such areas.

Fiber-optic "light pipes," used along with either remote or self-contained sensors, are purely passive, mechanical components of the sensing system. Since fiber optics contain no electrical circuitry and have no moving parts, they can safely pipe light into and out of hazardous sensing locations and withstand hostile environmental conditions. Moreover, fiber optics are completely immune to all forms of electrical "noise" and may be used to isolate the electronics of a sensing system from known sources of electrical interference.

**Sensing Modes**   The optical system of any photoelectric sensor is designed for one of three basic sensing modes: opposed, retroreflective, or proximity. The photoelectric proximity mode is further divided into four submodes: diffuse proximity, divergent-beam proximity, convergent-beam proximity, and fixed-field proximity. Ultrasonic sensors are designed for either opposed- or proximity-mode sensing.

**Opposed Mode**   Opposed-mode sensing is often referred to as "direct scanning" and is sometimes called the "beam break" mode. In the opposed mode the emitter and receiver are positioned opposite each other so that the sensing energy from the emitter is aimed directly at the receiver. An object is detected when it interrupts the sensing path established between two sensing components. (See Figure 7-42.)

**Retroreflective Mode**   The photoelectric retroreflective sensing mode is also called the "reflex" mode, or simply the "retro" mode. (See Figure 7-43.) A retroreflective sensor contains both emitter and receiver circuitry. A light beam is established between the emitter, the retroreflective target, and the receiver. Just as in the opposed-mode sensing, an object is sensed when it interrupts the beam.

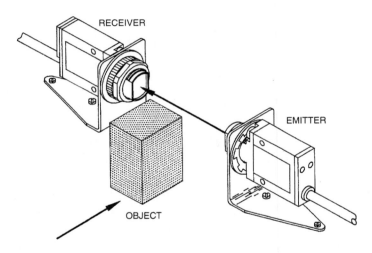

**Figure 7-42**   **Opposed sensing mode.** *(Courtesy of Banner Engineering Corporation.)*

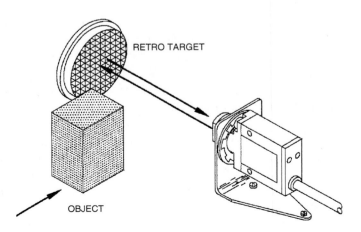

**Figure 7-43**   **Retroreflective sensing mode.** *(Courtesy of Banner Engineering Corporation.)*

**Proximity Mode**   Proximity-mode sensing involves detecting an object that is directly in front of a sensor by detecting the sensor's own transmitted energy reflected back from the object's surface. For example, an object is sensed when its surface reflects a sound wave back to an ultrasonic proximity sensor. Both the emitter and receiver are on the same side of the object, usually together in the same housing. In proximity sensing modes, an object, when present, actually "makes" (establishes) a beam, rather than interrupts the beam. Photoelectric proximity sensors have several different optical arrangements. They are described under the following headings: diffuse, divergent, convergent beam, and fixed-field.

*Diffuse Mode*   Diffuse-mode sensors are the most commonly used type of photoelectric proximity sensor. In the diffuse sensing mode, the emitted light strikes the surface of an object at some arbitrary angle. The light is then diffused from that surface at many angles. The receiver can be at some other arbitrary angle and some small portion of the diffused light will reach it. (See Figure 7-44.)

*Divergent*   To avoid the effects of signal loss from shiny objects, special short-range, unlensed divergent-mode sensors should be considered. By eliminating collimating lenses, the sensing range is shortened, but the sensor is also made much less dependent on the angle of incidence of its light to a shiny surface that falls within its range. (See Figure 7-45.)

*Convergent Beam*   Another proximity mode that is effective for sensing small objects is the convergent beam mode. Most convergent beam sensors use a lens system that focuses the emitted light to an exact point in front of the sensor and focuses the receiver element at the same point. This design produces a small, intense, and well-defined sensing area at a fixed distance from the sensor lens. (See Figure 7-46.)

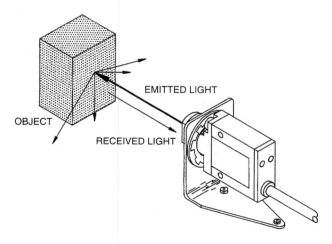

**Figure 7-44**   **Diffuse sensing mode.** *(Courtesy of Banner Engineering Corporation.)*

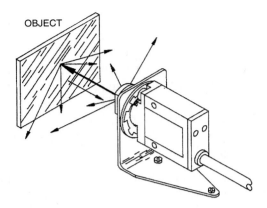

**Figure 7-45** **Divergent proximity sensing mode.** *(Courtesy of Banner Engineering Corporation.)*

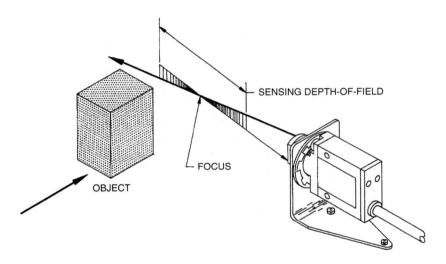

**Figure 7-46** **Convergent beam sensing mode.** *(Courtesy of Banner Engineering Corporation.)*

*Fixed Field* There is a photoelectric proximity mode that has a definite limit to its sensing range. Fixed-field sensors ignore objects that lie beyond their sensing range, regardless of the object surface reflectivity.

Fixed-field sensors compare the amount of reflected light that is seen by two differently aimed receiver optoelements. A target is recognized as long as the amount of light reaching receiver R2 is equal to or greater than the amount "seen" by R1. The sensor's output is cancelled as soon as the amount of light at R1 becomes greater than the amount of light at R2. (See Figure 7-47.)

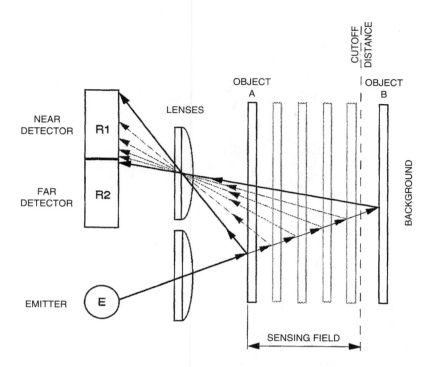

**Figure 7-47** Fixed-field proximity sensing mode. Object is sensed if amount of light at R1 is greater than the amount of light at R2. *(Courtesy of Banner Engineering Corporation.)*

## 7.19 Flow Sensors*

The ifm efector flow sensors (Figure 7-48) are rugged devices constructed of corrosion resistant material. These sensors measure the flow rate of fluids or gases and provide an analog or switched output.

The operating principle of a flow sensor is based on the calorimetric principle. These sensors use the cooling effect of a flowing fluid or gas to monitor the flow rate. The amount of thermal energy that is removed from the tip determines the local flow rate. As shown in Figure 7-49, the temperature differential will be the highest at low flow; therefore, a small increase in flow velocity will remove a large amount of heat from the tip. Tip temperature is lower at higher flow velocities, and so small changes in flow velocities will remove proportionally smaller amounts of heat from the tip. This temperature-based operating principle can reliably sense the flow of virtually any liquid or gas.

An example of how a flow sensor is used can be seen in Figure 7-50

*Courtesy of ifm efector, inc.

***Figure 7-48***   **Flow products.**  *(Courtesy of ifm efector, inc.)*

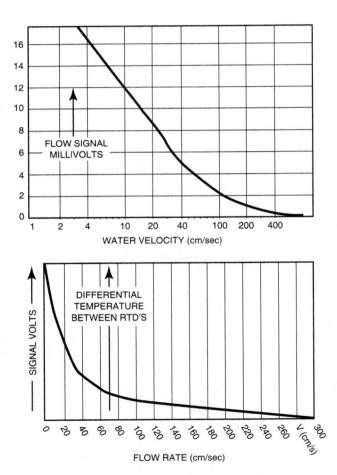

***Figure 7-49***   **Signal voltage versus flow velocity charts. Top graph shows velocity on log scale.**
*(Courtesy of ifm efector, inc.)*

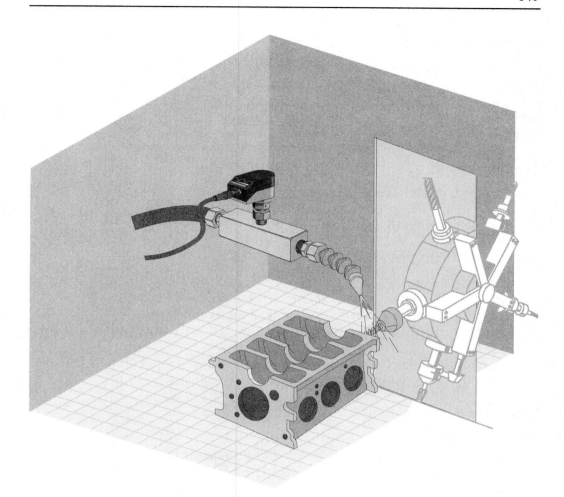

**Figure 7-50** **Flow sensors are used in machine tool coolant flow applications.** *(Courtesy of ifm efector, inc.)*

# Recommended Web Links

Students are encouraged to view the following Web sites as a supplement to the concepts presented in this textbook. Review and analyze the array of products that are available for electrical control applications. Many of these sites offer technical information that can help in converting the principles to practical applications. To view catalogs, your PC may require Adobe Acrobat Reader software.

## Motion Control

Motion Control.com
www.motioncontrol.com/products/level1.cfm
Review: Motion Controllers

## Mechanical Limit Switches

Siemens
www.sea.siemens.com/contrlbu/upld/files/p667-668.pdf
Review: Concepts and Application of Limit Switches

Square D
ecatalog.squared.com/catalog/pdf/9007CT9701R8-98.pdf
Review: Concepts and Application of Limit Switches

## Limit Switch Applications

Honeywell
content.honeywell.com/sensing/prodinfo/limitandenclosed/
Review: Industrial Applications That Use Limit Switches

## Position sensors

ATC
www.flw.com/atc/atc.htm
Review: Sensor Technology

## Proximity switches

Visible Machine Company
www.visiblemachines.com
Review: Proximity Switches

Baumer Electric
www.baumerelectric.com/baumer/boxe.html
Review: Proximity Switches and see applications provided for specific sensors

### Linear Displacement Transducer

Daytronic Corporation
www.daytronic.com/products/trans/lvdt/t-lvdt2.htm
Review: LVDT product electrical and displacement characteristics

Magtrol
www.magtrol.com/displacement/
Review: electrical and displacement characteristics

### Stepper Motors

Warner Electric
www.electrosales.com/warner/slosyn/
Review: Slo-Syn Step Motors and see the application information
available in pdf files

# Achievement Review

1. What are the three basic classes of industrial limit switches?

2. What is one use for the rotating-cam limit switch?

3. List some of the considerations when selecting a limit switch for operation in an electrical control circuit.

4. What is the difference between operating force and release force in a mechanical limit switch?

5. Draw a sketch to show the difference between pretravel and overtravel.

6. Discuss the accuracy and response time in a vane-type switch.

7. Draw the mechanical limit switch symbols for four different conditions.

8. Which of the following are methods used to achieve the operation of proximity limit switches?

   a. Capacitive fields
   b. Light rays
   c. Low current input

9. Which of the following determine the speed of response for the proximity limit switch using an oscillator in the output circuit?

   a. The location of the switch
   b. The speed of the object being sensed
   c. The oscillator frequency

10. Draw an electrical control circuit, using two push-button switches and a limit switch, relay, and solenoid operating valve. The solenoid operating valve is a single-solenoid, spring-return valve. The relay and solenoid are to be energized when the normally open push-button switch is operated. The relay and solenoid are to be deenergized when the limit switch is operated. The relay and solenoid can also be deenergized at any time during a cycle, before the limit switch is operated, by operating a second push-button switch.

11. Under what conditions would you use fiber optics in applying photoelectric control?

12. What thermal element is used to convert temperature differences into electrical signals in a flow monitor?

13. The flow monitor operates on what principle?
    a. Velocity of flow
    b. Type of medium being measured
    c. Thermal conductivity

14. Explain the operation of a pulse modulated photoelectric sensor as shown in Figure 7-39. Also, explain the impact on its operation if the emitter is moved further away from the receiver.

15. Explain the fundamental differences between an ultrasonic sensor and a photoelectric sensor. Also, in what type of industrial environment are each best suited.

16. In a multiple-axis controller, describe the role of the position decoder. Would the controller need to be programmed? What is the purpose of the software program?

17. Figure 7-38A illustrates a motion profile (speed and direction) of some device traveling along some axis. Describe this profile in terms of a cylinder rod in motion.

18. Figure 7-38B illustrates a motion profile (speed and direction) of some device traveling along some axis. Describe this profile in terms of a cylinder rod in motion.

# Chapter 8 — Pressure Control

## Objectives

After studying this chapter, you should be able to:

- Define terms used with pressure switches.
- Discuss tolerance in a pressure switch.
- List four types of pressure switches.
- Explain how the piston-type pressure switch operates.
- Discuss why the rated operating pressure of a Bourdon-tube pressure switch should not be exceeded during use.
- Describe the operation of the diaphragm pressure switch.
- Explain how the pressure transducer operates.
- Draw the normally open and normally closed pressure switch symbols.
- Draw an electrical control circuit showing how a normally closed pressure switch contact can be used to deenergize the circuit.

## 8.1 Importance of Pressure Indication and Control

Pressure indication and control are important in many electrical control systems in which air, gas, or a liquid is involved. The control takes several forms. It may be used to start or stop a machine on either rising or falling pressure. It may be necessary to know that pressure is being maintained.

The *pressure switch* is used to transfer information concerning pressure to an electrical circuit. The electrical switch unit can be a single normally open or normally closed switch. It can also be both, with a common terminal or two independent circuits. Usually it is easier to use a switch with two independent circuits in the design of a control circuit.

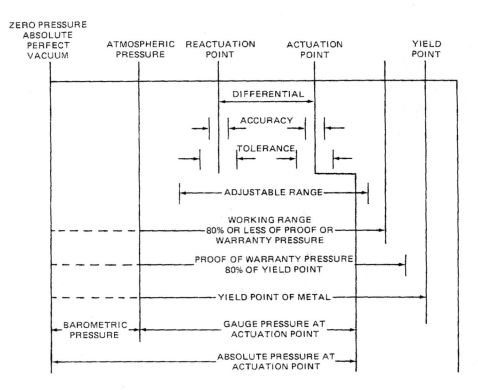

**Figure 8-1** Pressure chart.

Some terms associated with pressure switches are best explained in chart form. Figure 8-1 shows some of the terms we are concerned with and some of the most important points.

*Tolerance* in most switches is about 2%; *accuracy* is about 1%. These are expressed as percentages of the working range. In some cases it is necessary to know and use a differential pressure. As used here, the *differential* is the range between the *actuation point* and the *reactuation point*. For example, the electrical contacts operate at a preset rising pressure and hold operated until the pressure drops to a lower level. This differential may be a fixed amount, generally proportional to the operating range, or it can be adjustable.

The pressure switch symbol is shown in Figure 8-2. The NC contact opens and the NO contact closes on reaching preset pressures.

**Figure 8-2** Pressure switch symbols.

## 8.2 Types of Pressure Switches

There are several types of pressure switches available. Some of the most widely used types are:

- Sealed piston
- Bourdon tube
- Diaphragm
- Solid state

**8.2.1 Sealed Piston Type**   The sealed-piston type of pressure switch is actuated by means of a piston assembly that is in direct contact with the hydraulic fluid. The assembly consists of a piston sealed with an O-ring, direct acting on a snap-action switch (Figure 8-3). An extremely long life can be expected from this switch where mechanical pulsations exceed 25 per minute and where high over-pressures occur. The sealed piston saves the cost of installing return lines.

The piston-type pressure switch can withstand large pressure changes and high overpressures. These units are suitable for pressures in the range of 15 pounds per square inch (psi) to 12,000 psi.

**8.2.2 Bourdon-Tube Pressure Switch**   The pressure operating element in a Bourdon-tube unit is made up of a tube formed in the shape of an arc. Application of pressure to the tube will cause it to straighten out. Care should be taken not to apply pressures beyond the rating of the unit. Excessive pressures tend to bend the tube beyond its ability to return to its original shape.

The Bourdon-tube pressure switch unit is extremely sensitive. It senses peak pressures; for this reason, it may be necessary to use some form of dampening or snubbing of the pressure entering the tube.

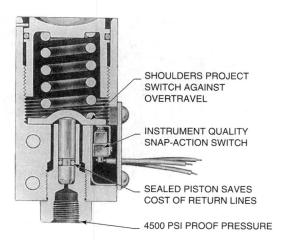

SHOULDERS PROJECT SWITCH AGAINST OVERTRAVEL

INSTRUMENT QUALITY SNAP-ACTION SWITCH

SEALED PISTON SAVES COST OF RETURN LINES

4500 PSI PROOF PRESSURE

*Figure 8-3*   **Piston-type pressure switch.** *(Courtesy of Barksdale, Inc., subsidiary of Crane Co.)*

This pressure switch is available for applications from 50 psi to 18,000 psi. It is generally used more in indicating service than in switching because of the inherent low work output of the unit.

Figure 8-4 shows a typical Bourdon-tube pressure switch.

**8.2.3 Diaphragm Pressure Switch**   A diaphragm pressure switch consists of a disk with convolutions around the edge. The edge of the disk is fixed within the case of the switch. Pressure is applied against the full area of the diaphragm. The center of the diaphragm opposite the pressure side is free to move and operate the snap-action electrical contacts. Units are available from vacuum to 150 psi.

Figure 8-5 shows a diaphragm pressure switch.

**8.2.4 Solid-State Pressure Switch**   A solid-state pressure switch is interchangeable with existing electromechanical pressure switches. Pressure sensing is performed by semiconductor strain gauges with proof pressures up to 15,000 psi acceptable. Switching is accomplished with solid-state triacs. An enclosed terminal block allows four different switching configurations:

1. Single-pole, single-throw (NC)
2. Single-pole, single-throw (NO)
3. Single-pole, double-throw
4. Double-make, double-break

See Figure 8-6 for a typical solid-state pressure switch.

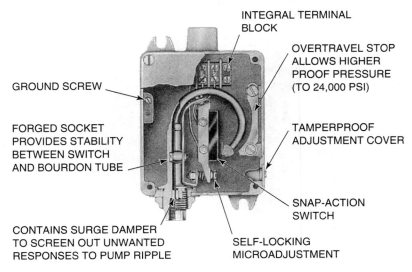

**Figure 8-4**   **Bourdon-tube pressure switch.** *(Courtesy of Barksdale, Inc., subsidiary of Crane Co.)*

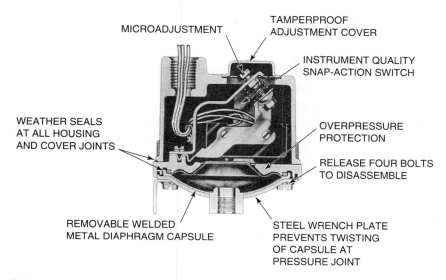

MICROADJUSTMENT

TAMPERPROOF
ADJUSTMENT COVER

INSTRUMENT QUALITY
SNAP-ACTION SWITCH

WEATHER SEALS
AT ALL HOUSING
AND COVER JOINTS

OVERPRESSURE
PROTECTION

RELEASE FOUR BOLTS
TO DISASSEMBLE

REMOVABLE WELDED
METAL DIAPHRAGM CAPSULE

STEEL WRENCH PLATE
PREVENTS TWISTING
OF CAPSULE AT
PRESSURE JOINT

**Figure 8-5** **Diaphragm pressure switch.** *(Courtesy of Barksdale, Inc., subsidiary of Crane Co.)*

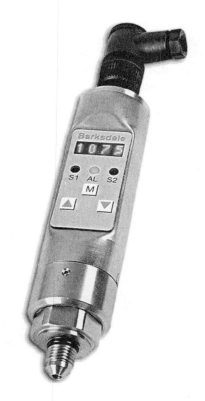

**Figure 8-6** **Solid-state pressure switch.** *(Courtesy of Barksdale, Inc., subsidiary of Crane Co.)*

**8.2.5 Pressure Transducers**   With the use of programmable controllers in pressure control applications in industry (covered in Chapter 15), the use of pressure has become important.

A pressure transducer, such as shown in Figure 8-7, utilizes semiconductor strain gauges that are epoxy-bonded to a metal diaphragm. Pressure applied to the diaphragm through the pressure port produces a minute deflection, which introduces strain to the gauges. The strain produces an electrical resistance change proportional to the pressure. Four gauges (or two gauges with fixed resistors) form a Wheatstone bridge.

**8.2.6 Pressure Sensors**   Figure 8.8 shows sensor solutions for hydraulic and pneumatic applications. The pressure family includes pressure sensors, pressure transmitters, and vacuum sensors. These sensors incorporate the same ceramic pressure-sensing technology found in high-accuracy pressure transmitters. Ceramic pressure sensors are the preferred choice because they can withstand higher overpressure and are more resistant to pressure spikes. The base of a ceramic sensor has a measuring electrode plus a reference electrode. A diaphragm containing a second measuring electrode is attached to the base with a glass solder layer. The two measuring electrodes form the plates of a capacitor, and with no pressure applied to the sensor, the electrodes are approximately 10 micrometers apart. As pressure is applied to the diaphragm, this distance decreases and thus increases the capacitance value and produces a signal that is proportional to the applied pressure. When combined with its mechanical mounting, this sensing element provides overall long term mechanical stability. See Figure 8-9 for an application

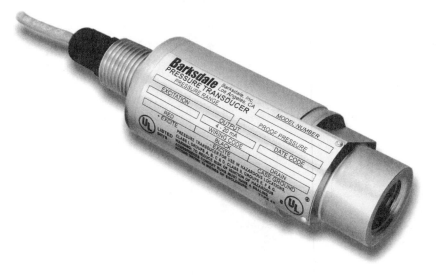

***Figure 8-7***   **Transmitter and transducer.** *(Courtesy of Barksdale, Inc., subsidiary of Crane Co.)*

*Figure 8-8*   **Pressure sensors.**  *(Courtesy of ifm efector, inc.)*

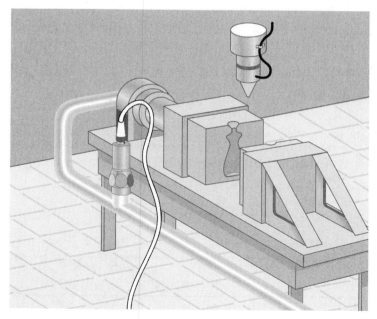

*Figure 8-9*   **Application of a pressure transmitter on an injection molding machine.**  *(Courtesy of ifm efector, inc.)*

of a pressure transmitter. These pressure transmitters have all the features needed for monitoring hydraulic clamping and ram pressures as, for example, those present in injection molding machines.

## 8.3  Circuit Applications

In application, the pressure switch contact is substituted for the limit switch contact. The basic circuit showing the limit switch as a means of reversing the piston is shown in Figure 7-5A. In Figures 8-10A and 8-10B, the pressure switch is connected into the fluid power pressure circuit. A normally closed contact is connected into the electrical circuit (Figure 8-10C).

   The function of this circuit is the same as that shown in Figure 7-5A, except now the piston reverses when a preset pressure is reached. Remember that a normally closed pressure switch contact opens on rising pressure.

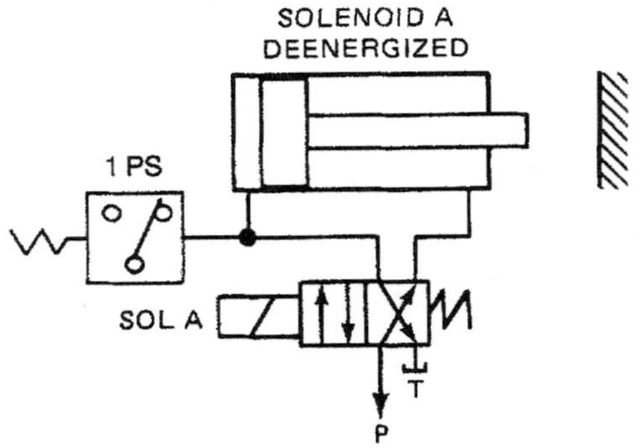

*Figure 8-10A*  **Hydraulic control circuit, piston retracted. P, pressure; T, tank.**

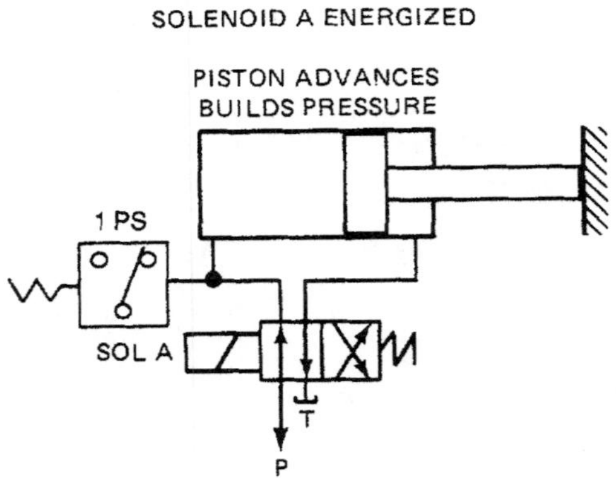

*Figure 8-10B*  **Hydraulic control circuit, piston extended (pressure switch activated). P, pressure; T, tank.**

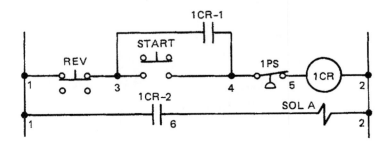

*Figure 8-10C*  **Control circuit for piston control.**

In the circuit shown in Figures 8-11A, 8-11B, 8-11C, and 8-11D, the double-solenoid valve is used. Remember that in the deenergized condition of both solenoids, pressure is allowed to bypass through the center of the valve spool to tank. The piston must be in position A for start conditions.

The sequence of operations proceeds as follows:

1. Operate the START push-button switch.
2. Relay coil 1CR is energized.
    a. Relay contact 1CR-1 closes, interlocking around the START push-button switch.
    b. Relay contact 1CR-2 closes, energizing solenoid A.

The valve spool now shifts, allowing pressure to enter the main cylinder area (left end). The piston moves forward. On reaching the workpiece, pressure builds against it. When a preset pressure is reached on pressure switch 1PS, its NC contact opens.

3. Relay coil 1CR is deenergized.
    a. Relay coil 1CR-1 opens, opening the interlock circuit around the START push-button switch.
    b. Relay contact 1CR-2 opens, deenergizing solenoid A.

When the NC pressure switch contact opens, the NO pressure switch contact closes. Since the cam on the piston is away from position A, the NC contact on limit switch 2LS is allowed to close. This action completes the circuit to relay 2CR.

4. Relay coil 2CR is energized.
    a. Relay contact 2CR-1 closes, interlocking around the NO pressure switch contact. (Note: This is necessary because the pressure drops when solenoid A is deenergized. The NO pressure switch contact opens.)
    b. Relay contact 2CR-2 closes, energizing solenoid B.

The valve spool now moves back, allowing pressure to enter the pull-back area (right end). The piston now returns to its initial position (A). On reaching this point, limit switch 2LS is operated, opening its NC contact.

5. Relay coil 2CR is deenergized.
    a. Relay contact 2CR-1 opens, opening the interlock circuit around the pressure switch 1PS contact.
    b. Relay contact 2CR-2 opens, deenergizing solenoid B.

Note the broken line in Figure 8-11D connecting the two limit switch contacts. This line indicates that they are two contacts in the same limit switch and operate at the same time. A similar explanation can be made for the broken line connecting the two contacts on 1PS. Both contacts are on the same pressure switch and operate at the same time.

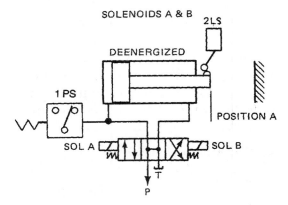

**Figure 8-11A**  Hydraulic control circuit, piston retracted (limit switch detector). P, pressure; T, tank.

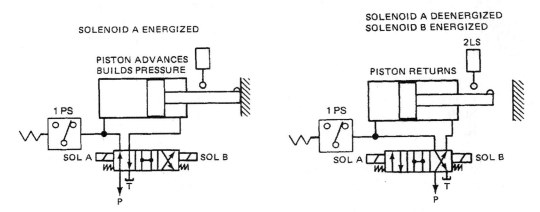

**Figure 8-11B**  Hydraulic control circuit, piston extended (pressure switch activated). P, pressure; T, tank.

**Figure 8-11C**  Hydraulic control circuit, piston retracting. P, pressure; T, tank.

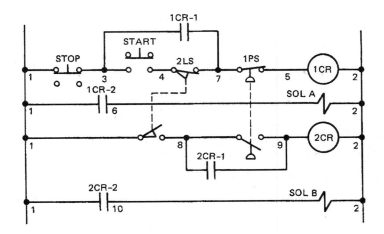

**Figure 8-11D**  Control circuit for piston control.

The circuits in Figures 8-12A, 8-12B, 8-12C, and 8-12D use two separate cylinder-piston assemblies. Each assembly is powered through a single-solenoid, spring-return operating valve.

The sequence of operations for these circuits is as follows:

1. Operate the START push-button switch.
2. Relay coil 1CR is energized.
   a. Relay contact 1CR-1 closes, interlocking around the START push-button switch.
   b. Relay contact 1CR-2 closes, energizing solenoid A.
3. The piston moves forward.

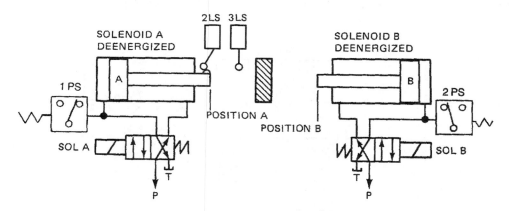

**Figure 8-12A**    Hydraulic circuit, two pistons, both retracted. P, pressure; T, tank.

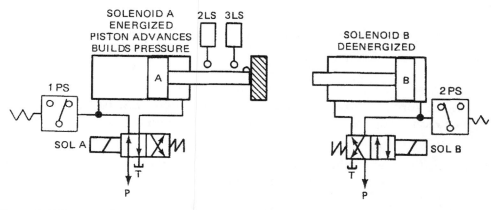

**Figure 8-12B**    Hydraulic circuit, piston A extended, piston B retracted. P, pressure; T, tank.

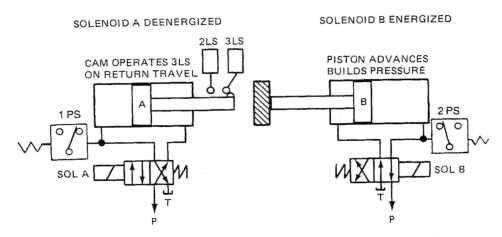

**Figure 8-12C**   Hydraulic circuit, piston A retracted, piston B extended. P, pressure; T, tank.

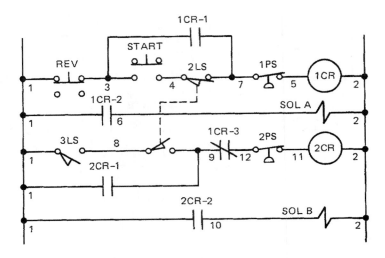

**Figure 8-12D**   Control circuit for two piston control.

On the forward stroke of piston A, limit switch 3LS, NO contact closes (cam is no longer operating 2LS), resulting in no circuit action as NC relay contact 1CR-3 is open.

   4. The piston reaches the workpiece and pressure builds to a preset amount on 1PS, operating the pressure switch contacts.
   5. The normally closed contact on 1PS opens, deenergizing relay coil 1CR.
      a. Relay contact 1CR-1 opens, deenergizing relay coil 1CR.

b. Relay contact 1CR-2 opens, deenergizing solenoid A.

c. Relay contact 1CR-3 closes.

Piston A now returns to position A. On the return travel, the cam on the piston operates limit switch 3LS. With relay contact 1CR-3, limit switch 2LS, NO contact, limit switch 3LS, and pressure switch 2PS are closed.

6.  Relay coil 2CR is energized.

    a. Relay contact 2CR-1 closes, interlocking around 2LS NC contact and 3LS NO contact.

    b. Relay contact 2CR-2 closes, energizing solenoid B.

Piston B now moves forward, meeting the workpiece and building pressure to an amount preset on 2PS.

7.  Normally closed 2PS contact opens, deenergizing relay coil 2CR.

    a. Relay contact 2CR-1 opens, opening the interlock circuit around 3LS and 2LS.

    b. Relay contact 2CR-2 opens, deenergizing solenoid B.

Piston B now returns to its initial position at B.

## Recommended Web Links

Students are encouraged to view the following Web sites as a supplement to the concepts presented in this textbook. Review and analyze the array of products that are available for electrical control applications. Many of these sites offer technical information that can help in converting the principles to practical applications. To view catalogs, your PC may require Adobe Acrobat Reader software.

Honeywell
content.honeywell.com/sensing/prodinfo/pressure/
Review: Select Application Notes

Entran Sensors and Electronics
www.entran.com/ptoc.htm
Review: Pressure Sensor Information and look at the application documents (pdf)

Omega Engineering
www.omega.com/toc_asp/sectionsc.asp?book=pressure&section=b
Review: The array of pressure transducers

# Achievement Review

1.  Draw the normally open and normally closed pressure switch symbols.

2.  Explain differential in pressure switch operation.

3.  List four major types of pressure switches.

4.  In a piston-type pressure switch, what is used to prevent oil leakage past the piston?

5.  Why is a form of dampening sometimes used with the Bourdon-tube pressure switch?

6.  What is the usual pressure operating range for the diaphragm-type pressure switch?

7.  Design and draw an electrical control circuit showing how the operation of a normally closed pressure switch contact can be used to deenergize an output solenoid. Use two push-button switches, a relay, pressure switch contact, and a solenoid.

8.  Which of the following can a pressure transducer use to transfer pressure information?

    a. A low resistance coil
    b. A specially designed plug-in relay
    c. Semiconductor strain gauges

9.  How is switching accomplished in the solid-state pressure switch?

10. List four different switching configurations available in the solid-state pressure switch.

11. How is the differential resistance in a pressure transducer measured?

    a. By applying a constant voltage to the bridge and deflection of the diaphragm
    b. Through calculation using data provided by the manufacturer
    c. By estimating the change on rising pressure

# Chapter 9 Temperature Control

## Objectives

After studying this chapter, you should be able to:

- Identify five important factors to consider when selecting a temperature controller.
- Explain the difference between controller sensitivity and controller operating differential.
- Describe how the thermocouple, thermistor, and resistance unit obtain temperature information that can be used in a control circuit.
- Explain how time proportioning is used in a temperature controller.
- List the advantages and disadvantages of the potentiometric-type controller.
- Discuss bandwidth, automatic reset, and rate control in pyrometers.
- Name two types of temperature switches or thermostats and explain how each operates.
- Show how a temperature switch contact can be used in an electrical control circuit to prevent the operation of the circuit unless the temperature of the part being sensed is at or above a preset temperature.

## 9.1 Importance of Temperature Indication and Control

The use of temperature control and indication is important in at least three areas:

1. Safety
2. Troubleshooting
3. Means of processing material

As a safety issue, it should be noted that excessive heat can cause many problems that lead to unsafe operating conditions, including overheated conductors and excessive heat in operating components, such as circuit breakers and other operating electromechanical components. Overheating is often the result of loose connections in a circuit carrying power current. Motor bearings may overheat due to improper lubrication or defective bearings. Overheating in a control cabinet can create false tripping of protective devices. More on this subject is covered in Chapter 19 on troubleshooting. Indication of overheated components or conductors can lead to their correction before trouble starts.

In processing materials, it is important that the correct temperature be used and maintained. Both control and indication are important to result in a good product, particularly in the molding of plastics and the diecasting of metals.

This chapter is devoted entirely to the use of temperature control and indication in the processing of materials.

## 9.2  Selection of Temperature Controllers

Temperature controllers basically consist of two parts: a sensing element that responds to temperature and a switch consisting of normally open and/or normally closed contacts. The contacts operate from the temperature-sensing element. The operating point is preset. The switch operation is usually accomplished when the temperature at the sensing point changes from any given level to the preset point. The main difference among temperature switches is the means by which the temperature information is transferred from the sensing element to the switching element.

There are several factors to consider when selecting a temperature controller:

- *Temperature range available*. Temperature range is the overall operating range of the controller. Not all controllers cover the entire temperature range used in industrial control. The controller is generally selected after the range is determined. The sensing element often becomes an important factor.
- *Type of sensing element*. There are three basic types: (1) electronic, (2) differential expansion of metal, and (3) expansion of fluid, gas, or vapor.
- *Response time*. Speed of response is a measure of the elapsed time from the instant the temperature change occurs at the sensing element until it is converted into controller action. The response to a temperature change may vary between one type of sensing element and another. Here the application may be the deciding factor. For example, if rapid response is not required, the slower-response and generally less-expensive types may be used.

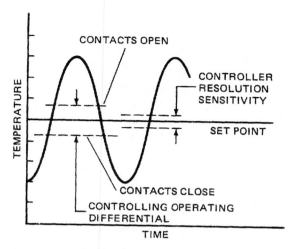

*Figure 9-1*  Temperature versus time curve.

- *Sensitivity.* Units vary with the amount of temperature change required to operate the switch. This amount is fixed in some switches; in others it is adjustable. It is usually desirable to have a controller that has a relatively high sensitivity. *Resolution sensitivity* is the amount of temperature change necessary to actuate the controller.
- *Operating differential.* Operating differential is the difference in temperature at the sensing element between make and break of the controller's contacts when the controller is cycled.

Figure 9-1 illustrates some of these operating characteristics in temperature controllers.

## 9.3  Electronic Temperature Controller (Pyrometer)

The electronic controller, or pyrometer, may have one of three different temperature sensing elements:

1. Thermocouple (approximate temperature range 200°F to 5000°F)
2. Thermistor (approximate temperature range 100°F to 600°F)
3. Resistance temperature detector (RTD) (approximate temperature range 300°F to 1200°F)

The *thermocouple* operates on the principle of joining two dissimilar metals, A and B, at their extremities (Figure 9-2A). When a temperature difference exists between the two extremity points, a potential is generated in proportion to the temperature difference. Combinations such as copper-constantan, iron-constantan, and chromel-alumel are used in thermocouple construction.

When the temperature at T1 is different than at T2, electromotive force (emf) exists. By the use of an indicating device that indicates emf or the flow of current, the temperature difference can be determined.

The relationship between temperature in degrees Fahrenheit and millivolts output for a type J thermocouple (iron-constantan) is shown in Figure 9-2B.

The *thermistor* is a semiconductor whose resistance decreases with increasing temperature. This element is connected into a null-reading-type ac bridge cir-

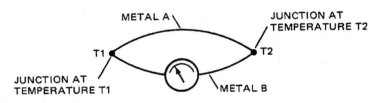

**Figure 9-2A**    **Thermocouple junction.**

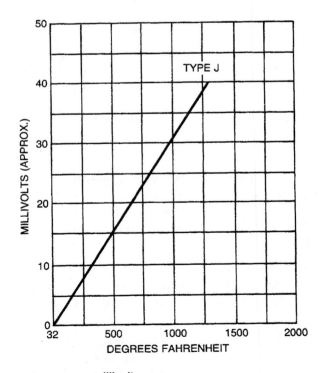

**Figure 9-2B**    **Temperature versus millivolt curve.**

cuit with the output fed to an amplifier. The output of the amplifier operates a relay that is used in the control circuit.

The *RTD unit* consists of a tube made of stainless steel or brass. It contains an element made in the form of a coil. The coil is generally wound of a fine nickel, platinum, or pure copper wire. As the temperature changes, the resistance of the coil changes proportionally. This change is converted to a voltage by its application in the electrical control circuit. Therefore, as the voltage changes with a temperature change, the voltage level is compared to a reference. Thus, an output proportional to the temperature change is obtained.

Three different physical arrangements of thermocouples are shown in Figures 9-3A, 9-3B, and 9-3C. Figure 9-3A shows the dual-element thermocouple. It can be used to sense temperature at two different physical levels. Figure 9-3B shows the single thermocouple. Both thermocouples are spring loaded. The single thermocouple uses an adaptor and small steel ferrule on a flexible armored cable to obtain the spring loading. The dual-element thermocouple uses a small steel plate that is secured to the deep couple with an adaptor and locking nut. The springs on both thermocouples are loaded against the steel plate. Figure 9-3C shows an adjustable-depth plastic melt-type thermocouple with a plug attachment.

Figure 9-4A shows an RTD unit. Figure 9-4B shows a line drawing of a thermistor.

The electronic temperature controller is usually the most expensive type. It is more sensitive and has a faster response. Generally, electronic controllers have a smaller sensing device. The sensing device can be remotely located with ease. Normal control accuracies of one fourth of 1% to 5% of full-scale reading can be expected.

The diversity of temperature control applications has brought about rapid advancement and refinement in electronic temperature controllers. Thus, some pertinent terms and expressions are explained on the following page.

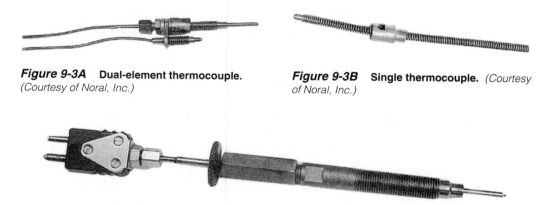

**Figure 9-3A**  Dual-element thermocouple. *(Courtesy of Noral, Inc.)*

**Figure 9-3B**  Single thermocouple. *(Courtesy of Noral, Inc.)*

**Figure 9-3C**  Adjustable-depth plastic melt-type thermocouple. *(Courtesy of Noral, Inc.)*

**Figure 9-4A**  **Resistance bulb unit (RTD).**  *(Courtesy of Noral, Inc.)*

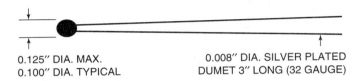

0.125″ DIA. MAX.           0.008″ DIA. SILVER PLATED
0.100″ DIA. TYPICAL        DUMET 3″ LONG (32 GAUGE)

**Figure 9-4B**  **Line drawing of a thermistor.**  *(Courtesy of Yellow Springs Instrument Co., Inc., Industrial Division.)*

*On/off* is the simplest and oldest method of temperature control. Assuming no time thermal lags in the object being heated or cooled, the on/off control produces a fluctuating temperature. The limits are within the heating or cooling band. Depending on the time lag, the amplitude of the temperature swing may exceed good operating control.

Time *proportioning* turns the heating elements on and off in accordance with the demands of the temperature set point. There is a point below the set point at which the power to the heating elements is on continuously. Likewise, there is a point above the set point at which the power is off. Between these two points the power is on at a proportional rate. For example, at a temperature 20% through the proportional band, the power may be 80% on; 50% of the way, it will be 50% on. This method differs from on/off control in that it does not require an actual temperature change to cycle.

To accomplish on/off and proportioning, pyrometers can be divided into two types. These are the millivoltmeter controller and the potentiometric controller.

The *millivoltmeter controller* is the oldest type on the market. It uses a millivoltmeter actuated directly from a temperature sensor such as a thermocouple. Setting the desired temperature positions either a photocell and light source or a set of oscillator coils. The indicating pointer carries a small flag. As the pointer approaches the set point, the flag moves between the photocell and light source or between the oscillator coils. Control action now starts, cutting back on heat.

One of the major sources of trouble with the millivoltmeter controller is the adverse effect of shock and vibration. Another problem involves the sensing device. For good indication accuracy, the external resistance of the thermocouple circuit should match that of the instrument. Some manufacturers offer controllers with adjustable external resistance circuits. Such instruments can be calibrated after they are installed.

The *potentiometric controller* differs from the millivoltmeter type in that the signal from the temperature sensor is electronically controlled and compared to the set-point temperature. This instrument can be supplied with or without indication because with this control, the indicating method is completely separate.

The potentiometric controller has the following advantages:

- No moving parts
- Does not have to be calibrated for external resistance
- Not affected by shock and vibration

The potentiometric controller has the disadvantage of more electronic circuitry that is not easily serviced; it is generally advisable to replace the entire plug-in controller. Figure 9-5 shows a potentiometric controller.

The millivoltmeter and potentiometric controllers both generally use a contactor to energize the heating element load. The contactor is energized and deenergized in response to a demand for heat. For example, when heat is required as sensed by a thermocouple, the contactor is energized by the controller output. Electrical power is then connected to the heating element load. When there are no heat requirements, the heating elements are completely isolated from the power source by the opening of the contacts on the contactor.

Mechanical contactors switch power in full ON and OFF cycles. Therefore, they should be used at cycle times of 15 seconds or longer for reasonable service

**Figure 9-5**   **Potentiometric controller.**  *(Courtesy of Barber-Colman Company.)*

life. Because of the full ON and OFF switching and the limited cycle time, contactor-controlled processes must have a higher tolerance for process temperature overshoot and undershoot (Figure 9-6).

Solid-state controllers do not use a contactor to energize and deenergize the heating element load. With this latest type of controller, the power remains connected to the load at all times. Just enough power is supplied to the heating elements to satisfy the temperature requirements. This type of controller is called a *stepless* or *proportional* controller and has a current output. It is generally used with silicon-controlled rectifiers (SCR) or triac solid-state power packs.

Two different outputs are available with this type of controller. Power is applied to the heating element load through:

- Phase control (phase-angle-fired)
- Zero crossing (zero-angle-fired)

With phase control, the output is continuous at the amount required for consistent temperature. Phase control is accomplished by "firing" the SCR at some point through the wave, depending on the amount of power required.

One problem with phase-fired control is the radio frequency interference (RFI) generated by the SCR switching at any place in the wave at an extremely fast rate. This switching is on the order of one-half microsecond (µs). Another problem is the resulting distorted wave shape, which creates a problem for the power company.

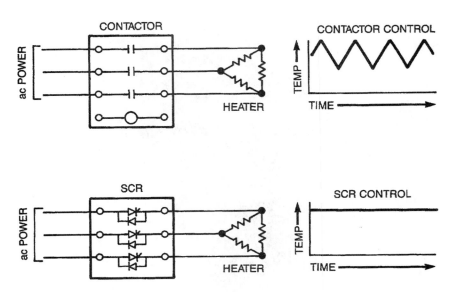

**Figure 9-6** **Contactor versus SCR control.** *(Courtesy of Chromalox.)*

These problems can be avoided by the use of zero-crossing SCR circuitry. Its characteristics are similar to the time-proportioning results. However, the power is always switched off and on when the power voltage is zero.

*Zero-crossing* or *zero-crossover-fired* power controllers are ideal for control of pure resistive loads that can accommodate rapid, full-power, ON/OFF cycling. Zero-crossover firing does not create RFI and will not adversely affect sensitive electronic equipment (computers, other SCR power controllers, logic controllers) located in the same area. Additionally, SCRs are protected from line voltage transients, making them more reliable in a wide variety of applications.

Zero-crossover-fired SCRs, when coupled with a time-proportioning control (firing package) will operate in a series of full ON and OFF cycles known as *time-proportional burst firing*. The time-proportioning control accepts the control output signal and converts it into a time-proportional signal, determining the amount of ON time and OFF time per duty cycle. The continuous, highly repetitious rate of full ON and full OFF produces a smooth power output to the load (heater) and a stable process temperature (Figure 9-7).

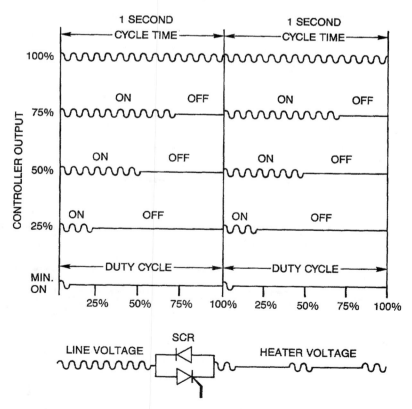

**Figure 9-7**   **Time-proportional burst firing.**   *(Courtesy of Chromalox.)*

A typical power control system consists of an RTD or thermocouple, temperature controller, firing package, and SCR power controller. Often the firing package is part of the temperature controller and is not a separate component. Figure 9-8 shows a schematic drawing of a typical system. These components work together to control the heating of the process, as follows:

- The temperature sensor provides a signal to the temperature controller.
- The temperature controller compares the sensor signal to the predetermined set point and generates an output signal that represents the difference between the actual process temperature and the set point.
- The firing package uses this control output to generate a time-proportional signal for the SCR power controller, switching the SCR on and off, thus regulating the power to the heater.

Figure 9-9 shows a complete SCR control panel.

The instrument shown in Figure 9-10 is a ramp/soak temperature and process controller. This controller has many features, including ON/OFF, proportional or PID control, ramp/soak function, dual display (process and set point), relay, triac (4–20 mA), and solid-state drive output modules. It has a nonvolatile memory, thus requiring no battery backup and eight event outputs to initiate remote activity such as valve actuation, operator annunciation, output to programmable controller, and status indication. See Figure 9-11 for a more detailed description of the front panel.

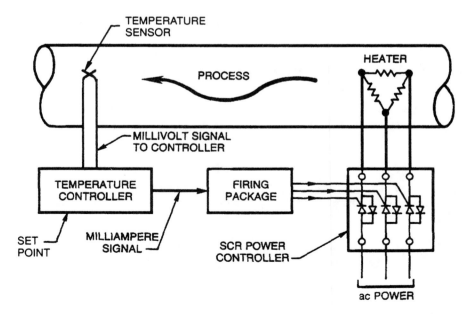

**Figure 9-8**   **Power control system.**  *(Courtesy of Chromalox.)*

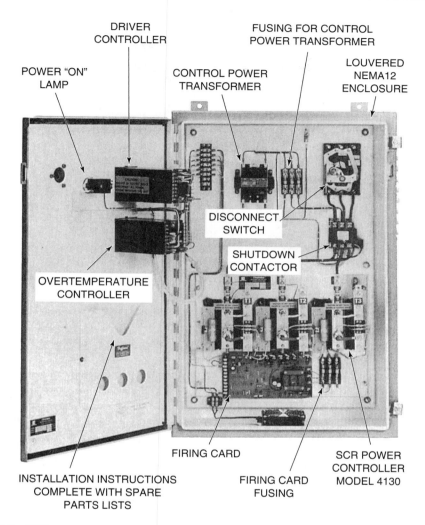

**Figure 9-9** **SCR control panel.** *(Courtesy of Chromalox.)*

One of a series of temperature controllers shown in Figure 9-12 has features of ON/OFF, proportional or PID control, ramp/soak, dual display (process and set point), relay, triac, (4–20 mA), and solid-state relay drive output modules. There is a nonvolatile memory requiring no backup battery. The front panel displays and push buttons are shown in Figure 9-13.

A drawing showing the remote set point and process signal output feature of this controller is shown in Figure 9-14A. A sketch showing the heat/cool feature of this controller is shown in Figure 9-14B.

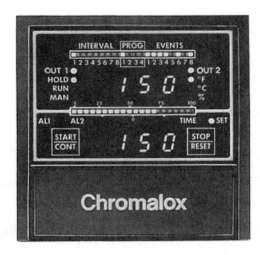

***Figure 9-10*** **Ramp/soak temperature and process controller.** *(Courtesy of Chromalox.)*

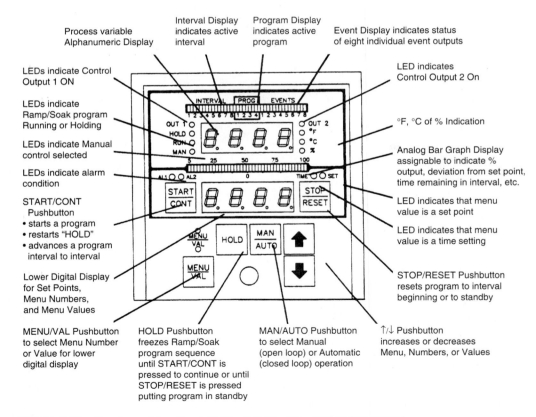

***Figure 9-11*** **Front panel functions of Figure 9-10.** *(Courtesy of Chromalox.)*

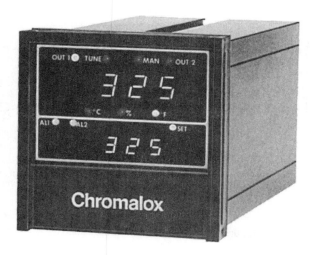

**Figure 9-12**  **Temperature controller.**  *(Courtesy of Chromalox.)*

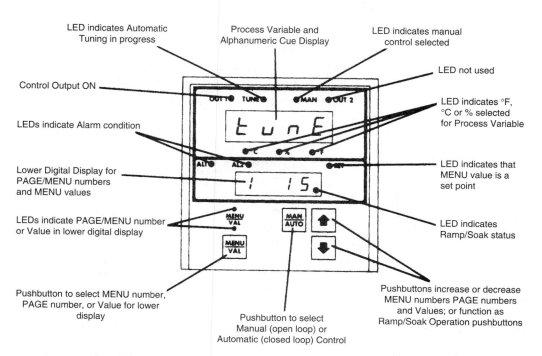

**Figure 9-13**  **Front panel display and push buttons for Figure 9-12.**  *(Courtesy of Chromalox.)*

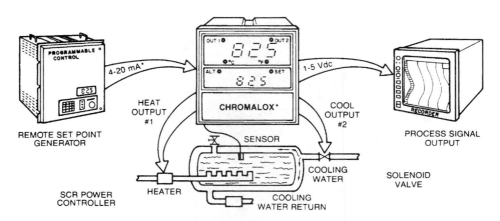

***Figure 9-14A*** **Remote set point and process signal output.** *(Courtesy of Chromalox.)*

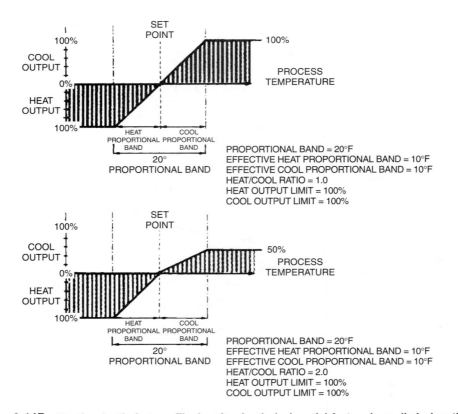

***Figure 9-14B*** **Heat/cool ratio feature. The heat/cool ratio (gain ratio) feature is applied when the cool output (#2) control mode is PID. It balances the heating and cooling capacities of the system, yielding stable process control. Note that the heat/cool ratio creates an "effective cool output limit" of 50% in the second illustration.** *(Courtesy of Chromalox.)*

## 9.4 Controller Outputs

The relay in pyrometer outputs generally is rated at 5–10 A. This rating means it can handle a small heating element load of approximately 500 watts (W) to 1000 W at 120 V. For larger loads, the relay is used to energize a contactor.

Solid-state controllers are now available with an output capacity capable of 20 A inrush and 1 A continuous. This control output can be used with load contactors up to sizes 2 or 3.

## 9.5 Additional Terms

In discussing the use of pyrometers in the control of temperature, the following terms apply:

- *Band width.* Sometimes referred to as the *proportioning band,* in most controllers band width is adjustable (Figure 9-15). A band width set for 3% is 24°F on an 800°F instrument. If the set point is at 450°F, the band extends from 437°F to 463°F. In controlling it is quite possible for the temperature to settle out at some temperature other than the set point. For example, assume the temperature settles out at 454°F. This temperature represents an offset of +4° and can be corrected by adjusting the set point at 446°F. Resetting of the control point in this way is called *manual reset.*
- *Automatic reset.* The addition of the auto reset function in a controller automatically eliminates the offset. Circuitry in the controller recognizes the offset. It will electronically shift the band up or down scale as required to remove an error. The auto reset feature is sometimes referred to as the *integral function.*
- *Rate.* Rate control is valuable in applications having rapid changes in temperature caused by external heating and cooling. It works in the opposite direction of reset and at a faster speed. For example, a sudden cooling causes the controller to turn the heaters full ON by means of an upward shift of the band. Rate is sometimes referred to as the *derivative* function.
- *Mode.* With reference to temperature controllers, mode refers to the operational functions that a controller has. A *two-mode controller* is one having proportioning and auto reset or proportional and auto reset. A *three-mode controller* is one having proportioning, auto reset, and rate or proportional, auto reset, and rate.
- *Analog and digital set point.* With an analog set point, the temperature is set on a scale. With the digital set point controller, the temperature is displayed in numbers (Figure 9-16).

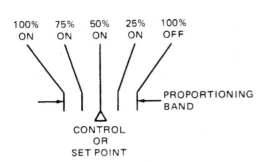

**Figure 9-15**  Control or set point.

**Figure 9-16**  Digital solid-state controller.
*(Courtesy of Barber-Colman Company.)*

## 9.6 Temperature Switches (Thermostats)

Temperature switches using differential expansion of metals may be of two different types. One uses a *mechanical link;* the other uses a *fused bimetal.*

The mechanical link has a temperature range of 100°F to 1500°F. The bimetal type has a temperature range of 40°F to 800°F.

The mechanical-link type has one metal piece directly connected or subjected to the part where the temperature is to be detected. The metal expands or contracts due to temperature change, producing a mechanical action that operates a switch.

The bimetal type operates on the principle of uneven expansion of two different metals when heated. With two metals bonded together, heat will deform one metal more than the other. The mechanical action resulting from a temperature change is used to operate a switch. In bimetal units the sensing element and switch are generally enclosed in the same container or are adjacent.

The bimetal type is usually the smallest and least expensive type of controller. It is not suited for high-precision work. However, its compact, rugged construction lends it to many uses. Accuracies of 1% to 5% of full scale can be expected. Figure 9-17 is a cutaway view of a bimetal unit.

**Figure 9-17**  Cutaway view of a bimetal-type temperature switch (thermostat). *(Courtesy of Fenwal Controls, a Division of Kidde, Inc.)*

With temperature switches using liquid, gas, or vapor, the sensing elements are generally located remote from the switch. The temperature ranges for these units are as follows:

- Liquid filled: 150°F to 2200°F (these may be self-contained)
- Gas filled: 100°F to 1000°F
- Vapor filled: 50°F to 700°F

These units operate on the principle that when the temperature increases, expansion of the medium (fluid, gas, or vapor) takes place. Thus, a force is exerted on a device that, in turn, operates a switch.

The liquid-filled type that is self-contained (switch and sensing element together) has a relatively fast response time.

A factor to consider in using units with a remote sensing head is the problem with the tube connecting the sensing bulb with the switching mechanism. The tube is easily damaged by mechanical abuse. Also, the response is slower with long lengths of connecting tube.

An example of a self-contained unit is shown in Figure 9-18A. A cross section through the unit is shown in Figure 9-18B. Figure 9-18C is an example of the unit using a remote sensing bulb.

The switch contact ratings on temperature switches (thermostats) vary from 5 to 25 A at 120 V. They generally consist of one normally open and one normally closed contact. They may have a common connection point, or they may have isolated contacts.

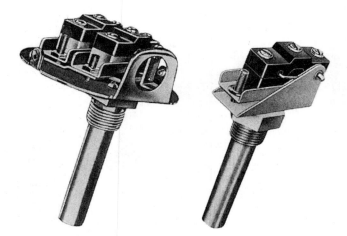

*Figure 9-18A*   **Dual-action, liquid-filled temperature controller with switch and sensing element together.** *(Courtesy of Fenwal Controls, a Division of Kidde, Inc.)*

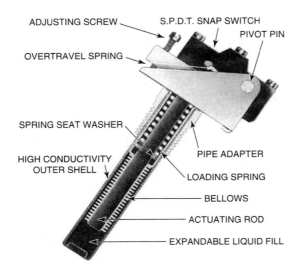

ADJUSTING SCREW
S.P.D.T. SNAP SWITCH
PIVOT PIN
OVERTRAVEL SPRING
SPRING SEAT WASHER
PIPE ADAPTER
HIGH CONDUCTIVITY
OUTER SHELL
LOADING SPRING
BELLOWS
ACTUATING ROD
EXPANDABLE LIQUID FILL

**Figure 9-18B** **Cross section of the temperature controller shown in Figure 9-18A.** *(Courtesy of Fenwal Controls, a Division of Kidde, Inc.)*

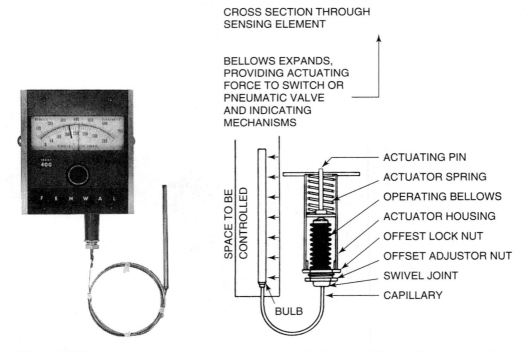

CROSS SECTION THROUGH
SENSING ELEMENT

BELLOWS EXPANDS,
PROVIDING ACTUATING
FORCE TO SWITCH OR
PNEUMATIC VALVE
AND INDICATING
MECHANISMS

SPACE TO BE CONTROLLED

ACTUATING PIN
ACTUATOR SPRING
OPERATING BELLOWS
ACTUATOR HOUSING
OFFEST LOCK NUT
OFFSET ADJUSTOR NUT
SWIVEL JOINT
CAPILLARY

BULB

**Figure 9-18C** **Indicating temperature controller with a liquid expansion, capillary-type remote sensing bulb.** *(Courtesy of Fenwal Controls, a Division of Kidde, Inc.)*

# 9.7 Temperature Sensors

Temperature monitoring systems utilizing sensors provide reliable feedback in temperature control applications. Such a system consists of an electronic control monitor and a variety of sensor probes (Figure 9-19). The control monitor is packaged in a compact stainless-steel housing that can be mounted either directly to a sensor probe or remotely using a cabled probe.

This temperature sensor utilizes a PT1000 RTD for temperature measurement (PT100 RTDs can also be used). The PT1000 RTD is a platinum resistor that exhibits a resistance of 1000 ohms at 0°C. Temperature is sensed by measuring the change in resistance of the RTD. Platinum is used because of its good linearity and stability with temperature. As temperature increases, the resistance of the RTD also increases.

The signal from the PT1000 RTD is evaluated by a microprocessor inside the temperature control monitor. This microprocessor along with other electronic components is mounted on a flexible, temperature-stable polyamid film.

Compared to thermocouples, RTDs are more repeatable and more stable. RTDs are also more sensitive, since the voltage drop across the RTD produces a larger signal than a thermocouple. RTDs have better linearity and do not require the cold junction compensation of a thermocouple. See Figure 9-20.

An example of an application for a temperature sensor is shown in Figure 9-21, in which a temperature sensor is used to monitor tank temperature to ensure a safe operating range.

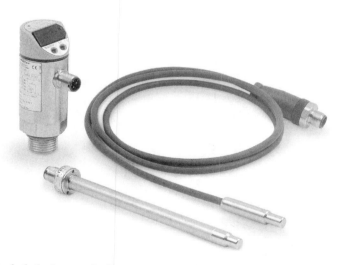

**Figure 9-19**   The imf efector monitoring system includes a control monitor and a variety of sensor probes. *(Courtesy of imf efector, inc.)*

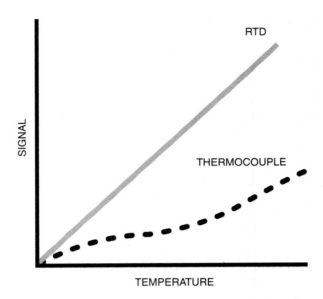

***Figure 9-20*** **Temperature-signal chart.** *(Courtesy of imf efector, inc.)*

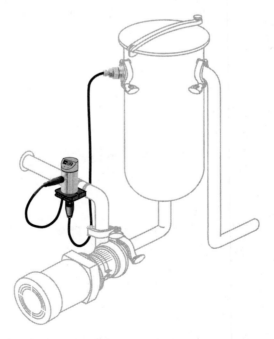

***Figure 9-21*** **The imf efector sanitary temperature sensors can be used to monitor tank temperature ensuring a safe operating range.** *(Courtesy of imf efector, inc.)*

## 9.8 Circuit Applications

Refer to Figure 4-11C, which shows a heating circuit with only heat OFF/ON push-button switches for control. These switches control the heat circuit through contactor 10CR.

In Figure 9-22A, a two-position selector switch is substituted for the two push-button switches in Figure 4-11C. A pyrometer is inserted in the circuit to control the contactor coil 10CR. Note that the symbol for the pyrometer shows only the relay contact in the pyrometer and the thermocouple connections. More details on internal circuit construction can be obtained from the manufacture of the pyrometer.

When the heat OFF/ON selector switch is operated to the ON position, the circuit is complete to the NO relay contact. If the temperature of the part being sensed is below the set point on the pyrometer, the relay contact will be closed, completing the circuit to the coil of contactor 10CR. The coil of contactor 10CR is energized. Contactor contacts 10CR-1, 10CR-2 and 10CR-3 close, energizing the heating elements at power line voltage.

When the temperature of the part being sensed by the thermocouple reaches the set point, the pyrometer relay contact opens, deenergizing the coil of contactor 10CR. Contactor contacts 10CR-1, 10CR-2, and 10CR-3 open, deenergizing the heating elements. The pyrometer relay contact continues to close and open, depending on the thermocouple input. The rate of opening and closing depends on the type of pyrometer.

The temperature switch (thermostat) is generally used at lower temperatures than the pyrometer. It is also used where accuracy and sensitivity are not important factors in temperature control. For example, a tank containing a fluid of some type (oil, asphalt, or brine) is to be heated to a given temperature. Depending on the size of the tank, it may take considerable time to heat it safely to a given temperature. If the preset temperature is 200°F, it generally is not important whether the temperature is actually 204°F or 196°F.

Another use of the thermostat is to indicate and/or control the temperature of fluid used in a machine operation. For example, it may be required that the fluid in a hydraulic system be at a minimum temperature of 70°F for the machine to operate, and the temperature should not exceed 125°F. Figure 9-22A and B illustrate this example.

In Figure 9-22B, a temperature switch is added to the circuit shown in Figure 4-3. The sequence of operations for the circuit in Figure 9-22B is as follows:

1. Temperature below set point of 125°F, circuit ready to operate.
2. Operate the START push-button switch.
3. Energize relay coil 1CR.
   a. Relay contact 1CR-1 closes, interlocking around the START push-button switch.

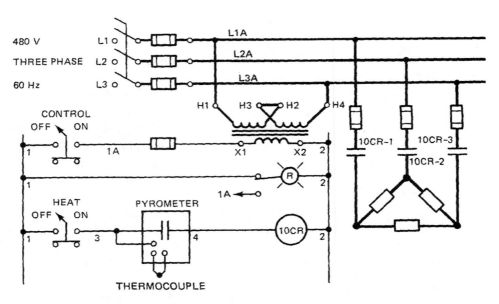

**Figure 9-22A**   Three-phase heating element control circuit with thermocouple. Conductors carrying load current at line voltage are denoted by heavy lines.

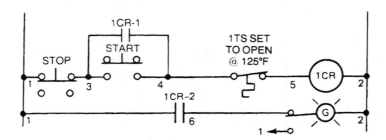

**Figure 9-22B**   Three-phase heating element control circuit modified to add latching circuit controlled by temperature switch.

     b. Relay contact 1CR-2 closes, energizing the green pilot light.

4. Temperature exceeds 125°F, NC temperature switch contact opens.

5. Relay coil 1CR deenergized.

     a. Relay contact 1CR-1 opens, opening the interlock circuit around the START push-button switch.

     b. Relay contact 1CR-2 opens, deenergizing the green pilot light.

    The temperature must drop below 125°F, allowing the NC 1TS contact to close before relay coil 1CR can again be energized.

Figure 9-23 shows the control for two heating elements. With the temperature of the medium being heated (oil, asphalt, brine) below 80°F:

1. Contactor coil 10CR is energized.
2. Contactor contacts 10CR-1 and 10CR-2 close, energizing the two heating elements.
3. Contactor auxiliary contact opens, preventing the energizing of relay coil 1CR through the operation of the START push-button switch.
4. The heated medium now reaches 80°F, deenergizing 10CR contactor coil by the opening of 1TS.
5. Contactor contacts 10CR-1 and 10CR-2 open, deenergizing the two heating elements.
6. Contactor auxiliary contact closes.
7. The START push-button switch can now be operated, energizing relay coil 1CR.
   a. Relay contact 1CR-1 closes, interlocking around the START push-button switch.

Relay coil 1CR now remains energized until such time as the temperature of the heated medium drops below 80°F. At that time contactor coil 10CR will

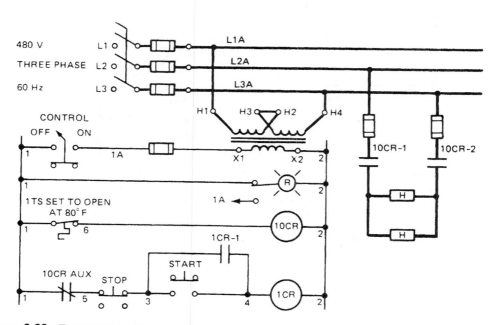

**Figure 9-23** Two-phase heating element control circuit with latching circuit and contactor. Conductors carrying load current at line voltage are denoted by heavy lines.

again energize, energizing the heating elements and opening the 10CR auxiliary contact. Relay coil 1CR will be deenergized.

In the circuit shown in Figure 9-24 hot bearing control is added to the circuit as shown in Figure 9-22B.

Assume that a machine has four bearings. The machine should not operate with any of the bearings at a temperature in excess of 130°F. A thermostat can be placed on each of the four bearings. The thermostat contacts will be normally open, set to close (operate) at 130°F.

1. With bearing temperature below 130°F, relay coil 2CR not energized.
2. When the temperature of one or more of the bearings exceeds 130°F, relay coil 2CR energized.
   a. Normally closed relay contact 2CR-1 opens.
      a-1. Relay coil 1CR deenergized.
      a-2. Relay contact 1CR-1 opens, opening the interlock circuit around the START push-button switch.
      a-3. Relay contact 1CR-2 opens, deenergizing the green pilot light.
   b. Relay contact 2CR-2 closes, energizing the hot bearing light (red).

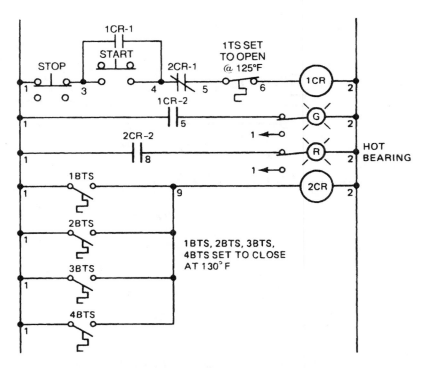

*Figure 9-24*  Heating element control circuit with four bearing temperature switches.

When the bearing problem has been corrected and all of the normally open thermostat contacts are open, the 1CR relay circuit can be energized through the START push-button switch.

## Recommended Web Links

Students are encouraged to view the following Web sites as a supplement to the concepts presented in this textbook. Review and analyze the array of products that are available for electrical control applications. Many of these sites offer technical information that can help in converting the principles to practical applications. To view catalogs, your PC may require Adobe Acrobat Reader software.

### Temperature Sensors

www.temperatures.com/
Review this interactive repository and guide to temperature sensors on the Web

Honeywell
content.honeywell.com/sensing/prodinfo/temperature
Review: Application Notes

### Temperature Controllers

Barber-Colman
www.eurotherm.com/products.htm
Review: Barber-Colman

Chromalox
www.chromalox.com
Review: Industrial Solutions

## Achievement Review

1. What are the three basic types of temperature sensing devices?

2. What is speed of response in a temperature controller?

3. Explain the difference between controller resolution sensitivity and controller operating differential.

4. Explain the thermocouple operating principle.

5. Explain the function of time-proportioning control.

6. What size of loads can be handled with a relay output controller? Can this load capacity be increased? If so, how?

7. If the band width in an 800°F temperature controller is set for 2%, what is the band width in degrees?

8. Explain how the mechanical-link temperature switch operates.

9. What are the disadvantages or problems found in the use of a liquid-filled sensing element as used with a temperature switch?

10. Draw the symbols for a normally open and normally closed temperature switch contact.

11. What is the approximate millivolt output of a type J thermocouple at 500°F?

   a. 10
   b. 15
   c. 20

12. On what principle does the bimetal type of temperature switch operate?

   a. Chemical reaction of the two metals
   b. Uneven expansion of two different metals when heated
   c. Expansion of air, gas, or vapor

# 10 Time Control

## Objectives

After studying this chapter, you should be able to:

- Describe the differences between a timer and a time-delay relay.
- Explain where the timer has a definite use.
- List three major types of timers discussed in this chapter.
- Explain the arrangement of the contacts in a reset-type timer under the conditions of reset, timing, and timed out.
- Draw simple control circuits showing (1) clutch and motor energized at the same time, (2) clutch energized before the motor, and (3) motor energized before the clutch.
- Explain how the multiple-interval timer operates.
- Discuss the operation of the repeat-cycle timer.
- Describe some of the design features of the solid-state timers.

## 10.1 Selected Operations

Many types of timers are available for use on industrial machines. Their major function is to place information about elapsed time into an electrical control circuit. The method of accomplishing this function varies with the type of timer used. Time range, accuracy, and contact arrangement vary among types of timers. A few suggestions are made here for the proper use of timers.

It is true that time elapses during nearly every action taking place on a machine. This fact does not necessarily mean that a timer is always the best means of control.

For example, a machine part is to move from point X to point Y, a definite distance. It is observed that this motion requires approximately 5 seconds to complete. If the position of the machine part is important, a position control device (such as a limit switch) not dependent on elapsed time should be used to indicate that the part did arrive at point Y. Due to variables in the machine, one cycle could require 5.0 seconds, the next 5.1 seconds, and a third cycle 4.9 seconds. Thus, if a timer is used the machine part will not always stop at point Y.

In Figure 10-1 the timer is set for 5 seconds. In case A, time runs out before the part reaches point Y. In case B, time runs out after the part has passed point Y.

A similar situation may exist when a timer is substituted for a pressure indication. In one case the pressure may build to a preset value before a preset time has elapsed. Also, time may elapse before the preset pressure is reached. Similar conditions may arise when a timer is substituted for a temperature controller to obtain temperature information.

The important point is to determine the critical condition to be met. Is it time, position, pressure, or temperature? Timers are very useful tools, but they must be applied properly.

Another point to clarify is reference to timers and time-delay relays. Generally, *time-delay relays* are devices having a timing function after the timer coil has been energized or deenergized. Time-delay relays have a normally open and a normally closed timing contact. In some cases the contacts are isolated. In other cases they have a common terminal. Sometimes time-delay relays have several NO and/or NC instantaneous contacts that operate immediately when the time-delay coil is energized.

When reference is made to *timers,* the time function may start on one or more of the contacts at energization, or at any time after energization during a preset time cycle. Likewise, the timing function may stop on one contact or more during the cycle after timing has been started on the particular contact(s).

In general, a timer opens or closes electrical circuits to selected operations according to a timed program. Instantaneous contacts are also available on some timers.

Timing functions will start from an electrical signal, initiated through any one of several components: a push-button switch, relay contact, temperature switch, pressure switch, limit switch, etc. Once an electrical signal is available to the timer, the timing function can, in general, be one of two types: ON delay (tim-

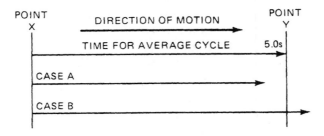

*Figure 10-1*   Timing for two different cases.

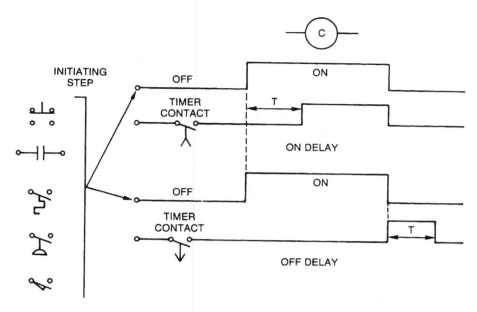

*Figure 10-2*   **Basic timing functions.**

ing function after energizing) or OFF delay (timing function after deenergizing), as shown in Figure 10-2.

Other forms of this basic timing function are found with the solid-state timer, including pulse and pulse-and-repeat operations. In some timers an inhibit function is available. This function is accomplished by applying a voltage to specific terminals. The timing cycle then stops and holds its outputs in the last state without resetting the circuit. Timing will continue and the outputs will be allowed to change according to their programmed operating modes when voltage is removed from the terminals.

## 10.2 Types of Timers

In grouping timers according to their method of timing, there are two types that are applied most frequently on industrial machines:

1. Synchronous motor driven
2. Solid state

Four other types of timers that have industrial applications are the dashpot, mechanical, electrochemical, and thermal. Since these have only limited use on industrial machines, they are not covered in this text.

# 10.3 Synchronous Motor-Driven Timers

The *synchronous motor-driven timer* can well be called the workhorse of the timer field. Its many applications make it a very useful tool. It is available in reset, repeat-cycle, and manual-set types.

### 10.3.1 Reset Timers

The *reset timer* depends on the clutch and the synchronous motor for its operation. The symbols for the clutch and synchronous motor are shown in Figure 10-3.

To provide an output for the timer to control electrical circuits, most reset timers have a set of three contacts. Each contact can be adjusted to perform in the same way or in a different way, as required. Figure 10-4 shows that there are three conditions for each contact during a complete cycle:

1. Reset or deenergized—clutch and motor deenergized
2. Timing—clutch energized; motor may or may not be energized, depending on its circuit application
3. Timed out—motor deenergized; clutch may or may not be deenergized, depending on its circuit application

When the contact is open, the o symbol above the contact is shown. When the contact is closed, the symbol is shown as ×. For example, in a given timer contact arrangement, assume contact 1 is open in the reset condition, closed in the timing condition, and open in the timed-out condition. The complete symbol appearing above that contact would be ○ × ○.

Remember that the contact changes from its reset condition to the timing condition when the clutch is energized. The contact changes from the timing con-

Figure 10-3   Symbols for timer clutch and motor.

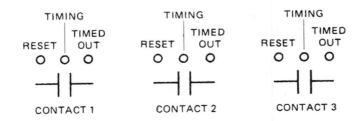

Figure 10-4   Timing symbols.

| | RESET | TIMING | TIMED OUT |
|---|---|---|---|
| SEQUENCE NUMBER | Clutch deenergized; motor deenergized | Clutch energized; motor energized | Clutch engaged until solenoid is deenergized; motor deenergized |
| 1 | O | O | X |
| 2 | O | X | O |
| 3 | X | X | O |
| 4 | X | O | X |
| 5 | O | X | X |
| 6 | X | O | O |

O CONTACT OPEN    X CONTACT CLOSED

*Figure 10-5*   Time sequence chart.

dition to the timed-out condition when the motor has run for the preset amount of time. The contact then changes back to the reset condition when the clutch is deenergized.

There are several arrangements available for each of the three contacts. A few are shown in Figure 10-5.

In some reset timers, the motor connections are not brought out for external connection. In this case they are generally wired so that the motor is energized from a normally open circuit contact that closes when the clutch is energized. On other timers, all connections are brought out. In this way, the motor as well as the clutch can be energized from an external source at any time during a cycle.

The circuit in Figure 10-6A illustrates the normal usage conditions in which the motor is energized directly from the clutch being energized. Figure 10-6B illustrates the motor energized at a later time, after the clutch is energized. Increased timing accuracy is possible with this arrangement if the timing period is relatively short in an overall long cycle. A third condition is shown in Figure 10-6C, in which the motor is energized before the clutch is energized. Increased accuracy can be obtained if the timing period is nearly that of the overall cycle.

Figure 10-7 shows an example of the reset timer.

Some of the simpler reset timers have only a snap-action switch that operates at the end of a preset time period. When the clutch is deenergized, the timer resets and the switch is released.

Another form of the reset timer is often called a *multiple-interval timer.* It is used extensively in programming control. Multiple time periods can be controlled by the adjustment of each individual ON and OFF setting.

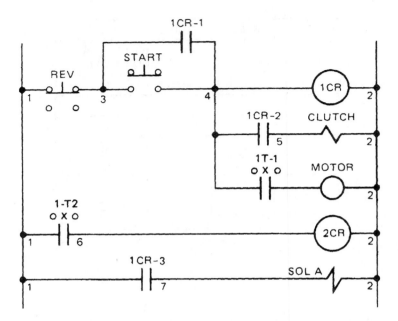

***Figure 10-6A***   Control circuit in which the motor is energized directly from the clutch being energized.

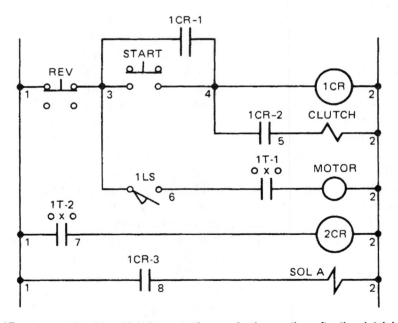

***Figure 10-6B***   Control circuit in which the motor is energized some time after the clutch is energized.

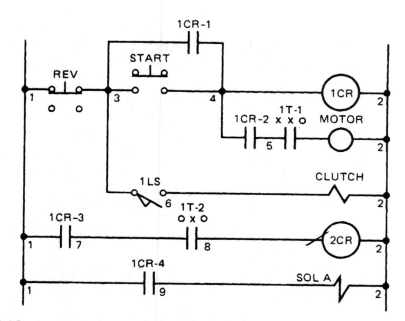

**Figure 10-6C** Control circuit in which the motor is energized before the clutch is energized.

**Figure 10-7** Reset timer. *(Courtesy of Eagle Signal Controls.)*

**Figure 10-8**  Multiple interval timer.  *(Courtesy of Eagle Signal Controls.)*

The timer displayed in Figure 10-8 consists of a series of contacts. Each contact has two fingers that ride on a cam plate. During the timing interval, a synchronous motor drives the plate downward through a solenoid-operated clutch. The contacts are closed and opened in the sequence set for the fingers to drop off the edge of the cam plate. When the clutch solenoid is deenergized to disengage the clutch, the cam plate is reset by a spring. Simultaneously, the contact fingers are lifted to their normally open or normally closed RESET position.

**10.3.2 Repeat-Cycle Timers**  The *repeat-cycle timer* is used to control a number of electrical circuits in a predetermined sequence.

The timer shown in Figure 10-9 consists of a synchronous motor driving a cam shaft. The cam shaft rotates continuously as long as the motor is energized. Adjustable cams determine the point of closing and opening a switch during each cam-shaft revolution.

In the unique split-cam design, each side of the cam is separately screwdriver-adjustable in either direction. Either side determines the precise instant

**Figure 10-9** **Precision switch cam programmer.** *(Courtesy of Automatic Timing and Controls Company, Inc.)*

during the cycle when the switch will actuate; the other side determines how long the switch remains actuated. Adjustments are easy and precise. One quarter turn of the adjusting screw equals .5% of cycle time. A setting disc, calibrated in 1% increments, facilitates program setup and indicates cycle progress.

**10.3.3 Manual-Set Timers**  The *manual-set timer* requires manual operation of the timer to start operation. The timer then runs a selected time and stops automatically.

In the timers illustrated in Figures 10-10A and 10-10B, timing begins when the START switch is closed. At the same time, the timing LED goes on. A relaxation oscillator starts to run at a rate determined by the set point. The timer times out and the timing LED turns off when the oscillator count is to the level set by the range switch. At time out, the load relay is energized, transferring its contacts, and the timing circuit is automatically deenergized. Reset occurs when the START switch is opened or when power is interrupted.

**Figure 10-10A** **Manual-set timer.** *(Courtesy of Eagle Signal Controls.)*

**Figure 10-10B** **Manual-set timers.** *(Courtesy of Automatic Timing and Controls Company, Inc.)*

# 10.4 Solid-State Timers

Great advancements have been made in the design of timer control units. A thorough review of the many types and variations of solid-state timers would more than fill this entire book. For our purposes, then, we limit our discussion to the following important design features:

- Microprocessor-based timers
- Digital set and optional digital readout
- Provision for external set

Examples of solid-state timers are pictured in Figures 10-11A and 10-11B. The Eagle CX200 (Figure 10-11A) is a microprocessor-based timer/counter. Time or count operation, time range, and standard or reverse start operation are

**Figure 10-11A** Microprocessor-based timer/counter. *(Courtesy of Eagle Signal Controls.)*

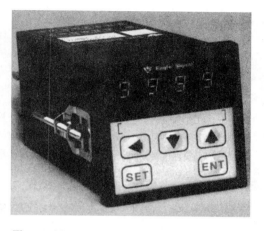

**Figure 10-11B** Microprocessor-based digital timer. *(Courtesy of Eagle Signal Controls.)*

selected via seven miniature rocker switches located inside the unit housing. Time or count set points are entered into the unit using a sealed membrane keypad on the front of the unit. Each digit in the set point is individually increased or decreased by pressing an appropriate keypad switch. Time or count set point and progress are displayed on the front of the unit by a 4½-digit liquid crystal display with 0.5-inch digits. Operational mode annunciators also appear in the display area on the front of the unit. The mode annunciator flashes when the unit is timing or counting.

The outputs can be programmed to operate in one of four load sequences: ○○×, ○×○, ○○× with pulse output, and ○○× pulse output with repeat-cycle operation.

The setting accuracy for count is 100%. The setting accuracy for time is ±0.5% or 50 milliseconds (ms), whichever is larger. The repeat accuracy for count is 100%. For time it is .001% or 35 ms, whichever is larger.

The SX200 series (Figure 10-11B) is a microprocessor-based digital timer. The programmable features include eight time ranges and eight output operating modes. These operating modes and all other setup functions are programmed with miniature rocker switches located on the back of the housing.

This timer uses a nonvolatile RAM memory (see Chapter 15) to retain the set point, actual time values, and program parameters. The expected life of data in memory is ten years.

The front panel of the unit is a sealed membrane keypad that provides excellent protection for most industrial environments. The keypad includes a special surface just below the display on which the function of the timer can be marked with a pen or pencil.

The SET and ENT keys on the front panel provide access to the set point and to the front panel programmed functions. Programming changes are entered using the increment and decrement keys.

A keypad "lock" function is built into the software of the unit, which allows the set point to be viewed but does not allow unauthorized changes.

The timing cycle progress is shown on four 0.3-inch red LED displays for easy readability. The front panel also has a flashing LED at the right side of the display to indicate that the unit is in the timing cycle and an LED at the left side of the display that lights when the programmed contacts are energized.

## 10.5 Circuit Applications

Refer again to Figure 8-10C. The ram advances until it reaches the work piece and builds pressure. On reaching a preset pressure, the normally closed pressure switch contact opens, and the ram reverses.

In Figure 10-12, a reset timer is added to the circuit shown in Figure 8-10C. The pressure switch contact is changed from normally closed to normally open. The sequence of operations is as follows:

1. Operate the START push-button switch.
2. Relay coil 1CR is energized.
   a. Relay contact 1CR-1 closes, interlocking around the START push-button switch.
   b. Relay contact 1CR-2 closes, energizing solenoid A. The piston advances to the work piece and builds pressure.
3. The normally open pressure switch contact closes, energizing the timer clutch.
   a. Timer contact 1T-1 closes, energizing the timer motor.
4. When a preset time elapses, timer contact 1T-1 opens, deenergizing the timer motor.
5. Timer contact 1T-2 closes, energizing relay 2CR.
   a. Relay contact 2CR-1 opens, deenergizing relay coil 1CR.
   b. Relay contact 1CR-1 opens, opening the interlock circuit around the START push-button circuit.
   c. Relay contact 1CR-2 opens, deenergizing solenoid A.
6. Pressure drops, opening pressure switch contact PS.
7. Timer clutch deenergized.
8. Timer contact 1T-2 opens, deenergizing relay 2CR.
9. Normally closed relay contact 2CR-1 closes, setting up the circuit for the next cycle.

In the circuit shown in Figure 10-13, an additional cylinder-piston assembly is added with an additional pressure switch contact 2PS.

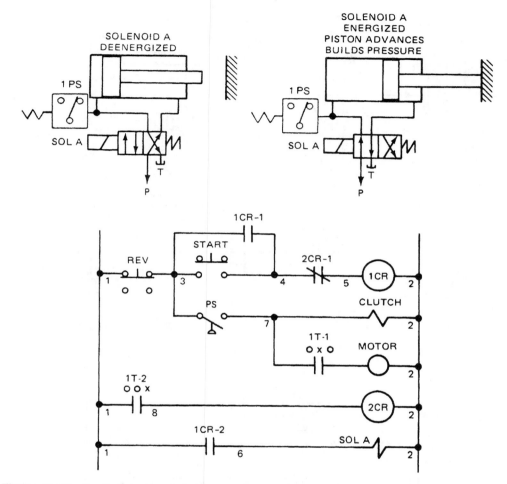

***Figure 10-12*** **Control circuit—timing how long piston is extended. P, pressure; T, tank.**

The function of this circuit is to start piston #2 with a time delay after piston #1 has built pressure to a preset level on the work piece. The sequence of operations proceeds as follows:

1. Operate the START push-button switch.
2. Relay coil 1CR is energized.
   a. Relay contact 1CR-1 closes, interlocking around the START push-button switch.
   b. Relay contact 1CR-2 closes, energizing solenoid A.

Piston #1 advances and builds pressure on the work piece. When a preset pressure is reached, pressure switch contact 1PS operates.

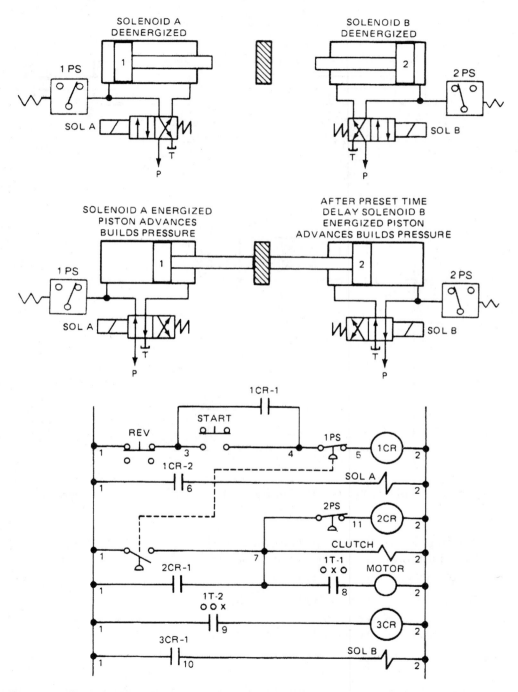

**Figure 10-13** Two-piston control circuit. Timer controls delay of piston B extension until piston A has been extended. P, pressure; T, tank

3. Normally closed pressure switch contact 1PS opens, deenergizing relay coil 1CR.
   a. Relay contact 1CR-1 opens, opening the interlock around the START push-button switch.
   b. Relay contact 1CR-2 opens, deenergizing solenoid A.

   Piston #1 now returns to its initial start position.

4. Normally open pressure switch contact 1PS closes, energizing relay coil 2CR.
   a. Relay contact 2CR-1 closes, interlocking around 1PS pressure switch contact.
5. Timer clutch energized.
   a. Timer contact 1T-1 closes, energizing the timer motor.
6. Timer times out.
   a. Timer contact 1T-1 opens, deenergizing the timer motor.
   b. Timer contact 1T-2 closes, energizing relay coil 3CR.
7. Relay contact 3CR-1 closes, energizing solenoid B.

   Piston #2 advances and builds pressure on the work piece.

8. Normally closed pressure switch contact 2PS opens, deenergizing relay coil 2CR.
   a. Relay contact 2CR-1 opens, deenergizing timer clutch.
9. Timer contact 1T-2 opens, deenergizing relay coil 3CR.
   a. Relay contact 3CR-1 opens, deenergizing solenoid B.

   Piston #2 now returns to its initial start position.

## Recommended Web Links

Students are encouraged to view the following Web sites as a supplement to the concepts presented in this textbook. Review and analyze the array of products that are available for electrical control applications. Many of these sites offer technical information that can help in converting the principles to practical applications. To view catalogs, your PC may require Adobe Acrobat Reader software.

Omron
www.omron-ap.com/admin/e-catalog/catalog.asp?catlvl=167
Review: Timers

Allen-Bradley
www.ab.com/industrialcontrols/products/relays_timers/index.html
Review: Product Line

General Electric
www.geindustrial.com/cwc/products?pnlid=3&famid=32&catid=104&id=r-iectim
Review: Specifications and Applications (pdf)

Automatic Timing and Controls
www.flw.com/atc/atc.htm
Review: Timer Products

## Achievement Review

1. Describe the differences between a time-delay relay and a timer.

2. List the two types of timers that are applied most frequently in the industrial field.

3. What are the two major components responsible for the operation of the reset timer?

4. List the conditions of the two major components in a reset timer under the following conditions:

   a. Reset or deenergized
   b. Timing
   c. Timed out

5. In a control circuit using two push-button switches, two relays, a reset timer, and a solenoid, show how the clutch and motor of the reset timer can be energized at the same time. The cycle starts with the operation of a push-button switch.

6. What is the major use of the multiple-interval timer?

7. What controls the opening and closing of a switch in the repeat-cycle timer?

8. What four sequences can be programmed to operate in the Eagle CX200 timer/counter? Explain the × and ○ notation.

9. In the control circuit shown in Figure 10-12, which of the following describes when the timer clutch is energized?

   a. Relay 1CR is energized
   b. Relay 2CR is deenergized
   c. Pressure builds, operating pressure switch 1PS

10. In the control circuit shown in Figure 10-13, which of the following describes when solenoid B is deenergized?

   a. Relay 1CR is deenergized
   b. The timer times out
   c. Pressure builds, operating pressure switch 2PS

## Objectives

After studying this chapter, you should be able to:

- Explain the basic difference between an electromechanical reset timer and an electromechanical reset counter.
- Discuss how the clutch and count motor operate through one complete cycle.
- Use the counter contact symbol in each of three conditions: deenergized, counting, and counted out.
- Follow through a typical control circuit using an electromechanical reset counter.
- Discuss some of the features of a solid-state counter.

## 11.1 Preset Electrical Impulses

The electromechanical control counter is similar to the electromechanical reset timer, except that the synchronous motor of the timer is replaced by a stepping motor. The stepping motor advances one step each time it is deenergized. After a preset number of electrical impulses, a contact is opened or closed.

The electromechanical counter requires a minimum of approximately 0.5 seconds OFF time between input impulses to reset.

As in the electromechanical timer, there are three output contacts. Each contact can be arranged for a sequence. The conditions for each contact can change as shown in Figure 11-1.

With the clutch deenergized, the contacts are in the reset condition. When the clutch is energized, the contacts go to the counting condition. When the count motor has received the same number of count impulses as set on the dial, the contacts go the counted-out condition. When the clutch is deenergized, the contacts return to the reset condition.

| CONTACT OPEN | RESET | COUNTING | COUNTED OUT |
|---|---|---|---|
| O | O | X | X |
|  | X | O | O |
| CONTACT CLOSED | O | X | O |
| X | X | O | X |

**Figure 11-1**   **Counter contact conditions.**

**Figure 11-2**   **Typical electromechanical counter.** *(Courtesy of Eagle Signal Controls.)*

Figure 11-2 shows a typical electromechanical counter. These devices are generally available with analog set and readout dials.

## 11.2 Circuit Applications

A typical control circuit using a reset counter is shown in Figure 11-3. In this example a single piston-cylinder assembly is used. The piston is to travel to the right as shown until it engages and operates limit switch 1LS at position P1. The piston returns to position P2, operating limit switch 2LS. The piston now travels to the right. This reciprocating motion continues until the preset number of counts has been reached. The counter is now in the counted-out condition. The piston continues to the right until a work piece is engaged and preset pressure builds to the setting of pressure switch 1PS. The piston now returns to the start position.

A single-solenoid, spring-return operating valve is used to supply fluid power to the cylinder. The circuit is arranged to return the piston to the start position at any time by operating the REVERSE push-button switch.

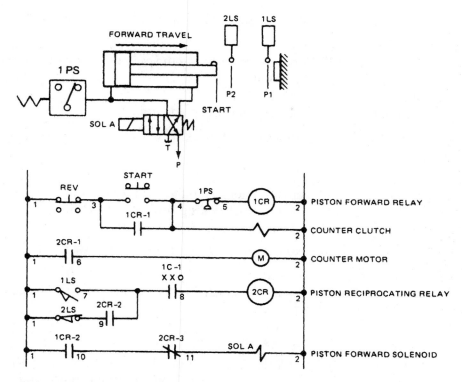

***Figure 11-3*** Control circuit using a reset counter.

The sequence of operation proceeds as follows:

1. Operate the START push-button switch.
2. Relay coil 1CR is energized.
   a. Relay contact 1CR-1 closes, interlocking around the START push-button switch and energizing the counter clutch.
   b. Relay contact 1CR-2 closes, energizing solenoid A.

The piston travels to the right. At position P2, limit switch 2LS is operated. No action results as relay contact 2CR-2 is open. The piston continues to travel to position P1, operating limit switch 1LS.

3. Relay coil 2CR is energized.
   a. Relay contact 2CR-1 closes, energizing the counter motor.
   b. Relay contact 2CR-2 closes, forming an interlock circuit with normally closed limit switch contact 2LS around normally open limit switch contact 1LS.
   c. Relay contact 2CR-3 opens, deenergizing solenoid A.

The piston travels back (to the left) until it operates limit switch 2LS.

4. Relay coil 2CR is deenergized.
   a. Relay contact 2CR-1 opens, deenergizing the counter motor. The counter has now completed one count.
   b. Relay contact 2CR-2 opens.
   c. Relay contact 2CR-3 closes, energizing solenoid A.

The piston now travels forward (to the right) until it operates limit switch 1LS.

5. Relay coil 2CR energized.
   a. Relay contact 2CR-1 closes, energizing the counter motor.
   b. Relay contact 2CR-2 closes, interlocking limit switch 1LS.
   c. Relay contact 2CR-3 opens, deenergizing solenoid A.

The piston travels back (to the left) until it operates limit switch 2LS.

6. Relay coil 2CR is deenergized.
   a. Relay contact 2CR-1 opens, deenergizing the counter motor. The counter has now completed two counts.
   b. Relay contact 2CR-2 opens.
   c. Relay contact 2CR-3 closes, energizing solenoid A.

The piston continues to shuttle between limit switches 1LS and 2LS until the counter has counted the number of preset counts. When the counter has counted out, counter contact 1C-1 opens. The next time the piston advances and operates limit switch 1LS, relay coil 2CR is not energized.

7. Solenoid A remains energized.
8. Piston continues past limit switch 1LS, meeting the work piece and building pressure to a preset amount on pressure switch 1PS.
9. The normally closed pressure switch contact 1PS opens.
10. Relay coil 1CR is deenergized.
    a. Relay contact 1CR-1 opens, opening the interlock circuit around the START push-button switch and deenergizing the counter clutch.
    b. Contact 1CR-2 opens, deenergizing solenoid A.

The piston now returns to its initial start position.

# 11.3 Solid-State Counters

Considerable advancements have been made in solid-state counter design. Solid-state counters are available with digital set and digital readout. High-speed pulse operation with 100% accuracy is available.

Figure 11-4A shows a typical solid-state counter. It is a single preset counter with many programmable features. Two input channels (A and B) allow a variety of inputs. Three counting modes can be programmed (count directional, add/subtract, quadrature). This unit is able to count up from zero or count down from the set point. The counter can be programmed to give a pulse at count-out

**Figure 11-4A** **Typical solid-state counter.**
*(Courtesy of Eagle Signal Controls.)*

**Figure 11-4B** **Solid-state counter with CMOS integrated circuits.** *(Courtesy of Eagle Signal Controls.)*

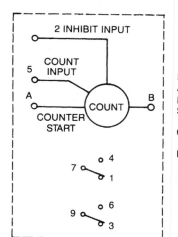

NOTE: RELAY OUTPUTS AND TRIAC OUTPUT ARE NOT AVAILABLE SIMULTANEOUSLY.
RELAY OUTPUTS—STANDARD
START INPUT—CLOSE TO START
                    OPEN TO RESET
COUNT INPUT—COUNTS ON SWITCH
                    OPENING
INHIBIT INPUT—CLOSE TO STOP
                    COUNT PROGRESS
                    OPEN TO RESTART
                    COUNT PROGRESS

**Figure 11-4C** **Wiring connections for counter shown in Figure 11-4B.** *(Courtesy of Eagle Signal Controls.)*

and automatically reset, or it can be programmed for single-cycle latch operation. The counter is programmed with six dip switches located on the side of the counter.

The counter shown in Figure 11-4B uses CMOS (complimentary metal-oxide semiconductor) integrated circuits for the counting function. Counter output action occurs when the count total indicated by front-mounted thumbwheel switches is reached.

Referring to the terminal arrangements shown in Figure 11-4C, the counter sets to the selected thumbwheel setting when power is applied to terminals A and B. Counts are applied to a count input terminal, and each count is registered on contact opening. When registered counts equal the set point, the output changes state. The output remains in this state as long as the line voltage is applied to terminals A and B.

A count inhibit is available with this counter. When line voltage is applied to the count inhibit from either terminals A or B, incoming count pulses are not counted. The counter remembers count total at the time that the inhibit is applied and resumes counting from that point after the inhibit voltage is removed. The count inhibits can be applied from either side of the power line.

Referring back to Figure 11-1, this counter has four usable output sequences: ○○×, ××○, ○×○, ×○×.

## Recommended Web Links

Students are encouraged to view the following Web sites as a supplement to the concepts presented in this textbook. Review and analyze the array of products that are available for electrical control applications. Many of these sites offer technical information that can help in converting the principles to practical applications. To view catalogs, your PC may require Adobe Acrobat Reader software.

Precision sales
www.precisionsales.com/kcount.html
Review: Product Line

Eaton/Durant
www.durant.com
Review: Count Controls, Electronic, Electromechanical, Mechanical

Automatic Timing and Controls
www.flw.com/atc/atc.htm
Review: Timer Products

## Achievement Review

1. How does the electromechanical reset counter differ from the electro-mechanical reset timer?

2. A given counter contact has been arranged to be open in the deenergized condition of the counter, closed in the counting condition, and closed in the counted-out condition. Show the contact and symbol for these conditions.

3. What accuracy can you expect to receive when using a solid-state counter?

4. Given the following counter contact symbol, explain the condition of the contact in the deenergized condition of the counter, the counting condition, and the counted-out condition.

5. In the electromechanical counter, what is the condition of the contacts with the clutch deenergized?

6. What accuracy can be expected in modern solid-state counters?

    a. 75%
    b. 90%
    c. 100%

7. What are some of the features available in the solid-state timer-counters?

8. In the circuit shown in Figure 11-3, what operational change would occur if the counter contact 1C-1 had a sequence of ×○× (closed-open-closed)?

9. In the same circuit (Figure 11-3), what is the function of the NC relay contact 2CR-3?

10. Under what condition does the stepping motor advance one step?

    a. Each time the motor is energized
    b. Each time the motor is deenergized

11. In a solid-state counter, when does the counter output action occur?

12. Explain the three output status conditions and the notation used.

# Chapter 12 Control Circuits

## Objectives

After studying this chapter, you should be able to:

- Explain how all complete control circuits progress through three basic areas: information or input, decision or logic, and output or work.
- Draw a sequence bar chart for an electrical control circuit.
- Demonstrate how various electrical components are used to gather information from a machine or system.
- Demonstrate how electrical components are used to make a decision as to how the gathered information is to be used.
- Demonstrate how the decision affects the output or work the machine is to accomplish.
- Draw simple electrical control circuits from a given set of requirements.
- Describe how an interlock circuit is used.
- Explain the function of the timer sequence symbol above each timer contact.

## 12.1 Placement of Components in a Control Circuit

Most of the components used in the electrical control of machines are discussed in Chapters 1 through 11 of this text. Figure 12-1 provides an outline of components and their usage. In this chapter, these components are placed in positions in a control circuit to show how to accomplish several different types of control.

Here are three examples that show the procedure as the control moves from the *input* (information) to the *output* (work):

1. A motor START push-button switch is operated, energizing a motor starter coil. The motor starter contacts then close, energizing the motor (Figure 12-2).

| INFORMATION<br>OR<br>INPUT | DECISION<br>OR<br>LOGIC | OUTPUT<br>OR<br>WORK |
|---|---|---|
| MANUAL<br>  Push-button switches<br>  Selector switches<br>  Drum switches | Relays | Lights |
| POSITION<br>  Limit switches | | Solenoids |
| PRESSURE<br>  Pressure switches | | |
| TEMPERATURE<br>  Pyrometers<br>  Thermostats | Contactors | Heating elements |
| TIME<br>  Timers | Motor starters | Motors |

*Figure 12-1*  **Chart showing division of control.**

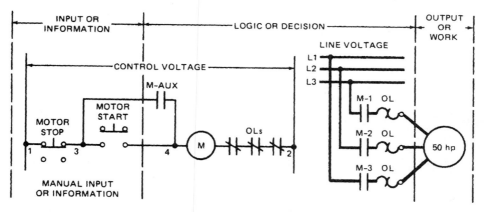

*Figure 12-2*  **Three-phase motor starter control circuit. Conductors carrying load current at line voltage are denoted by heavy lines.**

2. A normally open thermostat contact closes, energizing the coil of a contactor. The contactor contacts then close, energizing the heating elements (Figure 12-3).
3. A normally open limit switch contact may be held closed (operated) at the start of a cycle. This limit switch contact could be connected to energize the coil of a relay. A normally open relay contact then closes to energize a solenoid (Figure 12-4).

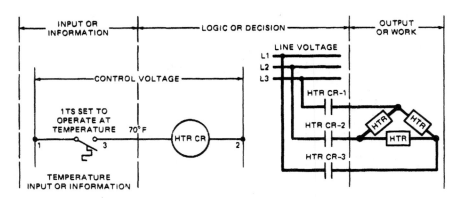

**Figure 12-3**   Three-phase heater control circuit. Conductors carrying load current at line voltage are denoted by heavy lines.

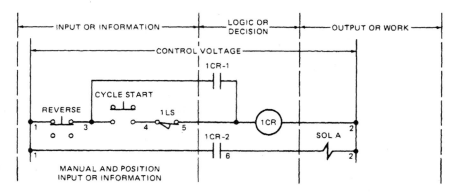

**Figure 12-4**   Solenoid control circuit.

In the eight circuit examples that follow, some or all of the input components are used in a control circuit to accomplish a given output. For each example, a *sequence bar chart* or analysis of the circuit is shown. The bar chart is the most efficient and simplest method of explaining the functioning of the control components from input to output.

The bar chart is arranged with the electrical components to be energized in a vertical column at the left-hand side. The components to be energized are the relays, time-delay relays, contactors, timers, and motor starters. The events that occur in sequence during a cycle of the machine are shown in a horizontal line at the bottom. These events are due to manual, position, pressure, temperature, or time inputs.

Each time a component is energized during a machine cycle from one event to a following event, a solid bar is shown on a horizontal line opposite to the component being energized. Thus, at any time during a machine cycle, the energized or deenergized condition of the component is observed by checking along a vertical line corresponding to the event that has occurred.

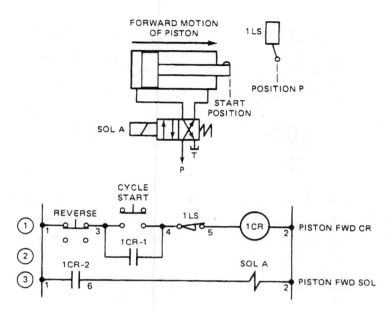

Figure 12-5A   Piston control circuit with extended piston detector.

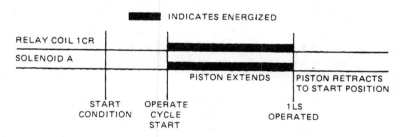

*Figure 12-5B*   Sequence bar chart for Figure 12-5A.

Figure 12-5A shows a typical control circuit. A pictorial drawing of the machine or equipment being used is included. In this circuit, at the position that limit switch 1LS is operated, both the coil of relay 1CR and solenoid A are deenergized. Figure 12-5B is a bar chart of this activity.

## 12.2 Control Circuit Examples

**Example #1—Figures 12-6A and 12-6B**   A three-phase motor is used to drive a fluid power pump that provides fluid power pressure. A fused disconnect switch provides a disconnecting means and short-circuit protection. The control transformer has fuses in the primary winding circuit. The transformer secondary is

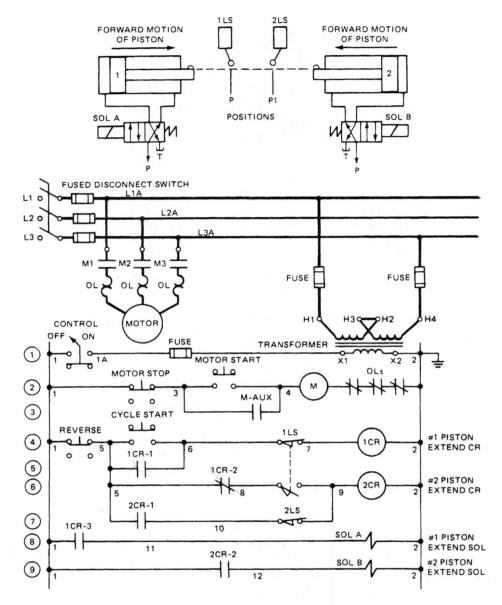

*Figure 12-6A*   Two piston control circuit with extended piston detection for Example #1. Three-phase heater control circuit. Conductors carrying load current at line voltage are denoted by heavy lines. P, pressure; T, tank.

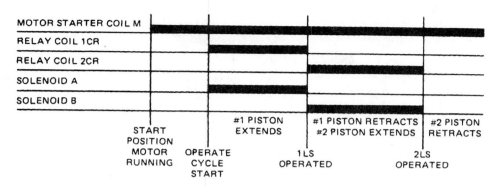

**Figure 12-6B** Sequence bar chart for Figure 12-6A.

grounded on the common side. The other side of the transformer secondary is fused. A selector switch is used for control disconnecting means. The operating cycle for the machine proceeds as follows.

The #1 piston moves from a start position to the right as shown. When the piston reaches a predetermined position P, it stops and reverses to its start position. At the same time that #1 piston reverses, #2 piston moves from a start position and moves to the left as shown. When #2 piston reaches a predetermined position P1, it stops and returns to its start position. Pressure can be available on the rod end of both pistons at the start positions. Provisions must be made to reverse either piston at any time in their forward strokes.

**Example #2—Figures 12-7A and 12-7B**   The same power and control circuit used for the motor and control transformer in Example #1 is used in this example. Therefore, in Example #2 only the ladder-type schematic circuit is shown.

As shown, the #1 piston moves from a start position to the right. The #2 piston moves from a start position to the left. The #3 piston moves from a start position up.

All three pistons will return to their start positions when any of the following conditions occurs:

- The #1 piston has engaged the work and built pressure to a predetermined setting
- A predetermined time setting on the #2 piston has been reached
- The #3 piston has reached a predetermined position P

Pressure can be available on the rod end of the pistons in their start positions. Provisions must be made to reverse all pistons at any time in their forward strokes.

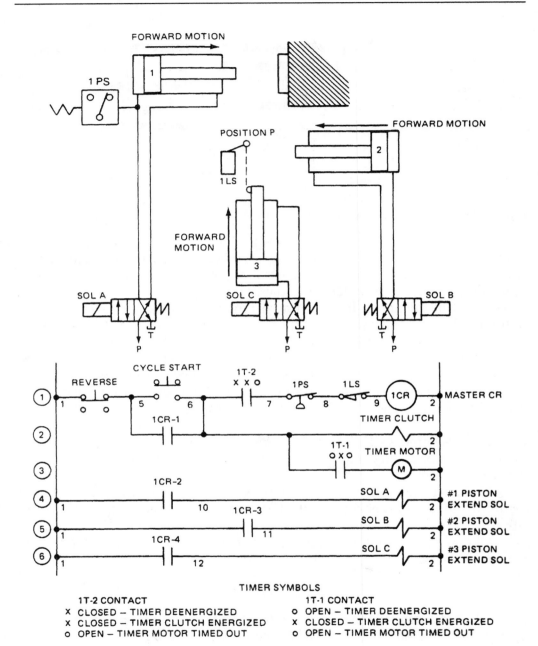

*Figure 12-7A*    Control circuit for Example #2. P, pressure; T, tank.

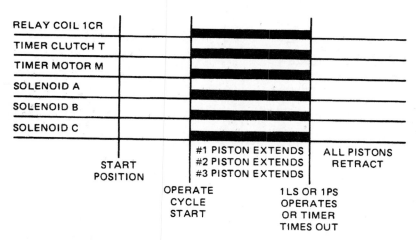

**Figure 12-7B** Sequence bar chart for Figure 12-7A.

One practical advantage of this circuit is that any one of the three inputs can be adjusted independently to control all three pistons. For example:

1. Adjust the pressure switch.
2. Change the timing.
3. Move the limit switch.

**Example #3—Figures 12-8A and 12-8B**  In this example the goal is to control the temperature of the operating fluid in the fluid power system. The motor should not be able to start unless the temperature of the operating fluid is more than 70°F and less than 100°F. The same power and control transformer circuit used in Example #1 is used here.

The #1 piston must be at position P1 for start condition. The #2 piston must be at position P4 for start condition. The #1 piston moves to the right as shown. At the same time, the #2 piston moves to the left. When the #1 piston reaches a predetermined position P2, it stops and reverses to position P1. When the #2 piston reaches a predetermined position P3, it stops and reverses to position P4. No pressure can be on the pistons at the start positions. Provisions must be made to reverse both pistons at any time in their forward strokes.

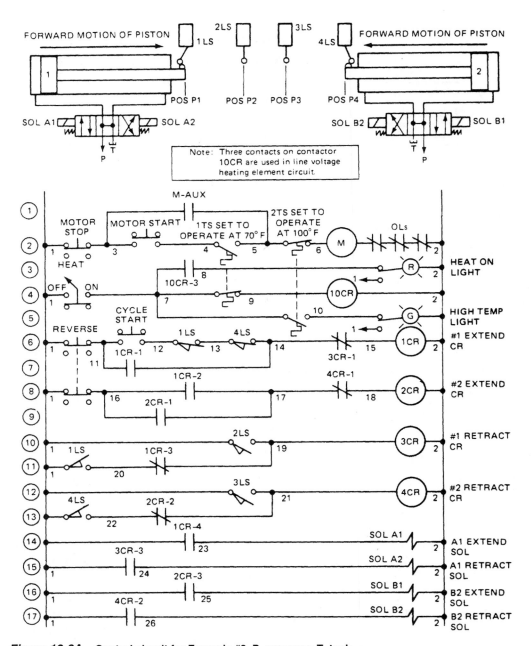

**Figure 12-8A**  Control circuit for Example #3. P, pressure; T, tank.

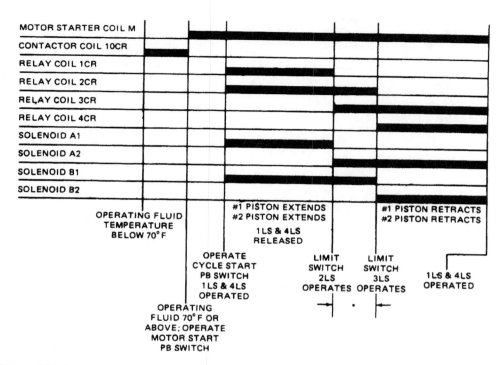

**Figure 12-8B** Sequence bar chart for Figure 12-8A.

**Example #4—Figures 12-9A and 12-9B** The same motor and control transformer circuit used in Example #1 is used here. Therefore, only the ladder-type schematic circuit is shown. Two cylinder-piston assemblies are used: #1 and #2. The #1 piston must be at position 1 and #2 piston must be at position 2 for start conditions.

When the cycle START push-button switch is operated, the #1 piston moves to the right and engages a work piece. A preset pressure is built on the work piece. When the preset pressure is reached, the #2 piston advances to the left. At the same time, the #1 piston returns to position 1. The #2 piston advances for a preset period of time. When the preset time expires, the #2 piston returns to position 2. A reset timer is used in this circuit.

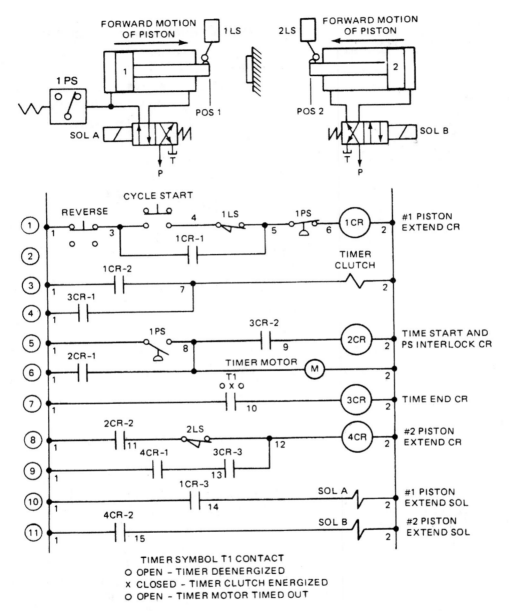

**TIMER SYMBOL T1 CONTACT**
O OPEN – TIMER DEENERGIZED
X CLOSED – TIMER CLUTCH ENERGIZED
O OPEN – TIMER MOTOR TIMED OUT

*Figure 12-9A*   **Control circuit for Example #4. P, pressure; T, tank.**

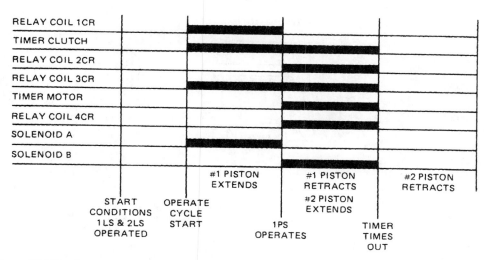

*Figure 12-9B*   Sequence bar chart for Figure 12-9A.

**Example #5—Figures 12-10A and 12-10B**   The same motor and control transformer control circuit used in Example #1 is used here. Therefore, only the ladder-type schematic circuit is shown.

One piston is used. The piston extends and contacts work. When the system reaches preset pressure, the piston reverses and operates limit switch 1LS. The cycle is repeated until the counter counts out. On the last forward stroke and after a preset time, the piston returns to the start position and stops.

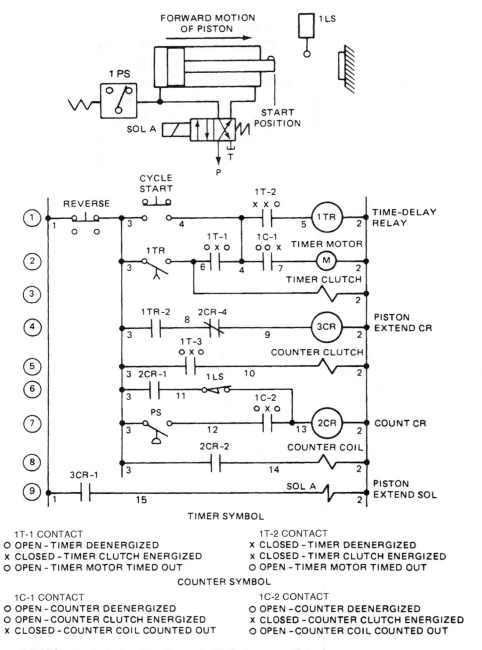

**Figure 12-10A**    Control circuit for Example #5. P, pressure; T, tank.

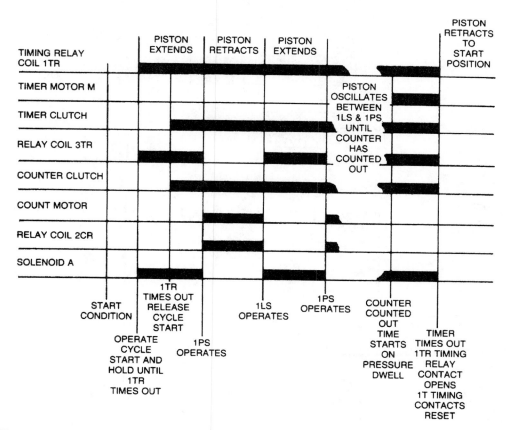

*Figure 12-10B*  Sequence bar chart for Figure 12-10A.

**Example #6—Figures 12-11A and 12-11B**  The same motor and control transformer circuit used in Example #1 is used here. Therefore, only the ladder-type schematic circuit is shown. This circuit is known as an *anti-tiedown*; that is, it is required that both START push-button switches be operated to start a cycle. Both switches must be released to the unoperated condition before the next cycle can be started.

One piston is used. The piston extends and stops in the extended position until a preset time has elapsed. The piston then returns to the start position.

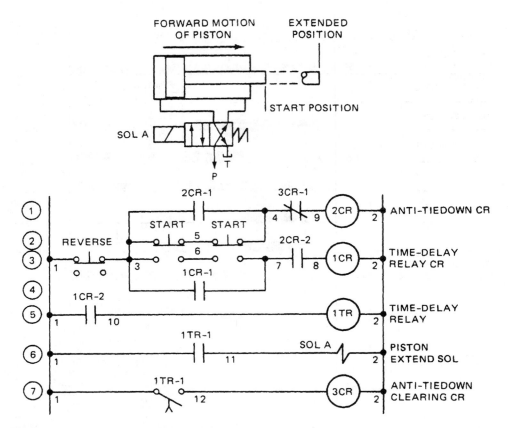

**Figure 12-11A**    Control circuit for Example #6. P, pressure; T, tank.

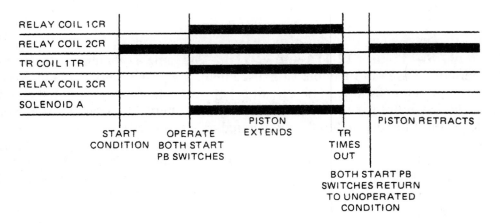

*igure 12-11B*    Sequence bar chart for Figure 12-11A.

**Example #7—Figures 12-12A and 12-12B**   The circuit in this example shows only a timed starting sequence circuit for two motor starters. The #1 motor starter coil is energized when the motor START push-button switch is operated, energizing motor #1. After a preset time delay, #2 motor starter coil is energized, energizing motor #2. Indicating lights are energized, showing that both motor starter coils are energized. The motor START push-button switches must be held operated until both motor starter coils are energized.

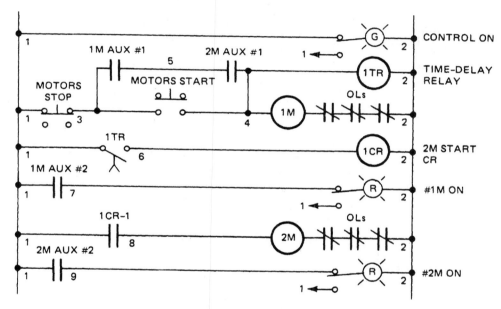

*Figure 12-12A*   Control circuit for Example #7.

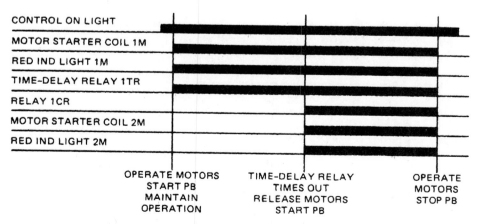

*Figure 12-12B*   Sequence bar chart for Figure 12-12A.

**Example #8—Figures 12-13A and 12-13B**   In this example, two separate three-phase motors are used. Each motor drives a fluid power pump. Fluid power pressure from #1 pump supplies fluid power pressure to the operating valve, using solenoids A1 and A2. Fluid power from #2 pump supplies fluid power pressure to the operating valve, using solenoids B1 and B2.

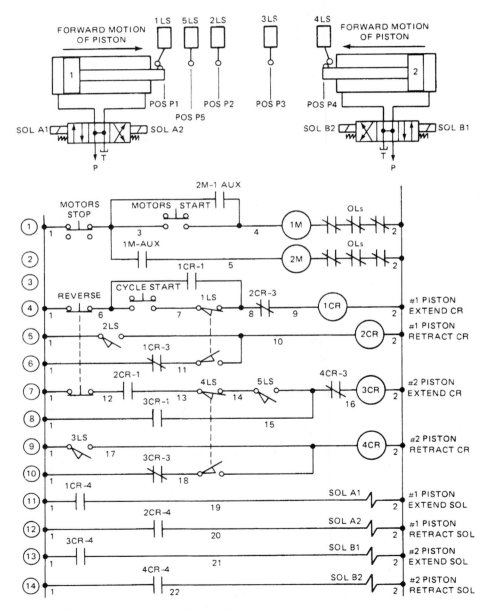

*Figure 12-13A*   **Control circuit for Example #8. P, pressure; T, tank.**

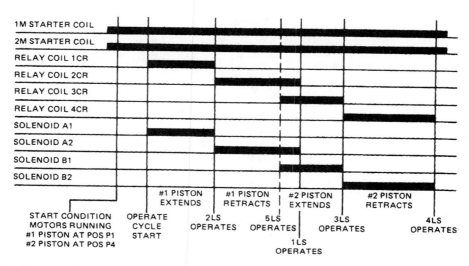

*Figure 12-13B*   Sequence bar chart for Figure 12-13A.

The motor start circuit uses only one push-button switch for starting. The #2 motor starter coil is energized through the closing of a normally open interlock contact on #1 motor starter. The operation of a single motor STOP push-button switch deenergizes both motor starter coils. An auxiliary contact on #2 motor starter is used to interlock the motor START push-button switch.

The #1 piston must be at position P1 for start condition. The #2 piston must be at position P4 for start condition. Piston #1 moves to the right as shown. When this piston reaches position P2, it stops and reverses to its start position P1. At a predetermined midpoint P5 in the travel of #1 piston on its return stroke, #2 piston moves to the left. When #2 piston reaches position P3, it stops and reverses to position P4. No pressure can be on the pistons in their start positions. Provisions must be made to reverse both pistons at any time in their forward strokes.

**Example #9—Batch Process Application**   This example is a typical application of a batch process involving several timed events initiated by some important action (Figure 12-14A). The circuit (Figure 12-14B) performs as follows.

The operator presses the PROCESS START switch to activate solenoid A, which allows product A (a liquid) to enter the tank. As the liquid level in the tank rises, the float switch 1FS activates, deenergizing solenoid A and starting 1TR to time and start the mixer motor. At the end of the timing cycle, 1TR-2 will stop the motor, deactivate 1CR, and activate solenoid B. As the liquid level in the tank empties, power to 1TR is removed and the timer is reset.

**Example #10—Fluid Control in an Off-Road Vehicle**   This example is an application of monitoring hydraulic fluid level and maintaining fluid temperature in both the primary and auxiliary fluid holding tanks of an off-road tractor. The circuit (Figure 12-15) performs as follows.

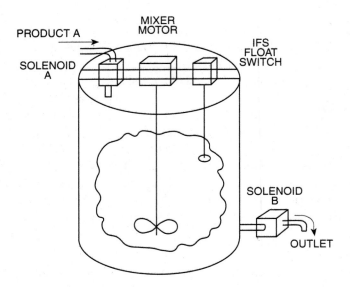

**Figure 12-14A**    Industrial control application: batch mixing process (Example #9).

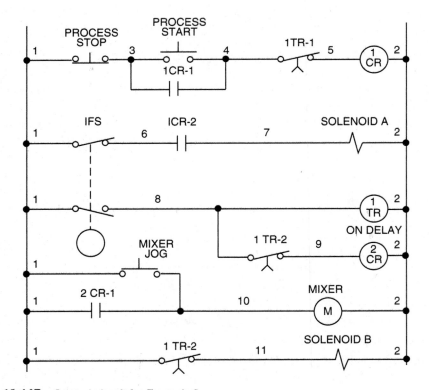

**Figure 12-14B**    Control circuit for Example 9.

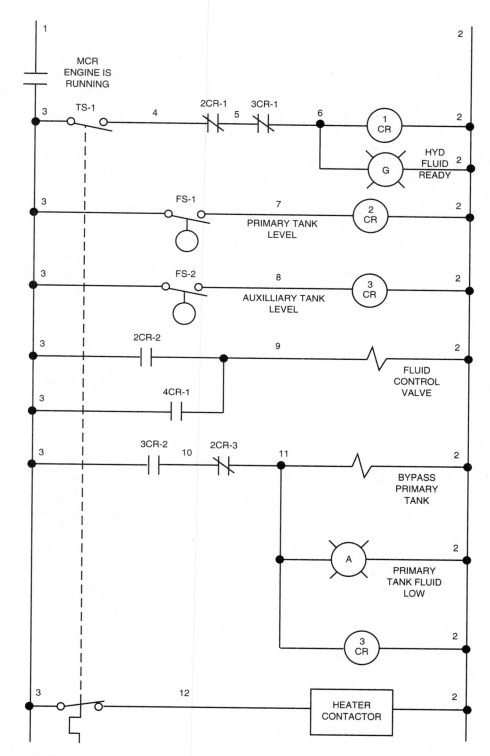

**Figure 12-15** Control circuit for Example #10: fluid control in an off-road vehicle.

As the motor is running, the master control relay (MCR) contact provides power to the circuit. If the fluid is too cold (below 45°F) a heater is activated through the thermal switch. As the temperature of the fluid increases to more than 45°F, 1CR and the hydraulic fluid ready light are energized. If the fluid level in the primary tank begins to drop below a designated level, and if the secondary tank has adequate fluid, the bypass primary tank solenoid is activated. If the levels in both tanks are low, then the fluid ready light will not be on.

**Example #11—Filling Process Control**   This application illustrates typical control of a filling process found in many material packaging operations (Figure 12-16A). The circuit (Figure 12-16B) performs as follows.

This process employs two motors: (1) a controlled motor that will stop and start in relation to the placement of the boxes below the filling chute, and (2) a noncontrolled motor that moves the filled boxes further down the process line.

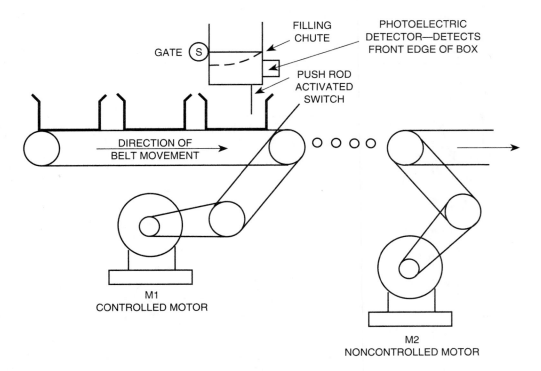

***Figure 12-16A***   **Industrial control application: Filling process control (Example #11).**

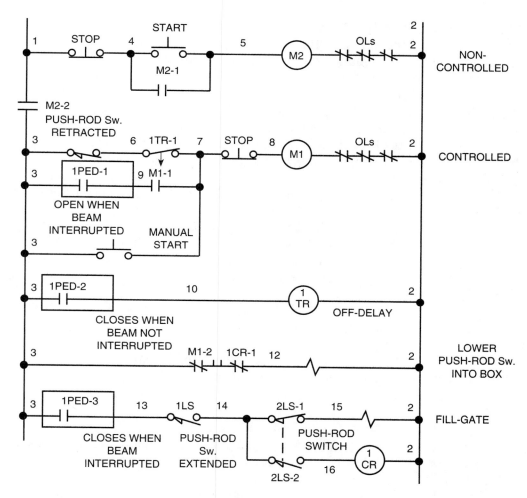

***Figure 12-16B*** **Control circuit for Example 11.**

The noncontrolled motor is controlled by the standard STOP-START control. The motor is expected to remain continuously on. Once the noncontrolled motor is turned on, contact M2-2 energizes the controlled motor circuit. Initially, if the push-rod is retracted, the controlled motor will start. When the photoelectric switch 1PED-2 senses the gap between the boxes, 1TR is energized, causing 1TR-1 to open immediately, deenergizing that latching leg of the circuit.

However, the contact 1PED-1 will be closed prior to the beam being interrupted by the box. Since the closed 1PED-1 is in series with the latching contact M1-1, the motor will remain running until the edge of the box is detected. When the next box enters the beam of 1PED-1, the contact will open, deenergizing the latch circuit and stopping the motor.

When the motor stops, M1-2 and 1CR-1 will both be closed, thus causing the push-rod to be extended into the box. Once the push-rod has fully extended into the box 1LS, the fill gate will open, causing the material to begin filling the box. As the material rises to activate the push-rod switch, 2LS-1 deenergizes the fill gate and 2LS-2 will deenergize the push-rod, causing it to rise out of the box. Once retracted, the controlled motor is started and the sequence is repeated.

**Example #12—Controlling the Rate of Speed and Position of an Electrohydraulic Proportional Valve**   This application example illustrates the use of control circuits to switch other electronic components. In this case, the circuit will switch potentiometers that control the acceleration and deceleration ramp rates. The circuit functions as follows.

Figure 12-17A shows the block diagram of the overall system. The function of the ramp module is to provide the hydraulic valve with a varying dc voltage. This voltage will cause the valve's solenoid to adjust the spool to a desired fluid flow rate. The module will accept a voltage developed from a resistance circuit, which creates a ramp rate (voltage change in time). The spool will, in turn, control the fluid flow to a cylinder. The cylinder's rod is the end-effector of this circuit.

Figure 12-17B shows the desired ramp rates as the valve's spool is moved. The ramp will determine the time it will take for the cylinder to reach the desired position. Process switches (S1, S2, and S3) will indicate the points where the cylinder's movement should change.

Figure 12-17C illustrates the input circuit to the ramp module. Note the selection of the potentiometers as inputs to the module. For example, this module will support three different ramp rates.

Figure 12-17D defines the ladder logic control circuit. When the START switch is activated, 1CR-4 will switch P1 to the ramp module. As the cylinder begins to move, 1LS will help latch 1CR. When the cylinder reaches position 2, 2LS will switch the ramp input from P1 to P2. Since P2 has a negative voltage, the ramp output will shift the valve's spool to the opposite direction. This shift causes the cylinder to retract at the designated ramp rate. Then 3LS is positioned to switch the ramp to deceleration into the home position, and stop.

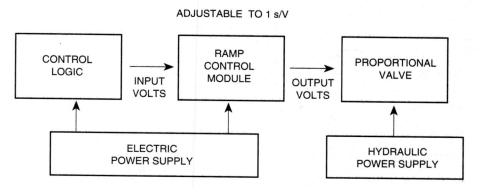

**Figure 12-17A** System block diagram. (Example #12).

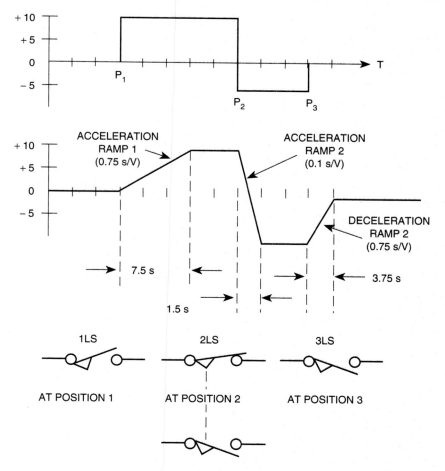

**Figure 12-17B** Control specifications for Example #12.

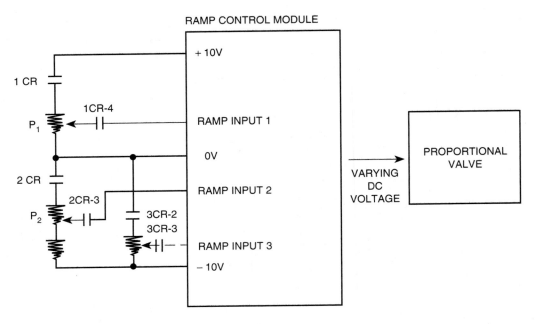

**Figure 12-17C** Control circuit for Example #12.

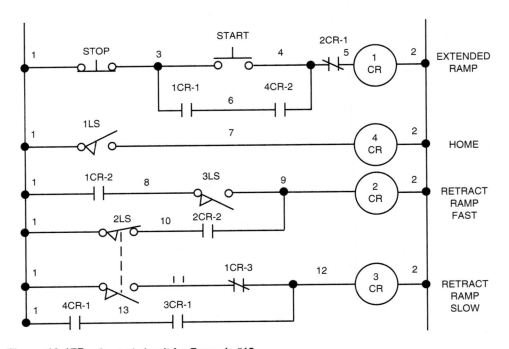

**Figure 12-17D** Control circuit for Example #12.

# Recommended Web Links

Students are encouraged to view the following Web sites as a supplement to the concepts presented in this textbook. Review and analyze the array of products that are available for electrical control applications. Many of these sites offer technical information that can help in converting the principles to practical applications. To view catalogs, your PC may require Adobe Acrobat Reader software.

Relay Ladder Logic
www.controleng.com
click on "Issue Archive"
click on "1998"
click on "April"
under "Departments" click on "Back to Basics"—Article: Relay Ladder Logic

Entertron Industries, Inc.
www.entertron.com/software.htm
Review: The information about ladder logic software

# Achievement Review

1. List five components that provide input (information) to an electrical control circuit.

2. Using electrical components, show how input, logic, and output progress through an electrical control circuit.

3. What electrical components should be shown on the left-hand vertical column of a bar chart?

4. In Example #1 (Figures 12-6A and 12-6B), what use is made of the fused disconnect switch?

5. Redesign the circuit shown in Figure 12-7 to allow a short time delay after operating the START push-button switch for the energizing of solenoid C. A time-delay relay may be used.

6. In the circuit shown in Figure 12-8, the operating fluid temperature being sensed by temperature switch 1TS is at 80°F. The heat OFF/ON switch has been turned on. What is the condition now of contactor coil 10CR?

   a. Energized
   b. Deenergized

7. In the circuit shown in Figure 12-9, what is the function of the normally open (held closed) limit switch contact 2LS?

8. In the circuit shown in Figure 12-10, when is the counter clutch energized? Explain the sequence of operations that follows the operation of the cycle START push-button switch.

9. In the circuit shown in Figure 12-11, what is the function of relay contact 2CR-2?

10. In the circuit shown in Figure 12-12, 2M motor has experienced an overload. This condition has opened the overload relay contacts on the 2M motor starter. Following this action, what is the condition of 1M motor starter?

    a. Deenergized
    b. Remains energized until the motor STOP push-button switch is operated

11. From Figure 12-14, what are the similarities and differences between a batch process and a machining operation?

12. From Figure 12-15, add a red alarm light to indicate when both tanks are low.

13. From Figure 12-16, add a counter to count the number of boxes being filled.

14. From Figure 12-17, what would need to be changed if the return profile is to be exactly the same as the extend profile? What are the purposes of the limit switch 2LS?

## Objectives

After studying this chapter, you should be able to:

- Name the important mechanical parts of an induction motor.
- Explain *slip* and its importance in the operation of a three-phase induction motor.
- Explain the major difference between a polyphase motor and a single-phase motor.
- List four different types of single-phase motors.
- Explain why the shaded-pole single-phase motor is different from the other single-phase motors.
- Know the basic types of dc motors
- Explain the characteristics of each type of dc motor
- Become acquainted with the operation of a brushless dc motor

## 13.1 ac Motors—Theory of Operation*

The principal parts of an electric motor are the stationary frame or *stator* and the revolving part or *rotor*. The rotary motion is produced by the interaction of the magnetic fields of the stator and rotor under the fundamental law of magnetism that like poles repel and unlike poles attract.

Figure 13-1 shows schematically the stator of an alternating, three-phase motor and indicates the manner in which the wires from each phase are wound around successive poles. Figure 13-2 shows in schematic form a three-phase, 60-Hz alternating current. The entire length of Figure 13-2 is one cycle, or 1/60

*Information courtesy of TECO-Westinghouse Motor Company.

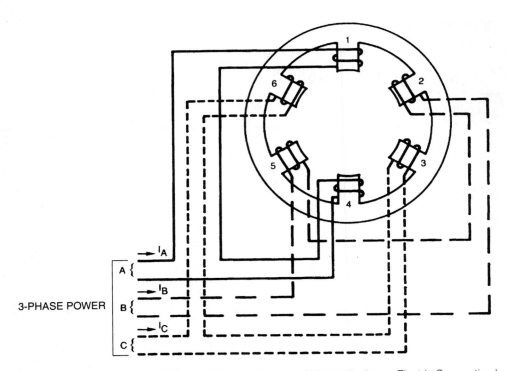

**Figure 13-1**   **Three-phase ac rotating field.** *(Courtesy of TECO-Westinghouse Electric Corporation.)*

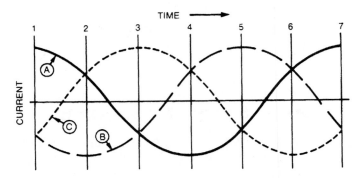

**igure 13-2**   **One complete three-phase (A, B, and C) ac cycle.** *(Courtesy of TECO-Westinghouse Electric Corporation.)*

of a second. Characteristically, the three phases of a three-phase power system maintain a relationship with each other as shown in this figure, each one offset from the other an equal amount.

It can be seen that phase A is at its greatest intensity at time 1 and is therefore producing the maximum magnetic field in motor poles 1 and 4. With current flowing as indicated at $1_A$, pole 1 will be a south pole and pole 4 a north pole. If the rotor should consist of a simple bar magnet mounted on a shaft, the north pole of the magnet would be attracted to the south pole 1 and the south pole of the magnet to the north motor pole 4.

Referring again to Figure 13-2, as phase A declines, the intensity of magnetism in motor poles 1 and 4 declines until at time 2 phase B has reached its peak in the negative direction and has excited motor pole 5 as a north pole. This new dominant north–south magnetic field will pull the simple bar magnet of the rotor to a new position to line up with poles 2 and 5.

At time 3, when phase 3 becomes maximum and causes poles 3 and 6 to become strong south and north poles, respectively, the rotor magnet will be pulled further clockwise. Continuing to the end of the cycle will cause the rotor magnet to complete one revolution. Since one cycle and one revolution took 1/60 second, the speed of this motor is evidently 60 × 60 or 3600 rpm. Each phase of this motor has two poles, so it is known as a "two-pole" motor. The formula for speed is:

$$S = \frac{120f}{p}$$

where $S$ is the synchronous speed in revolutions per minute, $f$ is the frequency in hertz, $p$ is the number of poles. Therefore:

$$S = \frac{120 \times 60}{2}$$

$$= 3600 \text{ rpm}$$

## 13.2 Polyphase Squirrel-Cage Induction Motors

The most common ac motor is the three-phase squirrel-cage induction motor shown schematically in Figure 13-3. The stator does not have solid poles as in Figure 13-1. Solid iron poles would overheat and give poor electrical efficiency. In actual practice the stator consists of a number of flat sheets called *laminations,* held together in a package and insulated from each other so as to reduce the circulating or eddy currents.

A typical stator lamination has a circular periphery and a round hole in the center. Notches or teeth are cut around the inner edge of the center hole. When the laminations are stacked in a package, these notches are lined up to form continuous channels called *slots.*

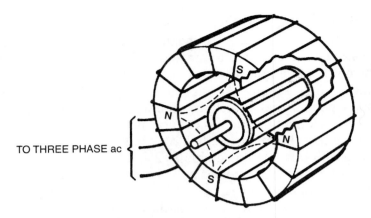

**Figure 13-3** **Schematic diagram of polyphase squirrel-cage induction motor.** *(Courtesy of TECO-Westinghouse Electric Corporation.)*

Insulated copper wires are coiled and laid in these slots in such a way that the ends of the wires can be connected to a three-phase circuit to bring about an electrical and magnetic arrangement similar to the schematic in Figure 13-1.

The small discs that were punched out of the center of the stator laminations are now available to become the rotor. These discs are notched around the periphery, insulated, and then pressed on the shaft in a package with the notches lined up to form slots. Copper or aluminum is shoved or cast in these slots, and the ends of the bars at each end of the rotor package are connected by fastening them to a ring of the same metal. The bars and rings give the appearance of a squirrel cage, hence the name of this motor.

The basic mechanical part is the frame into which the stator is tightly pressed. The frame has feet or other devices for firmly fastening the motor in place. Covers at each end of the frame carry the bearings that support the shaft and hold the rotor in proper relationship with the stator so as to allow smooth rotation. These covers are usually called *brackets* or *end bells*. Protection from mechanical damage, dirt, moisture, etc. is provided by various configurations of the frame and end bells while at the same time allowing necessary air circulation for cooling.

The electrical operation of the actual motor just described has some similarities to and some differences from the schematic design in Figure 13-1. The electrical operation of the stator is just like that in Figure 13-1. Current flowing through the coils of wire embedded in the slots creates definite north and south poles. These poles are strong or weak depending on the rise and fall of the current in Figure 13-2. As each phase of the ac cycle reaches its peak, the effective magnetism is rotated around the stator resulting in a rotating magnetic field.

The electrical operation of the rotor is somewhat different. The description of Figure 13-1 refers to a simple bar magnet fastened to a shaft. Actually the soft iron laminations of the real rotor have no permanent magnetism. It is the rotating magnetic field in the stator that induces a voltage in the squirrel cage of the rotor. This voltage causes a current to flow in the cage, which sets up effective north and south poles or magnetic fields in the rotor. It is the interaction of the stator and rotor fields that produces torque in the motor.

Continuous voltage will be induced in the rotor, provided the rotor's conductors continue to cut across the magnetic flux lines of the stator field. To bring this about, it is necessary for the rotor to travel at a speed somewhat less than the synchronous speed of the stator's rotating field. The difference between synchronous speed and actual rotor or motor speed is referred to as *slip* and is normally expressed as a percent of synchronous speed:

$$\% \text{ slip} = \frac{\text{Syn. rpm} - \text{Motor rpm}}{\text{Syn. rpm}} \times 100$$

For example, a four-pole motor might have a full-load speed of 1760 rpm, which indicates a slip of 40 rpm below synchronous speed of 1800 rpm. Using the formula, the calculation results in a slip of 2.2%.

When the number of poles and the percent slip are known, it is possible to calculate the full-load speed by reversing the preceding formula:

$$\text{Motor rpm} = \text{Syn. rpm}\left(1 - \frac{\% \text{ slip}}{100}\right)$$

Using the same example for a four-pole, 1800-rpm motor having a 2.2% slip:

$$\begin{aligned}
\text{Motor rpm} &= 1800\left(1 - \frac{2.2}{100}\right) \\
&= 1800\,(1 - 0.022) \\
&= 1800\,(0.978) \\
&= 1760
\end{aligned}$$

The magnitude of the torque produced by the interacting magnetic fields depends on several factors, but for any given motor the torque is mainly a function of the slip. Remember that any induction motor must have slip to produce torque.

## 13.3  Single-Phase Motors*

In polyphase motors, the correct mechanical placement of the multiple windings in the stator, together with the phase relationships between the supply voltages, produces a uniformly rotating stator magnetic field. With a single source of ac voltage connected to a single winding, a stationary flux field is created that pulsates in strength as the ac voltage varies but does not rotate. Consequently, if a stationary rotor is placed in this stationary stator field, it will not rotate. If the rotor is spun by hand, artificially creating relative motion between the rotor winding and the stator field, it will pick up speed and run. A single-phase, single-winding motor will run in either direction if started by hand, but it will not develop any starting torque (Figure 13-4).

What is needed is a second winding, with currents out of phase with the original or main winding, to produce a net rotating magnetic field, as was the case with the three-phase induction motor. These start windings, together with other components such as capacitors, relays, and centrifugal switches, make up the starting circuit. They have varying effects on motor starting and running performance.

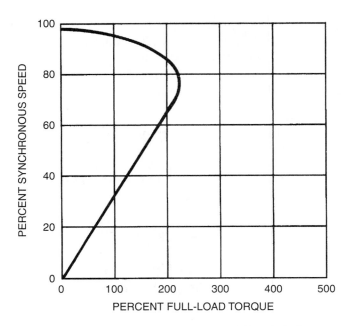

**Figure 13-4**   **Speed-torque characteristics of a single-phase, single-winding motor.** *(Courtesy of GE Motors [General Electric Company].)*

*Information courtesy of General Electric Company.

## 13.4 Resistance Split-Phase Motors

The winding arrangement of a typical distributed winding resistance, split-phase motor is shown schematically in Figure 13-5. If the main winding only is energized and the main winding current is recorded, the relationship between the supply voltage ($V_L$) and the main winding current ($I_M$) would be shown as in Figure 13-6. The main winding current lags behind the line voltage because the coils embedded in the steel stator naturally build up a strong magnetic field that slows the build-up of current in the winding.

The start winding is not wound exactly like the main winding, but its coils contain fewer turns of a much smaller diameter wire than that of the main winding coils. This difference in windings is required to reduce the amount the start winding current lags behind the voltage if it is connected to the line.

When both windings are connected in parallel across the line, the main and start winding currents will then be out of time phase by about 30 degrees. Being

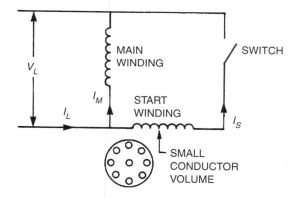

**Figure 13-5** Winding schematic—resistance, split-phase motor. $V_L$, load voltage; $I_L$, load current; $I_M$, main winding current; $I_S$, start winding current. *(Courtesy of GE Motors [General Electric Company].)*

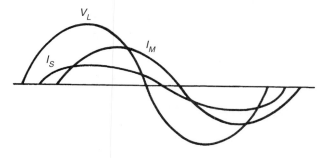

**Figure 13-6** Phase relationships of Figure 13-5. *(Courtesy of GE Motors [General Electric Company].)*

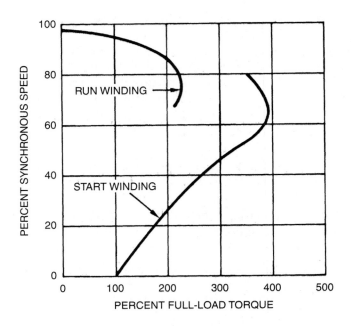

**Figure 13-7**   **General performance characteristics of the motor in Figure 13-5.**   *(Courtesy of GE Motors [General Electric Company].)*

out of phase produces a weak rotating flux field that is sufficient to provide a moderate amount of torque at standstill and start the motor.

The total current that the motor draws while starting is the vector sum of the main and start winding currents. Because of the small angle between these two vectors, the line current drawn during starting (inrush current) of a split-phase motor is quite high. Also, the small-diameter wire in the start winding carries a high current density, so it heats up very rapidly. A centrifugal switch and mechanism or relay must be provided to remove the start winding from the circuit once the motor has reached an adequate speed to allow running on the main winding only. The speed–torque relationship of a typical split-phase motor on both running and start connection is shown in Figure 13-7. The efficiency of the split-phase motor is moderate (50–65%). The starting torque is moderate with a high inrush current during starting.

## 13.5 Capacitor Start Motors

The capacitor start motor utilizes the same winding arrangement as the split-phase motor but adds a short time-rated capacitor in series with the start winding (Figure 13-8). The effect of the addition of this capacitor is shown in Figure 13-9. The

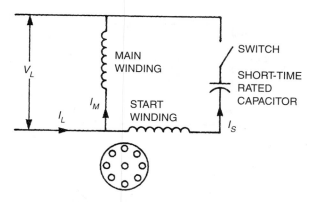

**Figure 13-8**    **Winding schematic for capacitor start motor.** *(Courtesy of GE Motors [General Electric Company].)*

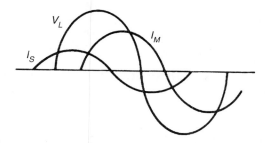

**Figure 13-9**    **Phase relationships of Figure 13-8.** *(Courtesy of GE Motors [General Electric Company].)*

main winding current ($I_M$) remains the same as in the split-phase case, but the start winding current is much different. Because of the addition of the capacitor, it now leads the line voltage rather than lagging as does the main winding. The start winding itself is also different. It contains slightly more turns in its coils than the main windings. It also utilizes wire diameters only slightly smaller than those of the main winding.

The net result is a time-phase shift that is much closer to 90 electrical degrees than with the split-phase motor. A stronger rotating field is therefore created, and starting torque is higher than with the split-phase design.

The start and running speed–torque characteristics of a typical capacitor start motor are shown in Figure 13-10. Again, a centrifugal switch and mechanism or relay must be used to protect both the start winding and capacitor from damage due to overheating. When running, the capacitor start motor performs identically to the split-phase motor. The efficiency of the capacitor start motor is moderate (50–65%). The starting torque is moderate to high. The capacitor controls inrush current.

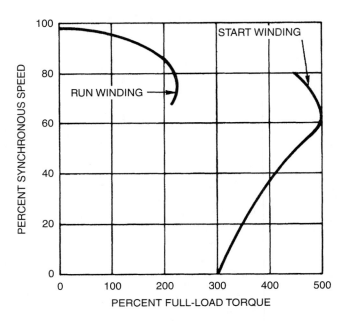

**Figure 13-10**   **General performance characteristics of the motor in Figure 13-8.** *(Courtesy of GE Motors [General Electric Company].)*

# 13.6  Permanent Split-Capacitor Motors

The windings of the permanent split-capacitor motor are arranged like those of the split-phase and capacitor start designs (Figure 13-11). The capacitor is capable of running continuously and replaces the intermittent-duty capacitor of the capacitor start motor. The centrifugal switch is also removed. Once again, the main winding current remains similar to the previous designs, lagging the line voltage (Figure 13-12). The start winding current and the start winding itself, however, are somewhat different from the capacitor start design.

Continuously rated capacitors are normally provided in small microfarad, high-voltage ratings. Therefore, the start winding is altered to boost the capacitor voltage to the correct level by adding significantly more turns to its coils than are in the main winding coils. Start winding wire size remains somewhat smaller than that of the main winding. The smaller microfarad rating of the capacitor produces more of a leading phase shift and less total start winding current. Starting torques will be considerably lower than with the capacitor start design.

The real strength of the permanent split-capacitor design lies in the start winding and the capacitor remaining in the circuit at all times. Therefore, the split-capacitor design results in better efficiency, better power factor, and lower

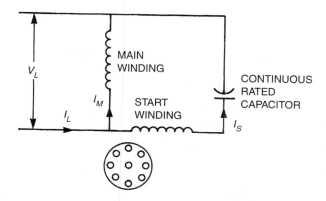

**Figure 13-11** **Winding schematic—permanent split-capacitor motor.** *(Courtesy of GE Motors [General Electric Company].)*

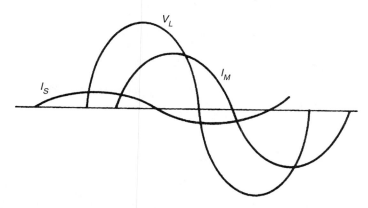

**Figure 13-12** **Phase relationships of Figure 13-11.** *(Courtesy of GE Motors [General Electric Company].)*

120-Hz torque pulsations than in the equivalent capacitor start and split-phase designs.

Figure 13-13 illustrates typical speed-torque curves for permanent split-capacitor motors. Note that varying the rotor resistance changes the shape of the speed–torque curve. Now various operating rpms can be utilized. With the addition of extra main windings in series with the original main windings, motors can be designed to operate at different speeds depending on the number of extra main windings energized. Note also that for a given full-load torque, less breakdown torque and, thus, a smaller motor is required with a permanent split-capacitor design than with any of the previously discussed designs. The efficiency is moderate to high (50–70%), and the starting torque is low to moderate (varies with rotor resistance).

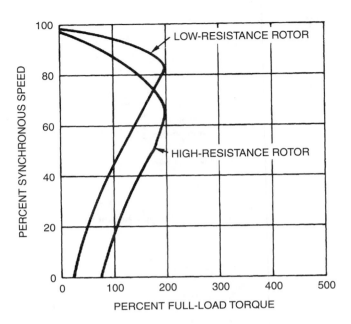

**Figure 13-13**   General performance characteristics of the motor in Figure 13-11.   *(Courtesy of GE Motors [General Electric Company].)*

## 13.7 Shaded-Pole Motors

The shaded-pole motor differs widely from the other single-phase designs. All the designs discussed to this point are distributed-wound motors having a main and start winding. They differ only in details of the starting method and corresponding starting circuitry. The shaded-pole motor is entirely different, both in construction and operation.

The shaded-pole motor is the most simply constructed and hence the least expensive of the single-phase designs. It consists of a run winding only plus shading coils that take the place of the conventional start winding. The construction of a typical four-pole shaded-pole motor is shown in Figure 13-14. The stator is of salient pole construction, having one large coil per pole wound directly in a single large slot. The shading coils are short-circuited copper straps wrapped around one pole tip of each pole.

The shaded-pole motor produces a very crude approximation of a rotating stator field through magnetic coupling between the shading coils and the stator winding. The placement and resistance of the shading coil is chosen so that as the stator magnetic field increases from zero at the beginning of the ac cycle to some positive value, current is induced in the shading coil. As previously noted, this

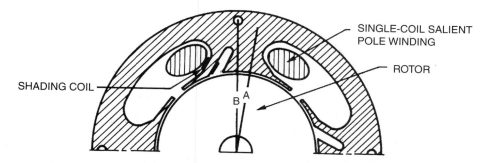

**Figure 13-14**  **Four-pole shaded-pole motor.** *(Courtesy of GE Motors [General Electric Company].)*

current will create its own magnetic field that opposes the original field. The net effect is that the shaded portion of the pole is weakened and the magnetic center of the entire pole is located at point A. As the flux magnitude becomes nearly constant across the entire pole tip at the top of the positive half-cycle, the effect of the shading coil is negligible and the magnetic center of the pole shifts to point B. As slight as this shift is, it is sufficient to generate torque and start the motor.

In Figures 13-15 and 13-16 the effect of the salient pole winding configuration on motor speed–torque outut is shown. A sinusoidal distribution of winding turns produces the best efficiency and lowest noise and vibration levels. The single coil winding is the crudest possible approximation of such a distribution. Thus, shaded-pole efficiency suffers greatly in the presence of winding harmonic content, particularly the third harmonic. This drop in efficiency produces a dip in the speed–torque curve at approximately 1/3 synchronous speed (Figure 13-17).

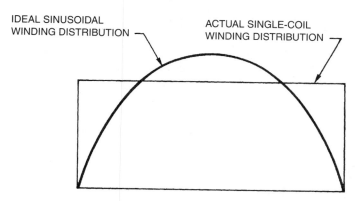

**Figure 13-15**  **Harmonic content resulting from single-coil winding distribution—ideal versus actual.** *(Courtesy of GE Motors [General Electric Company].)*

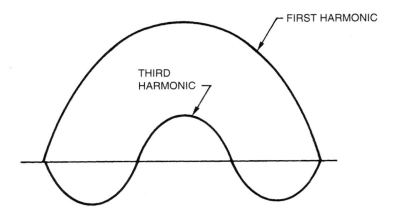

**Figure 13-16** **Harmonic content resulting from single-coil winding distribution showing first and third harmonic.** *(Courtesy of GE Motors [General Electric Company].)*

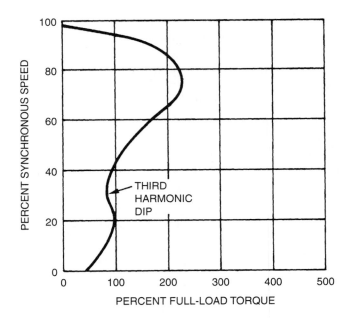

**Figure 13-17** **General performance characteristics of motor shown in Figure 13-14.** *(Courtesy of GE Motors [General Electric Company].)*

In addition, there are losses present in the shading coils. These factors combine to make the shaded-pole motor the least efficient and noisest of the single-phase designs. The efficiency is low (20–40%) and the starting torque low (plus there is the third harmonic dip).

# 13.8 dc Motors*

**13.8.1 dc Motor Classifications**    Dc motors are generally classified by motor output characteristic: constant torque, constant horsepower, or a combination of constant horsepower and constant torque.

Another general and common classification of motors is by the type excitation winding utilized. Standard type CD industrial dc motors are shunt wound. Stabilized shunt wound, series, and compound wound motors are available as a modification on selected frame sizes.

1. *Shunt wound* motors have the excitation winding connected in parallel with the armature or are excited from a separate power supply. With constant armature voltage and field excitation, a shunt wound motor will have a characteristic where speed tends to increase with increasing load. This speed increase with load will typically be small at full field and will increase at weaker field settings within the field-control range.

2. *Stabilized shunt exciting fields* have a small number of series turns. The purpose of the series winding is to provide additional flux as the load increases, which tends to offset the reduction in flux resulting from the effect of armature reaction. Essentially, constant main pole flux is maintained, providing less speed change with load. Most modern controls compensate for armature reaction effects and do not require or use the series field. Stabilized shunt motors do not provide any defined regulation.

3. *Compound wound* motors, as previously indicated, are available as a modification on selected frames only. When an application requires characteristics intermediate between a series and shunt motor, the standard type CD compound wound motor should be specified. Standard type CD compound wound motors will vary between a minimum and maximum speed regulation of 10–30% at full field. Very little speed range by field control is permissable on a compound wound motor as the speed versus load characteristic approaches that of a series wound motor when the shunt field is weakened, resulting in excessive speeds at light loads.

4. *Series wound* motors are available as a modification for those applications requiring very high starting torque and high speed regulation. Because excitation is entirely dependent on load, the application of series motors should be avoided where the load may drop below 25% rated torque or where the load is not solidly coupled, otherwise the motor may overspeed.

**13.8.2 dc Motor Speed Control Methods**    See Figure 13-18 for a cutaway of a typical dc motor. Commonly used methods of controlling the speed of a dc

---

*Courtesy of General Electric Company.

**Figure 13-18** **Cutaway view of a typical dc motor.** *(Courtesy of General Electric Company.)*

motor are armature voltage control, shunt field control, and a combination of armature voltage control and shunt field control.

**13.8.2.1 Armature Voltage Control.** As the name implies, for this type of speed control, the motor armature voltage is varied while maintaining the constant shunt field excitation. Output torque of a dc motor is proportional to the product of the main pole flux, armature current, and a machine constant that is a function of the armature winding. Thus,

$$T = K\varphi\, I_a$$

where    $T$   =   torque
         $K$   =   machine constant
         $\phi$   =   main pole flux
         $I_a$   =   armature current

Therefore, with armature voltage speed control and constant shunt field excitation, the torque is dependent on armature current only. A dc motor operated with armature voltage control and rated field excitation will develop rated torque at rated armature current independent of the speed. This configuration is commonly called constant torque operation.

Horsepower, at constant current, depends on applied armature voltage and, as a result, will increase as the armature voltage is increased.

Type CD motors will operate down to 1/50 base speed without excessive torque pulsation.

### 13.8.2.2 Shunt Field Control.

With speed control by field weakening, the voltage applied to the shunt field is adjusted by a variable resistance or rheostat in series with the shunt field circuit or by varying the voltage of the shunt field power supply.

Reducing the shunt field voltage decreases the field current, which, in turn, reduces the main pole flux and increases the speed of the motor. Field forcing or increasing the field voltage above rated voltage cannot be used to obtain speeds much below base speed, because the field will overheat at higher than rated current and saturation tends to limit the flux increase.

Speed range by field control for motors varies from 1:1 to 3:1 for standard motors and up to 6:1 for special motors.

As was indicated previously, torque is proportional to the product of flux, armature current, and the machine constant. Therefore, at constant armature current, the torque decreases as the flux is reduced to obtain higher speeds. Since horsepower is proportional to torque and speed,

$$HP = \frac{T \times rpm}{5252}$$

where   HP   =   horsepower
      rpm   =   revolutions per minute
      $T$    =   torque in lb-ft

The motor horsepower output at constant or rated armature current is essentially constant as the field voltage or flux is varied.

### 13.8.2.3 Combination of Armature Voltage Control and Shunt Field Control.

Utilizing both methods of speed control will give very wide speed ranges. Armature voltage control is typically used for speeds below base speed, resulting in a constant torque capacity, and shunt field control is used to obtain speeds above base speed, resulting in a constant horsepower capacity.

On shunt field control systems, if field weakening occurs before the armature voltage reaches rated voltage, motor horsepower will be less than rated even though the motor is operating at the rated base speed.

### 13.8.2.3.1 *Speed Regulation*

With constant line voltage and constant shunt field excitation, the change of speed with load depends on armature circuit resistance and the change in main pole flux with load. In classic dc motor theory, the speed change is predicted from only the voltage (IR) drop in the armature circuit, resulting in a straight-line regulation curve. However, with practical machines, the flux is not independent of armature current. In general, the main pole flux is reduced as armature current increases.

Speed regulation is defined as:

$$\frac{\text{No load rpm} - \text{Full load rpm}}{\text{Full load rpm}}$$

A decrease in main pole flux results in a reduced regulation or a tendency for motor speed to rise with increasing load. A stabilizing winding is used to maintain approximately constant flux by increasing the main pole excitation directly with load.

The reduction in flux is the result of two effects of armature current. The first is the distortion of main pole flux by armature reaction. The second is a compounding effect that depends on the commutating pole strength. In general, for motors, reducing pole strength results in a reduction of main pole flux with increasing armature current. The reduction may be considered independent of speed.

The percent IR drop will normally decrease as horsepower and base speed increase.

The slope of the saturation curve at the operating point also has a great effect on the regulation. A given number of main field ampere turns added or subtracted will have a much greater effect on speed at weak field than at full field conditions. A stabilized shunt wound motor with 15% regulation at top speed may have a 5% (or less) regulation at base speed.

For these reasons the regulation curve usually is not a straight line. It will not be identical for two motors due to slight physical variations and adjustments during manufacture. Typical speed regulation curves are shown in Figure 13-19.

### 13.8.2.4 Variation in Speed Due to Heating.

For Class F rated motors, the variation of rated speed from full load cold to full load hot shall not exceed 20% on ventilated motors and 25% on totally enclosed motors.

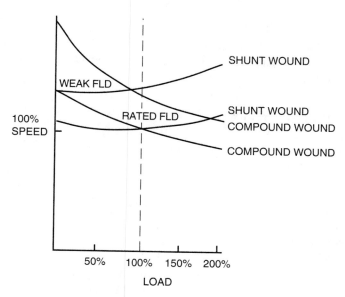

**Figure 13-19**   **Typical speed regulation.** *(Courtesy of General Electric Company.)*

Also at rated load hot and at rated field and armature voltage, motor speed will be 7.5% from rated base speed.

### 13.8.3 dc Motor Load Characteristics

**13.8.3.1 Constant Torque Drives.** Many industrial applications, such as conveyors, mixers, cement kilns, squeeze rolls, continuous processing machinery, and so on, require nearly constant torque over their operating range (see Figure 13-19). Dc motors operated with fixed field excitation and adjustable armature voltage have an approximately constant torque capacity over their speed range.

The load torque of a driven machine can be measured by wrapping a rope or cable around the input shaft or by using a torque arm and applying a steady pull through a spring scale. The horsepower can then be calculated, using the following relation:

$$HP = \frac{rpm \times Torque}{5252}$$

**13.8.3.2 Stalled Current Capability.** The continuous load section defines the torque capability at all speeds considering the thermal limits. While the curves may seem to indicate an intersection at zero speed, the data only applies at speeds other than zero, but may be very low, such as 10 rpm. At zero speed, localized heating of the segments under the brushes may produce segment distortion and result in brush vibration.

**13.8.3.3 Constant Horsepower Drives.** A common example of a constant horsepower drive is a center-driven winding wheel. The material is wound on the mandrel at constant speed and constant tension, using the following relation:

$$\text{HP} = \frac{\text{Linear speed (fpm)} \times \text{Tension (lb)}}{33,000}$$

The horsepower is constant. At the start of the winding process, the torque requirement is low because of the small radius and the high rotational (motor) speed. As the roll builds up, the radius increases, with a resulting increase in torque. The rotational speed must decrease in order to maintain constant linear speed.

The main drives of metal working machines require approximately constant horsepower, because an optimum cutting speed is maintained for particular types of material regardless of the diameter of the surface being machined. When machining small diameter stock, the torque requirement is low and the rotational speed is high. Stock of larger diameters requires higher torque and decreased rotational speed in order to maintain a constant cutting speed. Dc motors operated by field control and a constant armature voltage have a constant horsepower capacity over the speed range (see Figure 13-20).

In some applications, constant horsepower may be required over a wider speed range than is obtainable by armature voltage control. Horsepower and speed are approximately proportional to the applied armature voltage, and the horsepower requirements of the load must be available at the lowest operating speed obtained by armature voltage control.

**13.8.3.4 Combination of Constant Horsepower and Constant Torque.** Applications such as the center-driven winding reel often require a combination of constant torque and constant horsepower (see Figure 13-20). The horsepower required for a given linear speed of the material is constant during the build-up of the roll. However, it is often desirable to change the surface speed when reeling different material. Since horsepower is proportional to the surface speed, armature voltage control (constant torque) must be utilized to adjust the speed.

A dc motor with a suitable speed range by armature voltage control will provide for the surface speed adjustments, while field control will provide the constant horsepower requirements during the build-up of the roll.

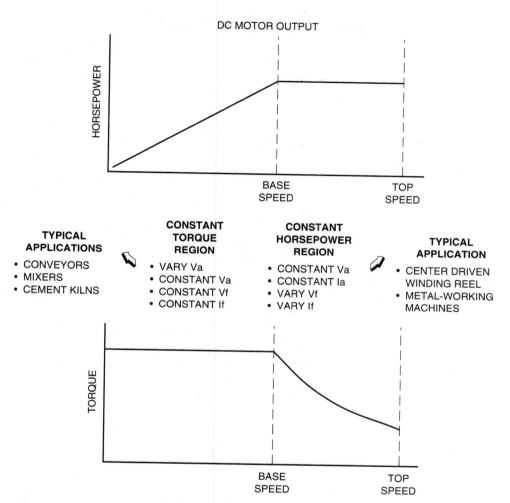

**Figure 13-20** Torque and hp characteristics of constant horsepower (top), constant torque load (bottom). *(Courtesy of General Electric Company.)*

# 13.9  Brushless dc Motors*

Three-phase ac plant power is converted to dc by the input side of a brushless dc motor control to charge up a bank of storage capacitors (called the *buss*), whose function is to store energy and supply dc power to the power transistors in the output bridge as power is required by the motor. The size of the buss (the number of capacitors) varies with the size of the motor.

*Courtesy of POWERTEC Industrial Motors.

This rectification is accomplished by six diodes, which may be in a single package or in several modules. The diodes are protected by the input fuses, which are chosen for their speed and interrupting capacity. An input choke in the dc leg of the diode bridge protects against line transients and limits the rate at which input current may increase or decrease. See Figure 13-21.

The input section is self-regulating. The highest voltage level possible on the capacitors is 1.4 times the line-to-line voltage (peak voltage). Initially, before the motor is turned on, the capacitors will charge up to this peak voltage. When the motor is started, it uses power from the buss to perform the work required. This usage partially discharges the capacitors and lowers the buss voltage. With three-phase input power, there are six periods in each cycle of ac when the line-to-line voltage is greater than the capacitor voltage. The capacitors will only draw current from the power lines when the capacitor voltage is lower than the instantaneous line-to-line voltage, and then it will only draw enough power to replenish the energy used by the motor since the last time the line-to-line voltage was greater than the capacitor voltage.

Torque in a motor is a function of current. Power is a function of speed and torque. Even though the current required by the motor to develop the torque may be large, the actual power used is small at low speeds. Because the energy drawn from the capacitors is the actual power used by the motor, the energy drawn from the input power lines is the actual power supplied to the motor. The brushless dc motor control is capable of running at very low speeds at very high torques while drawing very little current from the ac line. The result is that the rms current at the input of the brushless dc motor control is directly proportional to the output power of the motor rather than proportional to the motor load.

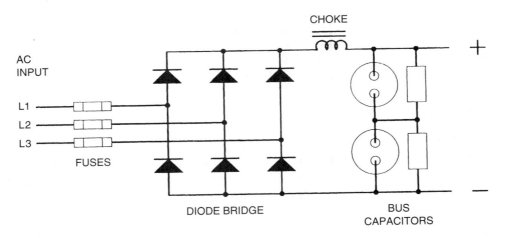

*Figure 13-21*   Simplified drawing of input power section of a brushless dc motor control.
*(Courtesy of POWERTEC Industrial Motors.)*

A brushless dc motor is wound like an ac induction motor, but it uses permanent magnets on the rotor instead of shorted rotor bars. There are three power carrying wires going to the motor. Each of these wires has to be, at synchronized times, connected to either side of the dc buss. This is accomplished with a six transister bridge configuration. See Figure 13-22.

Applying power to the motor requires turning on one transistor connected to the positive side of the buss and one transistor connected to the negative side of the buss, but never the two transistors in the same leg of the output. When two transistors are turned on, the entire buss voltage is applied to the windings of the motor through the two wires connected to those transistors and current will flow (if the CEMF of the motor is not greater than the buss) until the transistors are turned off. Due to the inductive nature of the motor windings, the current will not cease immediately when the transistors are turned off. It will decay quickly, but voltages in the bridge would rise dangerously if the snubber network were not present to prevent this from happening.

If two transistors were turned on and left on for any length of time, the current would build to very high levels too quickly, so the transistors are turned on for only brief intervals at a time. If the motor is lightly loaded, it will not take a lot of current (torque) to get it going. If the motor is heavily loaded, each turn-on interval will cause the current to build up until there is sufficient torque to turn the load. Once the motor has stated turning, the current supplied will be, in either

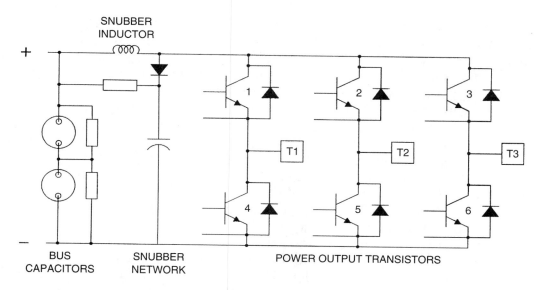

**Figure 13-22** The power output bridge consists of six power transistors and associated snubber components. *(Courtesy of POWERTEC Industrial Motors.)*

case, just enough to keep the motor and load turning. If a heavy load is taken off the motor, the current will quickly drop to the new level and an applied heavy load will be quickly picked up. Quick response toload changes accounts for the high efficiency of brushless dc.

Figure 13-23 is a schematic representation of the winding of the brushless dc motor. The connection shown in the drawing is a single-wye. There are three other ways the motor windings can be connected that will change the speed and horsepower (the torque remains constant in any given motor as the connections are changed), but these other connections are not made for different voltages as is the case with the ac induction motor. The standard connection conventions of the ac motor are followed in the assignment of markings on the motor leads, however.

Shown in Figure 13-24 are the major parts of the brushless dc motor in the off state. The stator windings are connected as in Figure 13-23 and the motor is operated from the power bridge shown in Figure 13-22. This figure is very simplified, showing only a two-pole motor for simplicity. Most brushless dc motors are four-pole or eight-pole motors.

Current is developed in the windings, producing torque by the interaction of the magnetic fields produced by the stator windings (with the power supplied from the control) and the fields of the permanent magnets mounted on the rotor.

The brushless dc motor control has an "electronic commutator" fed by an integral encoder mounted on the motor. This encoder tells the motor control which transistors should be turned on to obtain maximum torque from the motor at whatever position the motor shaft happens to be in at that point in time. Encoder data establish a communication between the motor and its control that is not present in ac motors and inverters, and that is not a part of the dc brush-type motor and its SCR control. The brushless dc motor control always knows where the motor shaft is in its rotation because the motor encoder is constantly monitoring it.

The control's power output bridge (Figure 13-21 consists of three "legs." Each leg has a power transistor from the positive side of the dc power buss to the output terminal (generally referred to as an *upper* transistor), and another transistor from the output terminal to the negative side of the power buss (herein referred to as a *lower* transistor). Each time a transistor turns on, it connects an output terminal to one of the sides of the dc power buss. Each output terminal also has a "free-wheeling" diode connected to each side of the buss to carry currents that the transistors cannot conduct.

Again, at whatever position the rotor happens to be, the encoder tells the drive which transistor should be turned on to deliver maximum torque from the motor. While this is actually done in an EPROM (an electrically programmable read-only memory integrated circuit), the simplified representation as a switch shown in Figure 13-24 will suffice for our purpose. Note that the longer

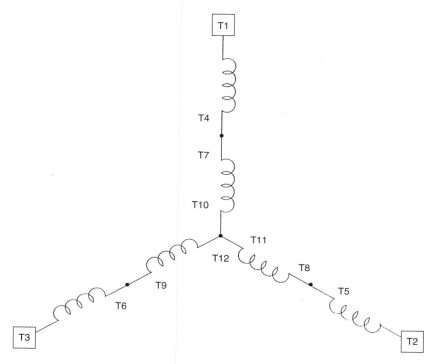

**Figure 13-23** **Single-wye connections for a brushless dc motor. Note: T indicates stator windings terminations.** *(Courtesy of POWERTEC Industrial Motors.)*

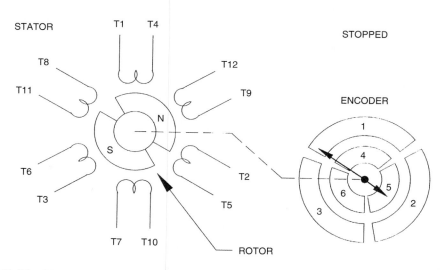

**Figure 13-24** **Simplified drawing of the major parts of the brushless dc motor. Note: T indicates stator windings terminations.** *(Courtesy of POWERTEC Industrial Motors.)*

arrow on the switch in the diagram governs the switching of the upper transistors, which are numbered 1, 2, and 3. The shorter arrow governs the switching of the lower transistors, 4, 5, and 6. The arrangement is such that each of the upper transistors may be operated with either of the lower transistors in the other two output legs, but an upper transistor may never be operated with the lower transistor in the same leg. To do so would produce a short circuit across the dc power buss and blow out the fuses, if not the transistors involved. The driver circuits are also logically interlocked to prevent the accidental turning on of opposing transistors.

The rotor in Figure 13-24 has two sets of magnets on it, a north pole and a south pole, which will interact with the electromagnetic fields and poles produced by the current in the stator windings (shown schematically in Figure 13-23 and in their relative positions around the rotor in Figure 13-24). Remember that the rotor is free to turn, but the stator windings are stationary.

When the control is turned on in the position shown in Figure 13-25, transistors 1 and 5 will turn on, allowing current to flow from the positive side of the buss through the number 1 transistor, out of the control terminal T1 into T1 winding of the motor, through T4 to T7, and through T10 to the center of the motor connection. Since the number 5 transistor is on, the current will flow through stator windings T11 to T8 to T5 and T2 in the motor to T2 on the motor control through the number 5 transistor to the negative side of the buss. The currents will

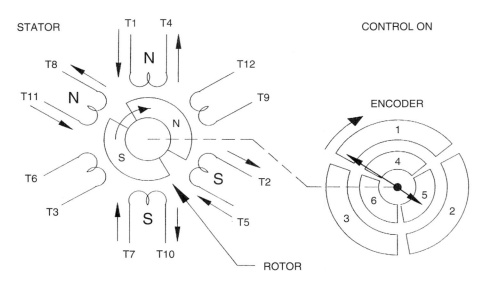

**Figure 13-25**   Current flows in the stator of the motor after the control is turned on in the position shown. Note: T indicates stator windings terminations. *(Courtesy of POWERTEC Industrial Motors.)*

be as indicated in Figure 13-25, setting up magnetic poles in the stator as shown (north poles at T1-T4 and T11-T8, and south poles at T7-T10 and T2-T5). The north poles in the stator attract the south pole of the rotor and the south poles in the stator attract the north pole of the rotor and repel the south pole. Rotation will occur in the indicated direction as torque is developed in the motor.

Since the load is heavily inductive, this current would build in a nearly linear way if the transistors were on continuously, but the switching of the lower transistors is pulse-width modulated, i.e., switched on and off at a relatively high frequency. The width of each pulse is determined by the torque required to turn the motor, which is under speed and, effectively, position control. The amount of torque required is determined by the load on the motor shaft. The lighter the load, the narrower the pulse width. As the load gets heavier, the pulses will get wider until they reach their maximum width.

With each pulse the current will build up a little until the end of the period of time that the two transistors are on, which will occur when the motor has turned far enough to turn off the 5 transistor. In this example, that condition will occur when the rotor has turned 60 degrees. In the actual motor, it will occur in 60 electrical degrees, not 60 physical degrees as in the illustration. There may be as many as 1440 electrical degrees per revolution in a motor. Then the next step will occur as the 5 transistor turns off, and the 6 transistor turns on. See Figure 13-26.

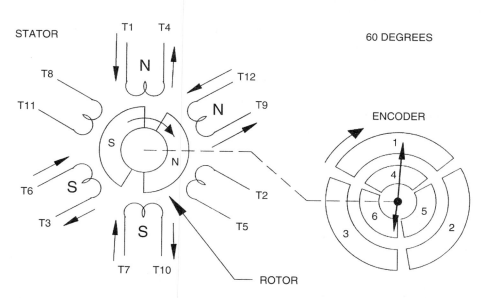

**Figure 13-26** When the motor rotates 60 degrees, the encoder switches the output transistors to rotate the field. Note: T indicates stator windings terminations. *(Courtesy of POWERTEC Industrial Motors.)*

Between the pulses and between stages of operation, the inductive current will want to continue to flow (since current cannot be created or destroyed instantaneously in an inductive circuit) if the energy is not used by the turning of the load, and the continued flow of current is allowed through the free-wheeling diodes, which are in inverse parallel connection with each output transistor. In this case the current, which is entering the motor at T1 and exiting at T2, will be forced to flow through the diodes around transistors 2 and 4. This current will decay rapidly, and it has the effect of slowing the motor down by loading it. It also has the effect of charging up the buss, but unless the motor has a large inertia on its shaft and is turning at a very high speed, it will have little effect.

After the motor has turned about 60 electrical degrees from the position shown in Figure 13-25, the 5 transistor will turn off and the current in the T11-T8-T5-T2 leg will die out; the 6 transistor will pick up the operation. Then the current flow will be through the T1-T4-T7-T10 leg to the center of the wye, and out through the T12-T9-T6-T3 leg. The stator magnetic poles will have shifted 60 degrees, causing the rotor to move in that direction.

Unlike the dc brush-type motor, which can only fire its SCR pairs every 2.6 milliseconds (.0026 second) at best, the brushless dc motor control can operate its transistors during the entire commutation cycle. A 2-kilohertz pwm (pulse width modulation) frequency allows the operation of a transistor every 500 microseconds (.0005 seconds). Frequencies from 2 kilohertz to 20 kilohertz or more are common. There is also no need to wait for the power line conditions to shut off the transistor; it may be cut off at any time.

As the field continues to move around the stator, the rotor follows it. Unlike the induction ac motor, in which all windings are continuously excited, the brushless dc stator is a dc excited field, and it moves because the windings are continuously switched in response to the movement of the rotor. The non-excited windings are carrying no current, since the windings in a brushless dc motor are either on or off. They do not go through the slow transitions that the ac motor windings go through. If the winding switching were to stop, the motor would come to a halt.

The switching of windings is controlled by a three channel position encoder mounted on the motor shaft. The Hall-effect transistors mounted on the feedback assembly are turned on and off by magnets in the encoder's *magnet wheel.* These magnets are aligned with the magnets on the rotor. The Hall-effect switches, which are noncontact (electrically isolated) devices, mounted on a four-pole motor in positions 60 degrees apart (they are 30 degrees apart on an eight-pole motor). It is not possible for the three switches to be all on or all off at the same time. There will always be two on and one off, or one on and two off at all times. See Figure 13-27.

The speed of the POWERTEC's four-pole, brushless dc motor is regulated by a two-channel, 30-pulse (60-pole), digital tachometer on the motor shaft,

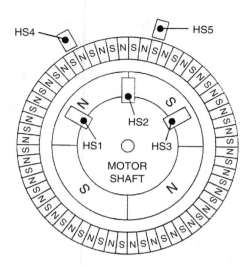

*Figure 13-27*   The encoder assembly—four magnets indicate position; outer magnets indicate speed and direction. Note: HS indicates Hall magnetic sensor.   *(Courtesy of POWERTEC Industrial Motors.)*

which indicates both speed and direction. Eight-pole motors (frames 280 and larger, except over 2500 rpm) have 120 magnetic poles around the outside of the magnet wheel. These poles alternately turn on and off two Hall-effect switches, which are connected to the control for speed control. There are two speed feedback channels offset by 90 electrical degrees, in quadrature.

These 30-pulse per revolution signals from the motor are electronically multiplied by 4 to supply a pulse every 3 degrees of shaft rotation (larger motors use a 60-pulse tachometer, yielding 240 pulses per revolution, a pulse every 1.5 degrees). Reference pulses are compared to pulses coming back from the motor's encoder. If the number of pulses from the motor (the absolute, not the frequency) does not equal the number of reference pulses applied, a position error count accumulates in an up/down counter on the speed control board. The number of accumulated pulses determines how much current is made available to the motor. If there are more reference pulses than feedback pulses, the accumulation is positive, and the current to the motor is positive, i.e., motoring current. This positive current will try to accelerate the motor to eliminate the pulse count. If there are more feedback pulses than there are reference pulses, the count is negative and the motor current will be shut off (the motor will coast along) until the accumulated count becomes positive again.

A motor running at no load will accumulate only a few pulses of position errors, but a fully loaded motor will accumulate 2/3 of the pulses necessary for current limit. Loads between no load and full load will accumulate some number between 0 and the full-load value. The maximum number of error pulses that may accumulate is about 100 (this is an adjustable "gain" function), at which time the brushless drive will be in current limit. When the maximum number of pulses is accumulated, the control is in current limit and further reference pulses will be ignored until the motor turns far enough to eliminate the excess pulses in the counter.

As long as the current limiting condition is avoided, the motor can track the reference frequency pulse for pulse. It is important to note that, short of current limit, this pulse accumulation does not affect speed. It only affects the shaft position relative to the no-load shaft position. Most motors classify speed regulation in terms of revolutions per minute difference between set speed and actual speed. A brushless dc motor set to run at 1750 rpm (with a frequency reference) will be running at 1750 rpm, period.

For motors operating at a steady speed and at a steady load, the pulse accumulation will not change, staying at the value necessary to drive that load at that speed. If the load should change, the pulse accumulation in the up/down counter will change to a value necessary to drive the new load. If the speed changes, the pulse accumulation will not change if the torque required by the load is the same at the new speed as the torque required at the old speed.

This introduction to the operation of brushless dc motors is brief and very simplified. It does not and cannot cover all the points necessary for a full or complete understanding. There are books available on the subject of brushless dc motors.

See Figure 13-28 for a cutaway view of a brushless dc motor.

*Figure 13-28*  **Cutaway view of a brushless dc motor.** *(Courtesy of POWERTEC Industrial Motors.)*

# Recommended Web Links

Students are encouraged to view the following Web sites as a supplement to the concepts presented in this textbook. Review and analyze the array of products that are available for electrical control applications. Many of these sites offer technical information that can help in converting the principles to practical applications. To view catalogs, your PC may require Adobe Acrobat Reader software.

### Motors, General

> General Electric
> www.geindustrial.com/cwc/products?famid=23
> Review: ac and dc Motors

> TECO-Westinghouse
> www.tecowestinghouse.com
> Review: Stock Motors

> US Motors
> www.usmotors.com/default_products.htm
> Review: Motors

### Brushless DC Motors

> POWERTEC Industrial Motors
> powertecmotors.com
> Review: Brushless Motors

# Achievement Review

1. How was the name *squirrel cage,* as applied to a three-phase induction motor, derived?

2. In a two-pole, three-phase, 60-hertz squirrel cage induction motor, what is the synchronous speed?

3. A four-pole, three-phase induction motor has a full-load speed of 1740 rpm. Calculate the percent slip from the synchronous speed.

4. Why does a single-phase motor have no starting torque if only a single winding is used?

5. Why is a centrifugal switch required in many single-phase motors?

6.  How does the addition of a capacitor in a single-phase motor create a rotating field?

7.  What is the difference between a capacitor start motor and a permanent split-capacitor motor?

8.  Why is the efficiency of a shaded-role motor low?

9.  How are dc motors generally classified?

10. Why should a series motor be loaded at all times?

11. What is the relationship between the output torque and main pole flux, armature current, and a machine constant in a dc motor?

12. What is the relationship between hp, torque, and rpm in a dc motor?

13. How is speed regulation determined in a dc motor?

14. In the description of a brushless dc motor in this text, how is the rectification of three-phase ac power accomplished?

15. What is the difference in the rotor in a brushless dc motor and the ac induction motor?

16. What are the Hall-effect switches used on a brushless dc motor, and how are they used on a four-pole motor?

17. In the brushless dc motor, what is speed regulation in terms of rpm difference between set-speed and actual speed?

# Chapter 14 Motor Starters

## Objectives

After studying this chapter, you should be able to:

- Discuss the difference between a contactor and a motor starter.
- Explain why overload relays are used on motor starters.
- Describe how the normally closed contact in the overload relay is connected into the motor starter control circuit.
- Explain how the ambient compensated overload relay operates.
- Draw the basic power and control circuit for a magnetic full-voltage motor starter.
- Explain why both mechanical and electrical interlocks are used on the reversing motor starter.
- Describe the relationship between applied voltage on a motor and the resulting torque.
- Draw the basic power and control circuit for a resistor-type, reduced-voltage motor starter and explain how it operates.
- Explain why a reduced-voltage motor starter should be used in some cases rather than a full-voltage motor starter.

## 14.1 Contacts and Overload Relays

The motor starter is similar to the contactor in design and operation. Both have one important feature in common: contacts operate when the coil is energized. The important difference is the use of overload relays on the motor starter.

*Manual motor starters* are available in which the contacts are closed by manual operation. The end results are energizing and protecting a motor. However, most machines today require additional control, safety, and convenience of remote control on one motor or multiple motors. Thus, very limited use is made of the manual motor starter on industrial machines.

The *magnetic motor starter* has three main contacts in the form used most frequently. These three contacts are normally open. This arrangement is used in the starting of three-phase motors. Most industrial plants in the United States have three-phase power available as standard. Sometimes two-phase, four-wire power is used. In such cases, four normally open contacts are supplied on the starter.

Like the power contactor, magnetic motor starters are available in many sizes. They start at approximately the relay size and extend up in capacity. Figure 14-1 lists motor starters, showing size, rated horsepower, and rated voltage.

The important concern with motor starters is the closing of contacts that connect the motor to the source of electrical power. Unfortunately, it is not always possible to control the amount of work applied to the motor. Therefore, the motor may be overloaded, resulting in serious damage. For this reason, overload

| NEMA SIZE | MOTOR STARTERS HORSEPOWER | | VOLTS |
|---|---|---|---|
| | SINGLE PHASE | TWO AND THREE PHASE | |
| 00 | 1 | 1½<br>2<br>2 | 200–230<br>460<br>575 |
| 0 | 2 | 3<br>5<br>5 | 200–230<br>460<br>575 |
| 1 | 3 | 7½<br>10<br>10 | 200–230<br>460<br>575 |
| 2 | 7½ | 15<br>25<br>25 | 200–230<br>460<br>575 |
| 3 | 15 | 30<br>50<br>50 | 200–230<br>460<br>575 |
| 4 | | 50<br>100<br>100 | 200–230<br>460<br>575 |
| 5 | | 100<br>200<br>200 | 200–230<br>460<br>575 |
| 6 | | 200<br>400<br>400 | 200–230<br>460<br>575 |
| 7 | | 300<br>600<br>600 | 200–230<br>460<br>575 |
| 8 | | 450<br>900<br>900 | 200–230<br>460<br>575 |

*Figure 14-1*   **Chart of motor starter horsepower.**

relays are added to the motor starter. Figure 14-2 shows a typical thermal over-load relay.

The goal is to protect the motor from overheating. The current drawn by the motor is a reasonably accurate measure of the load on the motor and thus of its heating.

Thermal protected overload relays today use a thermally responsive element. That is, the same current that goes to the motor coils (causing the motor to heat) also passes through the thermal elements of the overload relays.

The thermal element is connected mechanically to a normally closed contact. When an excessive current flows through the thermal element for a long enough time period, the contact is tripped open. This contact is connected in series with the control coil of the starter. When the contact opens, the starter coil is deenergized. In turn, the starter power contacts disconnect the motor from the line.

A motor can operate on a slight overload for a long period of time or at a higher overload for a short period of time. Overheating of the motor will not result in either case. Therefore, the overload heater element should be designed to have heat-storage characteristics similar to those of the motor. However, they should be just more sensitive enough so that the relay will trip the normally closed relay contact before excessive heating occurs in the motor.

The *ambient* (surrounding air) temperature in the location of the motor and starter also has some effect. It is necessary to specify the rating of a given temperature base plus the allowable temperature rise due to the load current. For example, an open motor rating is generally based on 40°C (104°F). The motor nameplate will specify the allowable temperature rise from this base.

The motor and starter are usually located in the same general ambient temperature. Thus, the overload heater elements are affected by the same temperature

OVERLOAD RELAY HEATERS INSTALLED HERE

*Figure 14-2*  **Typical thermal overload relay.**  *(Courtesy of TECO-Westinghouse Motor Company.)*

conditions. The overloads will open the motor starter control circuit either through excessive motor current, a high ambient temperature, or a combination of both.

*Ambient compensated overload relays* are used when the control is located in a varying ambient temperature and the motor that it protects is in a constant ambient temperature.

The ambient compensated overload relay operates through a compensating bimetal relay. The relay maintains a constant travel-to-trip distance, independent of ambient conditions. Operation of this bimetal relay is responsive only to heat generated by the motor overcurrent passing through the heater element.

All starter manufacturers list the size of overload heaters for a given starter application. The lists show the range of motor currents with which they should be used. These may be in increments of from 3% to 15% of full-load current. The smaller the increment, the closer can be the selection to match the motor to its actual work. Since the heater varies with the enclosure, this information is also included.

All overload relays should be *trip-free*; that is, it should be impossible for anyone to hold down or block a RESET button, resulting in damage to the motor. After a relay has tripped, the cause of the overload should be investigated. The problem should be solved before the RESET button is depressed to put the starter back into operation.

There is one point about overload relays that is often misunderstood. Overload relays are not intended to protect against short-circuit currents. Short-circuit protection is the function of fuses and circuit breakers.

Another feature of the motor starter is the auxiliary contact normally used in the starter control circuit. In some small starters, an additional load contact can be used. It is not generally used in large starters.

Auxiliary contacts may be normally open or normally closed and have a 10-ampere rating. In most cases, one NO interlock contact is supplied as standard. Additional contacts (up to four) can be obtained. These can be purchased with the original starter or can be added later. A kit is available from the manufacturer containing all hardware necessary to make the installation.

## 14.2  Across-the-Line (Full-Voltage) Starters

The *across-the-line (full-voltage) starter* is often referred to as the "simple" type. This starter has one set of contacts that close when the starter coil is energized. When the contacts close, the motor is connected directly to the line voltage. This type of motor starter has control that is relatively simple, inexpensive, and easy to maintain.

Figure 14-3A shows a typical across-the-line motor starter. Such items as the line terminals, load terminals, motor starter coil, overload relays, and auxiliary contact can be seen. Figure 14-3B is a complete circuit diagram of a typical full-voltage starter. Here, as in most machine control, the voltage for control is reduced to 120 V.

Figure 14-3C shows an across-the-line motor starter assembled in a NEMA 12 enclosure with a control transformer and disconnecting and protective devices. The NEMA 12 enclosure is designed to be oil-tight and dust-tight. (Enclosures are discussed in more detail in Appendix E.) Note that the disconnecting

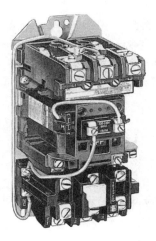

**Figure 14-3A** **Full-voltage starter.** *(Courtesy of Allen-Bradley, a Rockwell International Company.)*

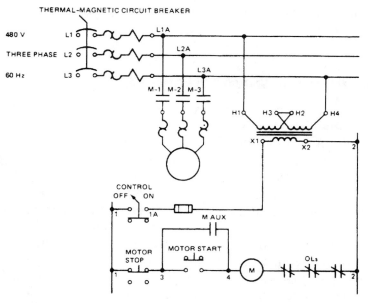

**Figure 14-3B** **Control circuit for a typical full-voltage starter.**

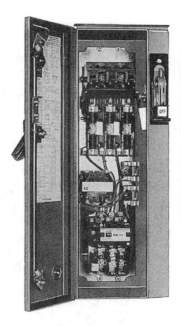

**Figure 14-3C**   **Size 1, NEMA 12 combination motor starter (fusible).**   *(Courtesy of Eaton Corporation.)*

device is mounted in the dead front of the enclosure to allow it to be locked in the open position with the door open or closed. This arrangement is referred to as a *combination starter.*

## 14.3  Reversing Motor Starters

Reversing and multispeed motor starters may be considered as special applications of across-the-line starters.

In the *reversing starter,* there are two starters of equal size for a given horse-power motor application. The reversing of a three-phase, squirrel-cage induction motor is accomplished by interchanging any two line connections to the motor. The concern is to properly connect the two starters to the motor so that the line feed from one starter is different from the other. Both mechanical and electrical interlocks are used to prevent both starters from closing their line contacts at the same time. Only one set of overloads is required as the same load current is available for both directions of rotation.

A reversing starter is shown in Figure 14-4A. Figure 14-4B is the circuit diagram for a typical reversing starter. Note the change of connections from the line to the contacts from one unit to the other (lines L2 and L3).

***Figure 14-4A*** **Reversing starter.** *(Courtesy of Allen-Bradley, a Rockwell International Company.)*

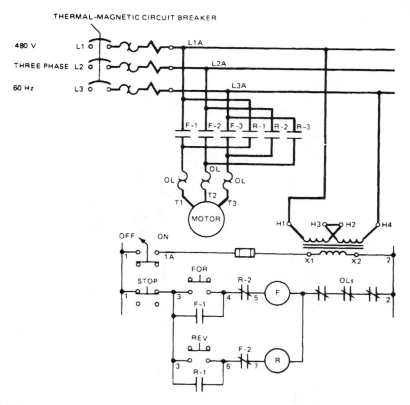

***Figure 14-4B*** **Circuit for a typical reversing starter. Conductors carrying load current at line voltage are denoted by heavy lines.**

## 14.4  Multispeed Motor Starters

Many industrial applications require the use of more than one speed in their normal operation. While many methods for adjustment of speed are available, the *pole changing* method is still used. For example, in a squirrel-cage induction motor, the speed depends on the number of poles. Pole changing is obtained by the design of the stator winding. At 60 Hz, which is standard in the United States, a two-pole motor operates at approximately 3600 rpm. A four-pole operates at 1800 rpm, a six-pole at 1200 rpm, an eight-pole at 900 rpm, and a ten-pole at 720 rpm. By having two or more sets of winding leads brought out into the terminal connection box, the number of effective poles can be changed. Also, one winding can be connected in more than one way.

With a change in speed, the horsepower will also change. For example, in a two-speed motor for which the slower speed is one-half the higher speed, the horsepower will also be one-half the horsepower of the higher speed. Therefore, two sets of overload relays must be used to provide adequate protection.

Figure 14-5A shows a two-speed reconnectable multispeed starter. The circuit of a typical two-speed motor and starter is shown in Figure 14-5B.

*Figure 14-5A*  **Two-speed reconnectable multispeed starter.** *(Courtesy of Eaton Corporation.)*

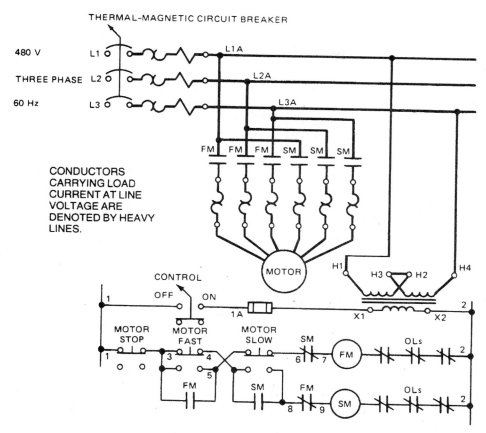

**Figure 14-5B** Circuit for two-speed starter. Conductors carrying load current at line voltage are denoted by heavy lines.

# 14.5  Additional Across-the-Line Starter Circuits

Additional across-the-line starter circuits are shown in Figures 14-6 through 14-12. To start this group of circuits, it is important to become acquainted with two more expressions in the electrical field: *no-voltage or low-voltage release* and *no-voltage or low-voltage protection.*

No-voltage or low-voltage release means that if there is a voltage failure, the starter will open its contacts or "drop out." However, the contacts will close again, or "pick up," as soon as the voltage returns. A typical wiring diagram showing no-voltage or low-voltage release is shown in Figure 14-6A. Note that this is a wiring diagram, not a schematic or elementary diagram. The elementary diagram is shown in Figure 14-6B.

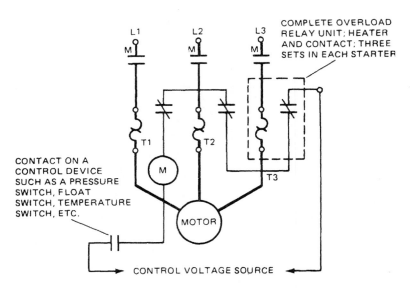

COMPLETE OVERLOAD
RELAY UNIT: HEATER
AND CONTACT: THREE
SETS IN EACH STARTER

CONTACT ON A
CONTROL DEVICE
SUCH AS A PRESSURE
SWITCH, FLOAT
SWITCH, TEMPERATURE
SWITCH, ETC.

**Figure 14-6A**    Line circuit incorporating no- or low-voltage release. Conductors carrying load current at line voltage are denoted by heavy lines.

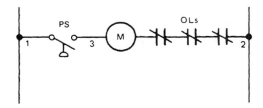

**Figure 14-6B**    Two-wire control circuit.

A *wiring diagram* shows the actual wiring on a group of components in a control center or system. In this case, it is an across-the-line motor starter. This approach to explaining the circuit is helpful when physically checking the wires and their connections. Most manufacturers of motor starters include a wiring diagram in the starter enclosure.

No-voltage or low-voltage protection means that if there is a voltage failure, the starter contacts will open or drop out. However, they will not reclose, or pick up, automatically when the voltage returns. Figure 14-7A is a typical wiring diagram showing this arrangement. The ladder diagram is shown in Figure 14-7B.

There are times when a motor starter must be energized and deenergized from several locations. Figure 14-8 shows such an arrangement, using four motor STOP/START push-button units. Note that each set of STOP/START push-buttons may be located remote from any of the other sets. Each has complete control of

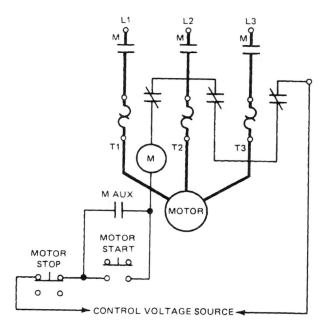

**Figure 14-7A** Line circuit incorporating no- or low-voltage protection. Conductors carrying load current at line voltage are denoted by heavy lines.

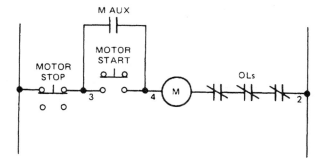

**Figure 14-7B** Three-wire control circuit.

the starter coil. For safety, the START push-buttons may have individual locks to prevent unauthorized operation of any units.

The use of a pilot light to indicate that a motor is energized is a convenience. Many times it is also a safety factor. For example, a motor may be located remote from its START/STOP push-button and therefore cannot be seen. Also, sometimes the shop noise level is so high that it is hard to know if a motor is operating. Figure 14-9A shows a motor starter control. The pilot light indicates its energized condition (Figure 14-9A). The motor is deenergized in Figure 14-9B.

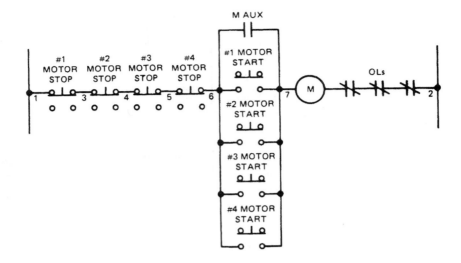

**Figure 14-8**  Control circuit utilizing for STOP/START push-buttons to allow operation from several locations.

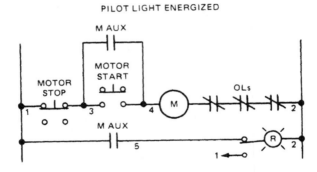

**Figure 14-9A**  Motor starter control circuit with pilot light; motor energized.

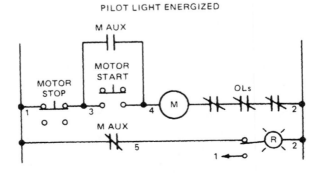

**Figure 14-9B**  Motor starter control circuit with pilot light; motor deenergized.

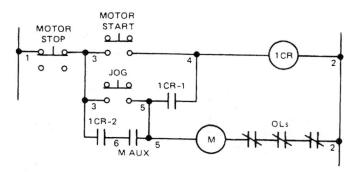

**Figure 14-10** Control circuit for jogging.

In addition to the start/stop function of control for the motor starter, there is also *jogging*. The starter is energized only as long as the JOG push-button is operated. The use of the relay in this circuit provides a safety factor to prevent the starter from "locking in" during a jogging operation. The circuit for the jogging operation is shown in Figure 14.10.

The sequence of operation for the jogging circuit is as follows:

**For normal motor operation:**

1. Operate the motor START push-button switch.
2. Relay coil 1CR is energized.
   a. Relay contact 1CR-1 closes, energizing the motor starter coil.
   b. Relay contact 1CR-2 and motor auxiliary contact M Aux. both close. Together with the closed 1CR-1 contact, they form an interlock circuit around the motor START push-button switch.
3. Operating the motor STOP push-button switch deenergizes the motor starter coil and relay coil 1CR.

**For jogging operation:**

1. Operate the JOG push-button switch.
2. The motor starter coil M will remain energized only as long as the JOG push-button switch is held operated. Note that relay coil 1CR is not energized in this operation.

Some machines have multiple motors. Due to power line conditions, it may not be advisable to start all the motors at the same time.

An arrangement of three motors starting from a single START push-button but energizing at different times is shown in Figure 14-11. Two time-delay relays are used. They can be set to start #2 and #3 motors at predetermined time intervals after #1 motor is energized. The STOP push-button switch will deenergize all

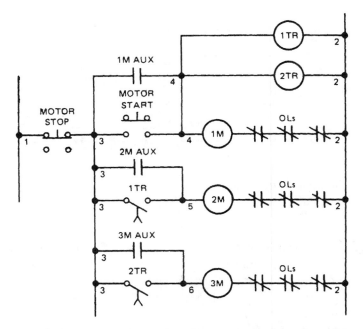

**Figure 14-11** Control circuit for starting three motors from a single START push-button but energizing at different times.

three motors. Additional motors can be started from this arrangement by adding time-delay relays for each motor starter. They should be added in the same way that #1 and #2 time-delay relays are used.

The sequence of operations for the circuit shown in Figure 14-11 is as follows:

1. Operate the motor START push-button switch.
2. Motor starter coil 1M is energized.
   a. Time-delay relay coils 1TR and 2TR are energized.
3. Motor starter auxiliary normally open contact 1M Aux. closes, interlocking around the motor START push-button switch.
4. After a preset time delay as set on timing relay 1TR, timing relay contact 1TR closes.
5. Motor starter coil 2M is energized.
   a. Motor starter auxiliary normally open contact 2M Aux. closes, interlocking around the 1TR timing relay contact.
6. After a longer preset time as set on timing relay 2TR, timing relay contact 2TR closes.
7. Motor starter coil 3M is energized.
   a. Motor starter auxiliary normally open contact 3M Aux. closes, interlocking around the 2TR timing relay contact.
8. All motors are now energized.

9. Operation of the motor STOP push-button switch deenergizes all motor starter coils and the timing relay coils.

A number of motors can be sequenced into operation using the auxiliary contact of the preceding motor starter. This method gives a short, fixed time delay of only the starter closing time. The opening of any of the motor starters through overload (opening overload contacts) will deenergize all the motors in sequence. All the motors will be deenergized through operation of the one common STOP push-button switch. This circuit is shown in Figure 14-12. Note that the START push-button switch must be held operated until #3 motor starter coil is energized and the 3M auxiliary contact closes.

The sequence of operations for the circuit shown in Figure 14-12 is as follows:

1. Operate the motor START push-button switch.
2. Motor starter coil 1M is energized.
   a. 1M motor starter auxiliary contact 1M Aux. closes.
3. Motor starter coil 2M is energized.
   a. 2M motor starter auxiliary contact 2M Aux. closes.
4. Motor starter coil 3M is energized.
   a. 3M motor starter auxiliary contact 3M Aux. closes, interlocking around the motor START push-button switch.
5. Operating the motor STOP push-button switch at any time deenergizes all motor starter coils.

Additional motors can be added to this sequence circuit. The auxiliary contact on the last motor starter is generally used to interlock the START push-button.

Occasionally it is necessary to prevent one motor from stopping until some time has elapsed after the stopping of another motor. This arrangement is used in fluid power applications in which a larger motor is driving a main pressure pump. If the motor driving the main pressure pump and a small motor driving a pilot pressure pump were deenergized at the same time, the small motor would come

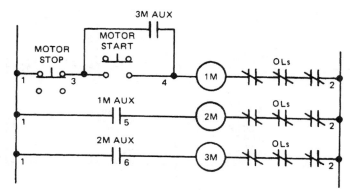

**Figure 14-12**   Control circuit for starting three motors in sequence.

to rest before the main pressure pump motor. This sequence would cause a loss of control pressure on the large main pressure pump during the time the large motor was coming to rest.

Figure 14-13 shows a circuit using two motor starters with a time delay. This arrangement prevents #2 motor starter from being deenergized until after a preset time has elapsed from the time #1 motor starter is deenergized. The sequence of operations for the circuit shown in Figure 14-13 uses two motor starters with a time delay. Motor starter coil 2M must be energized first.

1. Operate 2M motor START push-button switch.
2. Motor starter coil 2M is energized.
   a. 2M Aux. #1 contact closes, interlocking around the #2 motor START push-button switch.
   b. 2M Aux. #2 contact closes, setting up the start circuit for the #1 motor.
3. Operate the #1 motor START push-button switch.
4. Motor starter coil 1M is energized.
5. Time-delay relay coil TR is energized.
6. OFF time-delay contact TR closes, interlocking around the 2M motor STOP push-button switch.

   Motor starter coil 1M must be deenergized before 2M.

7. Operate the 1M motor STOP push-button switch.
   a. Motor starter coil 1M and time-delay relay coil TR are deenergized.
8. After a preset time delay as set on TR, the OFF time-delay contact opens.
9. Operate 2M motor STOP push-button switch.
10. Motor starter coil 2M is deenergized.

The circuit shown in Figure 14-14 is similar to that shown in Figure 14-11. The main difference is in the operation of the motor START push-button switch to initially energize the #2 motor starter coil 2M. In this circuit, the 2M motor starter

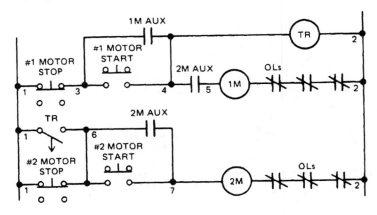

**Figure 14-13** Control circuit using two motor starters with a time delay.

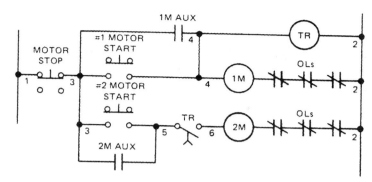

**Figure 14-14** Control circuit for starting two motors. Motor 2 cannot be started until some time after Motor 1 has started.

coil cannot be energized until the preset time on timing relay has expired. The sequence proceeds as follows:

1. Operate the #1 motor START push-button switch.
2. Motor starter coil 1M and timing relay coil TR are energized.
   a. 1M motor starter auxiliary contact 1M Aux. closes, interlocking around the motor START push-button switch.
3. After a time delay set on timing relay TR has expired, timing contact TR closes.
4. Operate the #2 motor START push-button switch.
5. Motor starter coil 2M is energized.
   a. 2M motor starter auxiliary contact 2M Aux. closes, interlocking around the #2 motor START push-button switch.

Note that in this circuit, 2M motor starter coil can be deenergized through the opening of an overload relay contact without deenergizing 1M motor starter coil. However, if 1M motor starter coil is deenergized through the opening of an overload contact, timing relay coil TR is deenergized. This action opens the circuit to 2M motor starter coil, deenergizing the motor starter coil 2M.

The sequencing of several motors is shown in Figure 14-15. The sequence of operations for that circuit is as follows:

1. Operate the motor START push-button switch.
2. Motor starter coil 1M is energized.
   a. Green pilot light is energized.
   b. 1M motor starter auxiliary contact 1M Aux. closes.
3. Motor starter coil 2M is energized.
   a. 2M motor starter auxiliary contact 2M Aux. closes.
4. Motor starter coil 3M is energized.
   a. 3M motor starter auxiliary contact 3M Aux. closes.

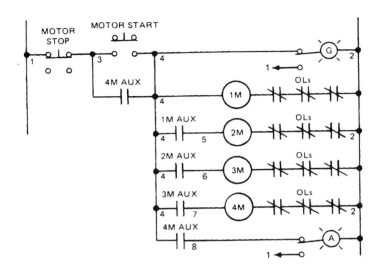

**Figure 14-15**   Control circuit for sequencing several motors.

5. Motor starter coil 4M is energized.
   a. Amber pilot light is energized.
   b. 4M motor starter auxiliary contact 4M Aux. closes, interlocking around
      the motor START push-button switch.
6. Operating the motor STOP push-button switch will deenergize all motor
   starter coils and pilot lights.

Note that the motor START switch must be operated until 4M Aux. closes.

   Note that this circuit is similar to that shown in Figure 14-12. The advantage
of the arrangement in Figure 14-15 is that the STOP push-button switch has complete control in opening the circuit (deenergizing) to all the starter coils. In the
case of the circuit in Figure 14-12, the starters drop out (deenergize) in sequence
from the preceding starter. The disadvantage of the circuit in Figure 14-15 is that
it may limit the number and/or size of starters used. The current used to energize
the starter coils comes through the push-button switches. Therefore, the start circuit is controlled to a degree by the rating of the switches.

   The use of a master motor STOP push-button switch can be arranged with
each motor starter circuit having an individual STOP push-button switch. The
master motor STOP push-button switch can be used as a convenience or as emergency safety. This switch is generally supplied with a mushroom head operator.
The circuit is shown in Figure 14-16.

   As a safety precaution, it is often necessary to provide an overstroke limit
switch to deenergize a motor starter coil. For example, a machine part such as a
moving table may travel a given distance under safe operating conditions. However, if this distance is exceeded, damage to equipment or injury to personnel

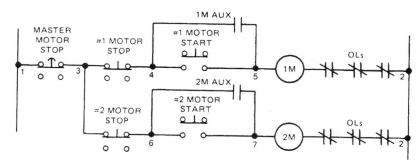

**Figure 14-16** Control circuit utilizing individual STOP and STOP push-button switches with master STOP.

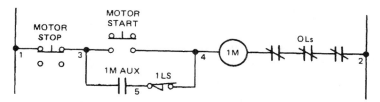

**Figure 14-17** Control circuit incorporating an overstroke limit switch.

may result. The limit switch is located to operate if the proper travel distance is exceeded. When the limit switch is operated, deenergizing the motor starter coil, the motor must then be energized on an emergency basis to back the operating cam off the limit switch. This action is accomplished by switching the operating circuit to reverse and backing the cam off the limit switch by inching the motor START push-button switch. This circuit is shown in Figure 14-17.

## 14.6 Reduced-Voltage Motor Starters

Reduced-voltage motor starters can be divided into several different designs:

- Autotransformer (or compensator)
- Primary resistor
- Wye (or star) delta
- Part winding

In connecting a motor directly across the line, the resulting current at start condition may be on the order of 4 to 10 times the full-load rating of the motor. This current is sometimes referred to as *locked rotor current*. The high current often causes line disturbances if the power supply is inadequate or the motor is large.

The basic principle of the reduced-voltage starter is to apply a percentage of the total voltage to start. After the motor starts to rotate, switching is provided to apply full line voltage. For example, in the autotransformer type, starting steps may be 50%, 65%, or 80% of full voltage.

A timer is provided in the starter control circuit so that the time between starting and running may be adjusted to suit actual operating conditions.

It should be pointed out that at reduced voltage, the torque available from the motor will be reduced. This reduction may cause concern when the motor is starting under load. Torque varies as the square of the impressed voltage. For example, starting on the

- 50% voltage tap, torque will be 25% of normal.
- 65% voltage tap, torque will be 42% of normal.
- 80% voltage tap, torque will be 64% of normal.

Each type of reduced-voltage motor starter is best explained by referring to its circuit diagram. Therefore, a circuit diagram and photo accompany the description of each type of reduced-voltage starter on the following pages.

Detail control may vary between manufacturers and with the use of the starter. Therefore, the circuit diagrams are not necessarily those of the devices shown in the photographs. The manufacturer lists the maximum horsepower for a range of 5 through 1000 for voltages of 200, 230, 380, 460 and 575. The NEMA size is given for each rating.

### 14.6.1 Autotransformer or Compensator (Figures 14-18A and 14-18B)

Two contactors are used in the control. One is a five-pole called the *start contactor*

***Figure 14-18A*** **Reduced-voltage autotransformer starter, closed transition.** *(Courtesy of General Electric Company.)*

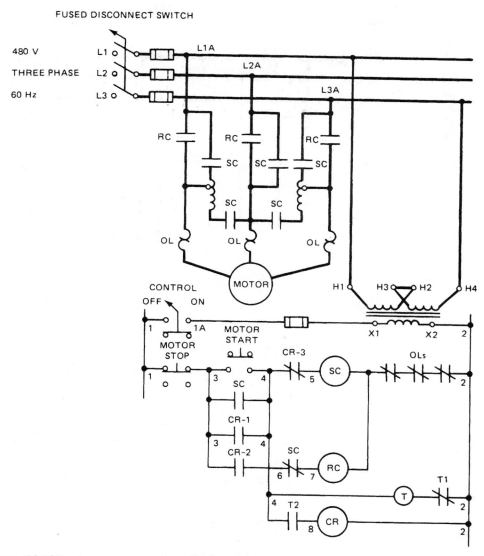

**Figure 14-18B** Control circuit for a typical reduced-voltage autotransformer starter, open transition.

(SC). The second contactor is a three-pole contactor, the *run contactor* (RC). A typical starter is shown in Figure 14-18A.

An autotransformer of sufficient size to handle the horsepower rating of the motor is shown connected to the line leads and motor through the open contacts. Taps are available for a starting voltage of 80%, 65%, or 50% of full line voltage.

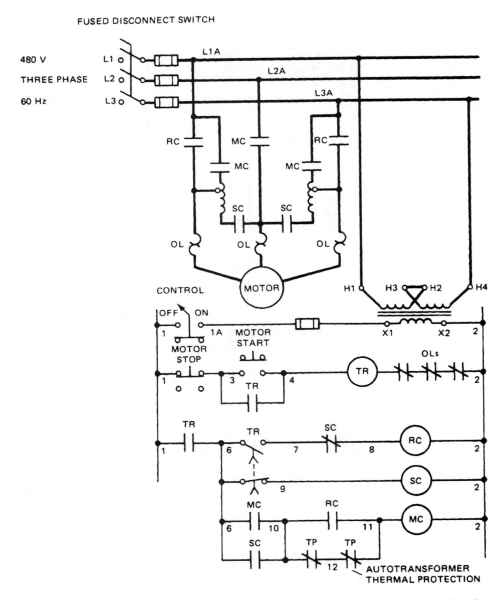

**Figure 14-18C**  Control circuit for a typical reduced-voltage autotransformer starter, closed transition.

On operating the motor START push-button switch, the start contactor is energized and the SC contacts close. This action connects the motor to the line through the selected autotransformer tap.

When time set on time $T$ runs out, the start contactor is deenergized and the run contactor is energized. This condition connects the motor to full line voltage.

This type of starting has one disadvantage. The motor is momentarily disconnected from the line during the changing from the start contactor to the run contactor. This occurrence is known as *open transition.* In some localities or under some conditions open transition may be objectionable. If in doubt, consult the local power company.

The circuit for open transition is shown in Figure 14-18B.

A more practical solution is to use a *closed-transition* autotransformer-type motor starter. The circuit for the closed-transition type is shown in Figure 14-18C. With closed transition, the motor is not disconnected from the line when changing from the start condition to the run condition.

**14.6.2 Primary Resistor (Figures 14-19A and 14-19B)**   Two contactors, each three-pole, are used in the control circuit. One is a *line contactor* (LC). The other is the *accelerating contactor* (AC). Three banks of resistors are used. These are shown connected to the line and the motor through the contacts. The resistors

*Figure 14-19A*   **Primary resistor-type reduced-voltage starter.** *(Courtesy of Square D/Schneider Electric.)*

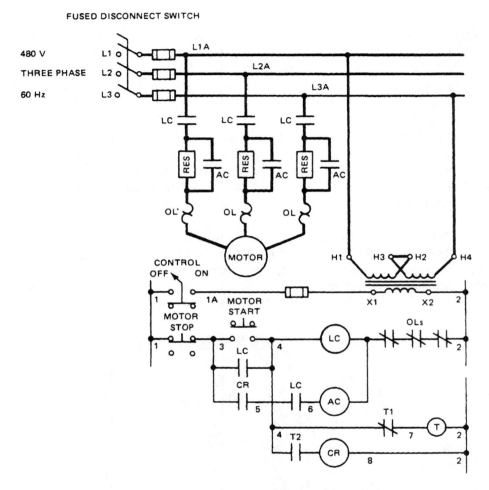

***Figure 14-19B*** **Control circuit for a typical primary-resistor reduced-voltage starter.**

must be of sufficient capacity to handle the horsepower rating of the connected motor. This type of starter is shown in Figure 14-19A.

On operating the motor START push-button switch, the line contactor is energized, closing the LC contacts and connecting the motor to the line through the resistors. After a preset time delay, the AC contacts are closed, connecting the motor directly to the line.

One advantage of the primary resistor-type starter is the smooth acceleration. As the motor accelerates, the current decreases. Thus, the voltage drop across the resistors is reduced, increasing the voltage applied to the motor. The disadvantage is inefficient operation, as the power dissipated in the resistors is comparatively high. The circuit for this type of starter is shown in Figure 14-19B.

### 14.6.3 Wye- (or Star) Delta Starter (Figures 14-20A, 14-20B, and 14-20C)

Figure 14-20B shows that to use this type of starting, all leads from each three-phase stator winding must be brought out for connecting to the starter, which provides 33% normal starting torque.

Three contactors are used: the LC, SC, and *run contactor* (RC). Operating the motor START push-button switch energizes the LC and the SC. The start contacts connect the motor in wye or star, putting approximately 58% of the line voltage across each motor phase. After a preset time delay, the start contacts open and the run contacts close, connecting the motor in delta and putting full voltage on the motor.

The cost of the control components for this type of starting is less than for the autotransformer and primary-resistor types. However, the motor must be supplied with all the leads brought out for this type of starter. Thus, wye-delta starters can be used only with wye-delta motors.

Closed transition versions are available that keep the motor windings energized for the few cycles required to transfer the motor windings from a wye connection to a delta connection. Such starters are provided with one additional contactor, plus a resistor bank. Figure 14-20C shows the circuit of a wye-delta closed transition motor starter.

*Figure 14-20A*  **Wye-delta reduced-voltage starter.**  *(Courtesy of Square D/Schneider Electric.)*

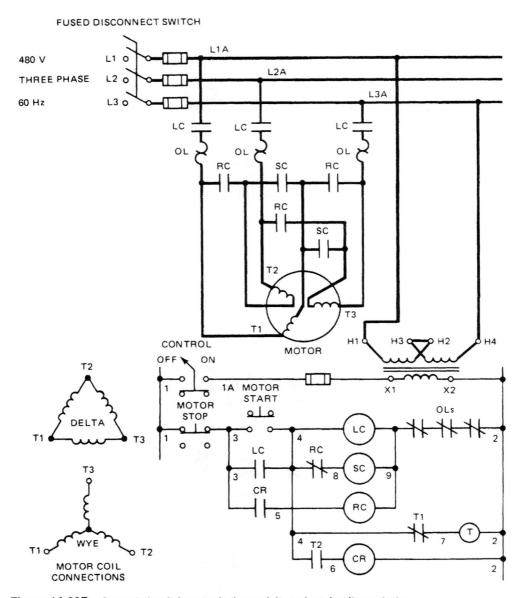

**Figure 14-20B**   Control circuit for a typical wye-delta reduced-voltage starter.

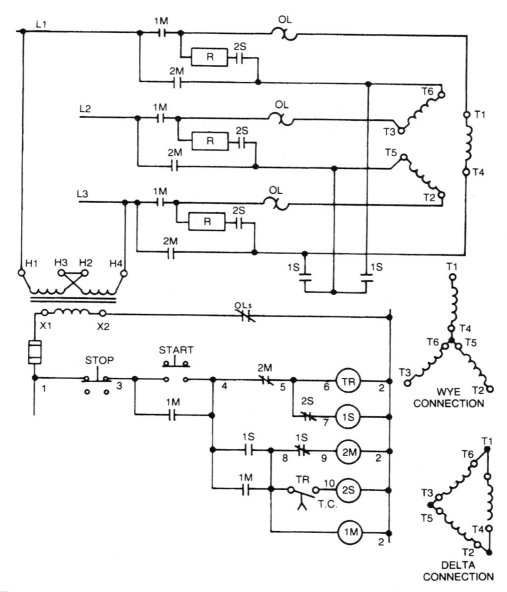

*Figure 14-20C*   **Control circuit for a typical wye-delta reduced-voltage starter, closed transition.**

## 14.6.4 Part Winding (Figures 14-21A, 14-21B, and 14-21C)   Like the wye-delta starter, the part-winding starter depends on the motor having the proper leads brought out for connection to the starter. The part-winding starter supplies 48% normal starting torque. Thus, while part-winding starting is not truly a

***Figure 14-21A*** **Part winding reduced-voltage starter.** *(Courtesy of Square D/Schneider Electric.)*

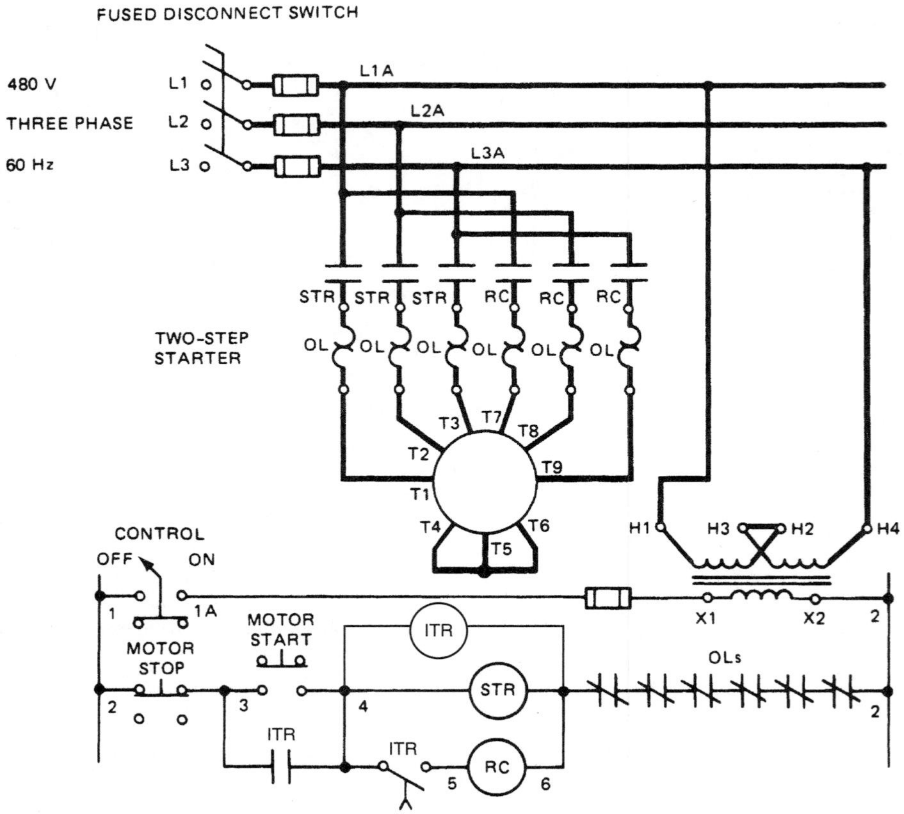

***Figure 14-21B*** **Control circuit for a typical two-step part-winding starter.**

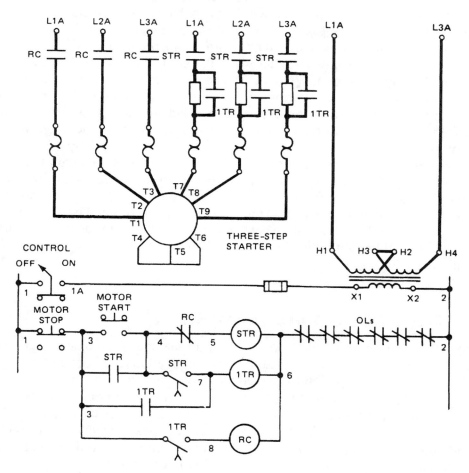

**Figure 14-21C** Control circuit for a typical three-step part-winding starter.

reduced-voltage means, it is usually so classified because of the resulting reduced current and torque.

If the starter windings of the motor are divided into two or more equal parts, with all terminals for each section available for external connection, then the motor may lend itself to what is known as part-winding reduced-voltage starting.

Part-winding starters are designed for use with squirrel-cage induction motors generally having two or three windings. A photo of a typical part-winding starter is shown in Figure 14-21A.

The circuit for the *two-step part-winding starter* is shown in Figure 14-21B. This starter is designed so that when the control circuit is energized, one winding of the motor is connected directly to the line. The winding draws about 60% of normal locked rotor current. It develops approximately 45% of normal motor torque. After about four seconds, the second winding is connected in parallel with the first. The motor is then fully across-the-line and develops its normal torque.

The *three-step part-winding starter* has resistance in series with the motor on the first step. The circuit for this starter is shown in Figure 14-21C. To start, one winding is connected to the line in series with the resistor. After a short interval, approximately two seconds, the resistor is shorted out. The first winding is connected directly to the line. About two seconds later the second winding is connected in parallel with the first winding. Part-winding starters provide closed transition starting.

## 14.7 Solid-State Motor Starters

Many applications require the lower starting torque and smooth acceleration offered by solid-state systems. Typical of such applications are conveyor systems, pumps, and compressors. Solid-state reduced-voltage controllers provide smooth, stepless acceleration of a motor through the use of silicon-controlled rectifiers. By controlling the conduction of the silicon-controlled rectifiers, voltage is gradually applied to the motor. This operation is sometimes referred to as providing a "soft start" to the motor. An adjustable current-limit feature limits current to 25–70% and starting torque to 6–49% of full voltage values.

Figure 14-22 shows the relationships among various types of reduced-voltage starters for the line current drawn from zero to full-load speed. It is also compared to full-voltage starting. Note that the entire range of most reduced-voltage starting is covered by solid-state reduced-voltage starters. This feature gives the solid-state type more flexibility over other types of reduced-voltage starting.

Solid-state reduced-voltage starters are available in voltage ratings of 200, 230, 460, and 575 volts for 60 Hz motors. The maximum horsepower ranges are approximately 10–1000 at 460 or 575 V. The maximum range for 200, 230 V drops to approximately 10–300 hp.

The SMC PLUS™ SMART motor controller shown in Figure 14-23 provides microcomputer-controlled starting for standard squirrel-cage induction motors. There are three standard modes of operation:

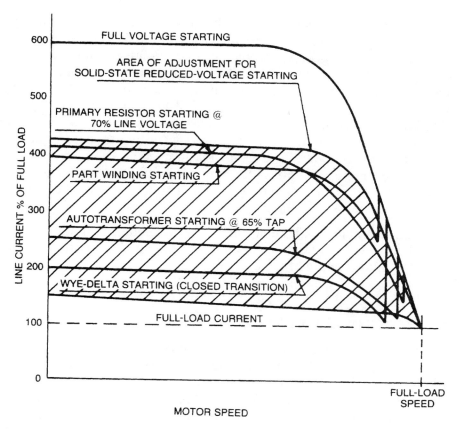

*Figure 14-22*   **Graph comparing starting currents.**

*Figure 14-23*   **Solid-state motor starter.** *(Courtesy of Allen-Bradley, a Rockwell International Company.)*

1. *Soft start with selectable kickstart.* This mode has the most general application. The motor voltage gradually increases during the acceleration ramp period, which can be adjusted from 2 seconds to 30 seconds. This setting is made for the best starting performance over the required load range. Kickstart is for high friction applications and is selectable from 0.4 second to 2 seconds. It provides a pulse of 500% of full-load amperage.
2. *Current limit.* This mode is used when it is necessary to limit the maximum starting current. The current is adjusted, according to starting current restrictions, from 50% to 500% of full-load amperage.
3. *Full voltage.* For applications requiring a full-voltage start, the acceleration ramp time is set to minimum ($\frac{1}{4}$ second), in effect allowing the controller to start the load across-the-line.

The microcomputer-based controller is self-calibrating. It adjusts itself for any line voltage level and frequency within its rating and any current value at or below the continuous rating of the controller.

## 14.8  Starting Sequence

A circuit showing the sequencing of two main drive motors using magnetic reduced-voltage starters is shown in Figure 14-24. The reason for the sequence starting is to reduce the line inrush current.

Associated with the two main drive motors are two auxiliary motors. They are used for pilot pressure and cooling or lubrication. These motors must be started first and must drop out (deenergize) the main motors in case either auxiliary motor deenergizes through an overload.

When the auxiliary motor starters are energized, the #1 and #2 auxiliary contacts close. The sequencing of the two main drive motor starters proceeds as follows:

1. Operate the motor's START push-button switch.
   a. The energizing of the control section of the #1 motor starter is now initiated.
      1a. Relay coil #1-1CR is energized.
      1b. Timing relay coil #1-1TR is energized.
      1c. Timing relay #1 times out, closing #1 TR contact.
      1d. Relay coil #1-2CR is energized.
      1e. Relay contact #1-2CR-1 closes, energizing the green pilot light.
      1f. Relay contact #1-2CR-3 opens, deenergizing relay coil #1-1CR.
2. The sequence for energizing the control section for #1 motor starter is now complete.

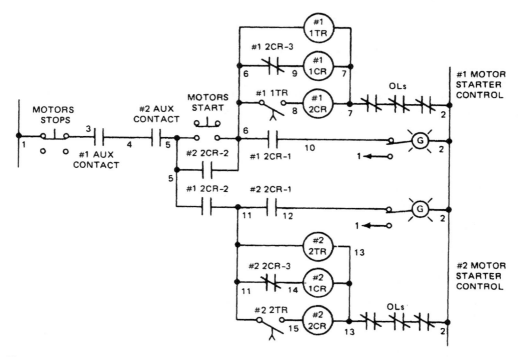

*Figure 14-24* **Control circuit for sequencing two main drive motors using magnetic reduced-voltage starters.**

3. Relay contact #1-2CR-2 closes.
   a. The energizing of the control section of the #2 motor starter is now initiated.
      3a. Relay coil #2-1CR is energized.
      3b. Timing relay coil #2-2TR is energized.
      3c. Timing relay #2 times out, closing #2-2TR contact.
      3d. Relay coil #2-2CR is energized.
      3e. Relay contact #2-2CR-1 closes, energizing the green pilot light.
      3f. Relay contact #2-2CR-3 opens, deenergizing relay coil #2-1CR.
4. Relay contact #2-2CR-2 closes, interlocking around the motor's START push-button switch.
5. The sequencing for the energizing of motor starter #2 is now complete.
6. Both motor starters are energized.
7. Both motor starters are deenergized at any time by the operating of the motor's STOP push-button switch or the opening of any of the overload relay contacts.

# Recommended Web Links

Students are encouraged to view the following Web sites as a supplement to the concepts presented in this textbook. Review and analyze the array of products that are available for electrical control applications. Many of these sites offer technical information that can help in converting the principles to practical applications. To view catalogs, your PC may require Adobe Acrobat Reader software.

## Motor Control Centers

Allen-Bradley
www.ab.com/mcc
Review: Motor Control Systems

General Electric
www.geindustrial.com/cwc/products?famid=22
Review: Motor Control Centers

TECO-Westinghouse
www.tecowestinghouse.com/Controls/controls.htm
Review: Controllers

## Motor and Line Protection

Allen-Bradley
www.ab.com/power/prodinfo/motorpro
Review: Starters and Protection Devices

# Achievement Review

1. Why are overload relays used in motor starters?

2. Explain how the normally closed relay contact in the overload relay is used to deenergize the motor starter coil. A basic circuit may be drawn.

3. How does a high ambient temperature in the location of the overload relay affect its operation?

4. Why should the overload relay be trip-free?

5. Draw the basic power and control circuit for a magnetic full-voltage reversing motor starter.

6. Explain how the reversing of an ac induction motor is accomplished. Why are mechanical and electrical interlocks used?

7. A reduced-voltage motor starter is set to apply 65% of full voltage to the motor for starting. What torque, in percent of full voltage, will be developed in the motor under these starting conditions?

8. Explain how the resistor-type, reduced-voltage motor starter reduces the voltage to the motor on starting. A basic power and control circuit should be drawn.

9. What is the difference between no-voltage or low-voltage release and no-voltage or low-voltage protection?

10. What is meant by *jogging* in motor starter control?

11. What protection is provided in the solid-state reduced-voltage starter to prevent a high-temperature condition due to overcurrent?

    a. Special fuses
    b. Current meter
    c. Thermal sensor

12. How is the effective voltage applied to a motor using a solid-state reduced-voltage starter controlled?

    a. By means of a timer
    b. Through controlling the conduction of the silicon-controlled rectifiers
    c. By the current drawn by the motor

13. What is the difference between open transition and closed transition in a reduced-voltage motor starter?

# Introduction to Programmable Control

## Objectives

After studying this chapter, you should be able to:

- Explain the three basic logic control functions.
- Describe electrical control devices that provide information or input to a control system.
- Describe electrical control devices that provide decision making or logical sequencing of a control system.
- Describe electrical control devices that provide work or output from a control system.
- Understand the programming language of a programmable logic controller (PLC).
- List the advantages of a PLC over hard-wired relay logic circuits.
- Explain the relationship of the PLC processor to input/output (I/O).
- Identify the input and output symbols of a PLC ladder logic diagram.
- Draw a simple motor starter circuit using a PLC ladder logic.
- List the different types of devices to "condition" the input signals to a PLC.
- Explain the relationship of PLC memory to I/O.
- Explain the types of computer memory used in PLCs.
- Describe the logical path of an input device through the input and output tables of a PLC.
- Draw a simple PLC control circuit.
- Convert an existing relay logic circuit to a PLC controlled circuit.
- Understand the role of the data communication highway.
- Describe the function of a complete PLC system utilized in a typical industrial application.

## 15.1 Primary Concepts in Solid-State Control

Before we begin discussion of programmable controllers outlined in Appendix H, it may be of some help to review the fundamental concepts of solid-state control. The same concepts we considered with electromechanical control are still applicable. For example, in electromechanical control, a contact is closed to energize a circuit or opened to deenergize a circuit. In solid-state control, these closed and open contact concepts are referred to simply as 1 and 0, respectively. In programmable controllers, the same logic is followed. The conditions of the inputs and outputs are always such as to allow a signal to pass or to prevent it from passing.

This chapter introduces the student to the basic logic and practical use of programmable controllers.

Figure 12-1 showed the control broken down into three sections. We repeat this explanation in Figure 15-1 to emphasize that where control is changed from electromechanical to solid-state programmable controllers, only the logic (decision-making) section is affected in most cases. The input (information) section and the output (work) section remain substantially the same. In a few cases contactors and motor starters shift from the logic section to the work section. The necessity for this shift generally depends on their size and function.

| INFORMATION OR INPUT | DECISION MAKING OR LOGIC | WORK OR OUTPUT |
|---|---|---|
| Push buttons | Relays | Solenoids |
| Limit switches | Reed relays | Heating elements |
| Proximity switches | | |
| Pressure switches | Magnetic amplifiers | Motors |
| Temperature switches | Solid-state elements | Contactors |
| Tape readers | Indicating lights | |
| Pulse generators | Motor starters | |
| Photo cells | | |

*Figure 15-1*  Chart showing division of control.

| RELAY LOGIC | PROGRAMMABLE CONTROLLER | BENEFITS OF PROGRAMMABLE CONTROLLERS OVER RELAY LOGIC |
|---|---|---|
| Relays | Solid-state components | High reliability |
| Hard wiring | Programming | Easy to implement<br>Easy to change |
| Large size | Small size | Save floor space |
| Timer/counters | Programmed functions | Easy to change presets |
| | Microcomputer based | Greater functionality |
| | I/O interface | Adjust to most field devices |
| | Diagnostic indicators | Simplify troubleshooting |

*Figure 15-2* **Chart of benefits of programmable controllers versus relay logic.**

## 15.2 Introduction to Programmable Logic Controllers

Electromechanical control is covered in Chapters 1 through 13, which show how the logical principles of electromechanical control can be adopted to any type of control device that can simulate the logical sequence required of a control circuit. However, the utilization of computer-based control devices has been proven to be less expensive to implement than a hard-wired relay logic circuit. Figure 15-2 shows the typical benefits of programmable controllers over "hard relay logic." Therefore, programmable controllers are used extensively in industrial machine control applications.

## 15.3 Programmable Logic Controller Concepts

The National Electrical Manufacturers Association (NEMA) standard ICS 3—1988, Part ICS 3—304.01 defines a programmable controller as a digital operating electronic apparatus that uses a programmable memory for the internal storage of instructions for implementing specific functions such as logic, sequencing, timing, counting and arithmetic to control, through digital or analog input/output modules, various types of machines or processes. A digital computer that is used to perform the function of a programmable controller is considered to be within this scope. Excluded are drum and similar mechanical type sequencing controllers.

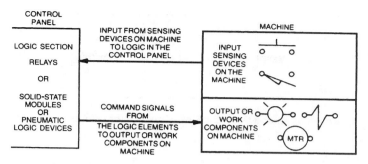

*Figure 15-3* **Electromechanical or solid-state control.**

| INFORMATION OR INPUTS | DECISION MAKING OR LOGIC | WORK OR OUTPUT |
|---|---|---|
| Pushbuttons and selector switches | Central processor unit (consisting of processor, memory, and power unit) | Solenoid valves (analog) |
| Limit switches or linear transducers | | Relays |
| | | Contactors |
| Pressure switches or hydraulic transducers | | Lights |
| | | Horns |
| Temperature switches or thermocouples | | Heating elements |
| | | Fans |
| Motor starter contacts | | Motor starter |
| | | Alarm messages |
| Relay contacts | | Status information |
| Photoelectric eyes | | Signal amplifiers |

*Figure 15-4* **Three sections of control with the use of PLC.**

To avoid any confusion in abbreviating programmable controller, we use the term *programmable logic controller* (PLC) to differentiate it from personal computer (PC).

In earlier discussions of electromechanical control, it was shown that all control is divided into three sections (Figure 15-3). This division into three basic parts does not change with the use of the PLC (Figure 15-4).

The relationships among the three sections of control in Figure 15-3 and Figure 15-4 are shown in block form in Figure 15-5.

In Figures 15-6 and 15-7, a simple motor starter circuit is shown in electromechanical and PLC formats. Note the use of the familiar electromechanical symbol in the PLC circuit.

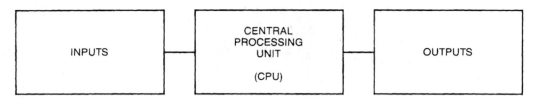

*Figure 15-5*   **Relationships among three sections of control (box form).**

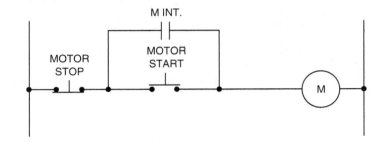

*Figure 15-6*   **Motor starter circuit—electromechanical.**

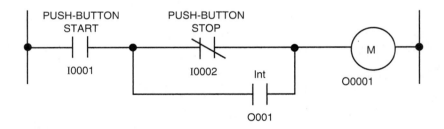

*Figure 15-7*   **Motor starter circuit—programmable.**

## 15.4 Input/Output (I/O)

As we have seen in reviewing electromechanical circuits, all control must depend on initial and continuing information from the machine and/or the process. This information comes from discrete inputs such as push-button switches, limit switches, transducers, thumbwheels, etc. (Figure 15-8). The input module then communicates the status of the discrete inputs to the processor.

The signal from the discrete input must be prepared for access to the processor. Field-supplied voltages from the input may cover a range of 24–240 V ac/dc. The input signal must be converted to a low-level dc logic voltage. This voltage will vary with different manufacturers, but it is in the area of 5–12 V. In case the

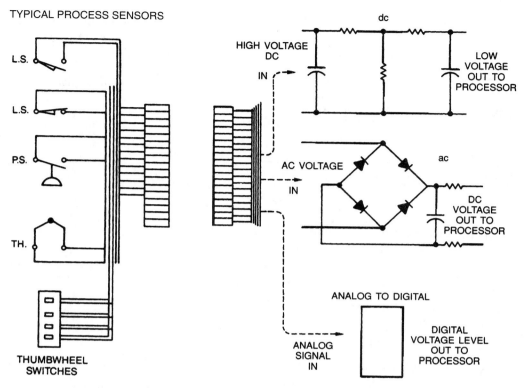

**Figure 15-8**   Discrete inputs.     **Figure 15-9**   Typical voltage interface modules.

input voltage is ac, a bridge rectifier converts the ac to dc, and resistors drop the voltage to the required level. If dc, only resistors are used to drop the voltage. In case the signal is of an analog nature, such as from temperature, speed, pressure, etc., the signal is converted to a binary-coded decimal (BCD) through an analog to digital converter (Figure 15-9).

Due to "noise" or large voltage spikes in the input signal, it is necessary to isolate this incoming signal before it enters the processor. Three devices are used:

1. Optoisolator (Figure 15-10A)
2. Reed relay (Figure 15-10B)
3. Transformer (Figure 15-10C)

While any of the three can be used successfully, the most common is the optoisolator. In Figure 15-10A, an LED is shown on the left side and a photodiode on the right side. A current is set up through the LED from the source voltage and a series resistor. When the light from the LED hits the photodiode, a reverse current is set up in the output. The output voltage equals the output supply voltage

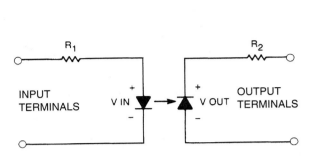

*Figure 15-10B*  Reed switch.

*Figure 15-10A*  Optoisolator.

*Figure 15-10C*  Transformer.

minus the drop across the resistor. As the input voltage varies, the light from the LED varies; the output voltage thus keeps in step with the input voltage.

In the input module there is a light to indicate that a valid signal has been detected. Again, there are three devices that have been used for this purpose:

1. LED (light-emitting diode)
2. Neon lamp
3. Incandescent lamp

The symbols for these are shown in Figure 15-11. While all three of these can be used, the most common is the LED due to its long operating life.

The output module has a function similar to the input module except in reverse order. It functions to allow the processor to communicate with the work components, such as solenoid valves, motor starters, alarm horns, LED displays, etc. As with the input signals, all the work components are terminated. A fuse is used in the work component circuits to protect the electronic components from excessive inrush currents in the work components and from short circuits. Other considerations include the transforming of low-power logic signals of the processor to a higher power level, providing visual indication of both the fuse and functional status of the output points, electrical isolation, and electronic circuitry that indicates the processor's decisions and actions (Figure 15-12).

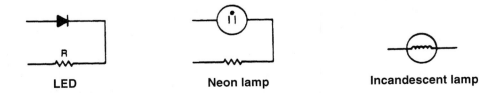

LED              Neon lamp              Incandescent lamp

*Figure 15-11*   Symbols for PLC indicator lights.

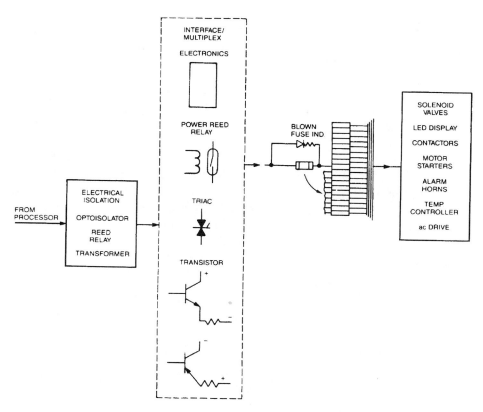

**Figure 15-12**  **Output in block form.**

# 15.5 Processor

The processor module is considered the brain of the programmable logic controller. This unit consists of a state-of-the-art microprocessor and memory chips. There are also circuits necessary to store and retrieve information from the memory and communication circuits required for the processor to interface with the programmer, printer, and other peripheral devices.

The processor module is designed to support a fixed number of discrete, analog, or specialty I/O ports. Also, processors determine variations in the available program instructions, fixed mathematical functions such as PID control, and communication link with other processors.

The processor serves to monitor the status (ON/OFF) of the input and output devices. It also scans and solves the logic of the user program. Scanning, then, is the process of examining inputs—solving logic and determining output status.

## 15.6 Memory

The memory section of the processor serves two important functions:

1. Remembers information that the processor may need to make decisions. This part of the memory is sometimes referred to as *storage* or *data table*. It is in this area that the status (ON/OFF) (I/0) of all the discrete inputs and outputs is stored. Numeric values of timers and counters may also be stored in this memory.

2. Remembers the instruction given by the user, telling the PLC what to do. This section of the memory may be referred to as the *user program memory* and contains ladder-diagram instructions. It is generally many times larger than the storage or data table. Instructions are placed in the memory via a programming device, magnetic tape, or systems computer. Figure 15-13 shows a programmable controller with a handheld programmer.

Figure 15-14 shows the location of the data table and the program storage in the memory.

There are two general classes of memory:

1. Volatile
2. Nonvolatile

The following describes the various types of memory within these two classes. It is important to learn how their characteristics affect the manner in which programmed instructions are retained or altered.

*Volatile memory* will lose its programmed contents if operating power is removed or lost. It is therefore necessary to have battery backup power at all times. One type of volatile memory is RAM, which means read/write, solid-state

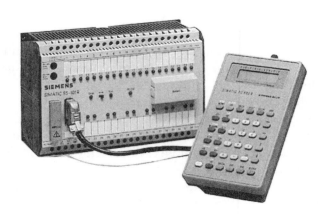

**Figure 15-13**  SIMATIC S5-101R programmable controller with handheld S5-605R programmer.
*(Courtesy of Siemens Energy and Automation, Inc., Programmable Controls Division.)*

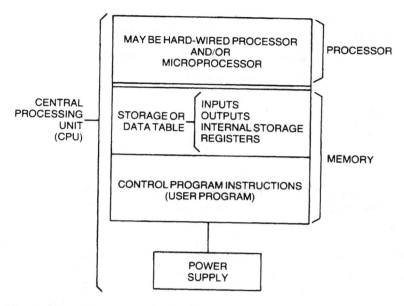

**Figure 15-14**   Location of units within CPU (block form).

*random access memory.* Information stored in RAM can be retrieved or "read." "Write" indicates that the user can program or write information into the memory. "Random access" refers to the ability of any location in the memory to be accessed or used.

There are several types of RAM memory; two of them are:

1. MOS (metal-oxide semiconductor)
2. CMOS (complimentary metal-oxide semiconductor)

*Nonvolatile memory* retains its information when power is lost. It thus does not require a battery backup. A common type of nonvolatile memory is ROM (*read-only memory*).

As with volatile memory there are several types of ROM:

- PROM
- EPROM
- EAROM
- EEPROM

PROM means *programmable read-only memory.* It is a special type of ROM and is generally programmed by the manufacturer. It has the disadvantage of requiring special programming equipment and once programmed cannot be erased or altered.

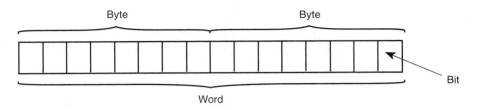

*Figure 15-15*   **Basic information unit—16-bit word.**

EPROM means *erasable programmable read-only memory*. A program can be completely erased by the use of an ultraviolet light source. After the program chip is completely erased, program changes can be made.

EAROM means *electrically alterable read-only memory*. An erasing voltage applied to the proper pin of an EAROM chip will completely erase the program.

EEPROM means *electrically erasable programmable read-only memory*. While it is a nonvolatile memory, it offers the same programming flexibility as RAM. The program can easily be changed with the use of a PC with EEPROM software or an EEPROM programming unit or manual programming unit.

Programmable logic controllers may have different memory sizes. The actual size will depend on the user's application. Sizes are generally expressed in K (1024) values (2K, 4K, 16K, etc.). These numbers represent the number of words available. The range may be from 256 words for small PLCs to 64K for large PLCs. A word is usually 8, 12, or 16 bits (Figure 15-15).

## 15.7  Power Supplies*

The power source in most industrial plants is 120-240-480 V. It is important that the source be as free as possible of heavy loads such as welding equipment, large horsepower motors, or heavy heating loads. It is advisable to have the source as free as possible of noise and voltage fluctuations. The manufacturer will generally supply the proper-size power-supply unit for the PLC. In some cases additional power supplies may be necessary where remote I/Os are required.

A satisfactory power supply will take the incoming line power and rectify and filter. In some power supplies the high-voltage dc will be switched at a high (ON/OFF) rate to the filter section. The filter will remove any switching transients that may be on the dc. Many manufacturers will include additional electronics to monitor the incoming line voltage.

*Refer Back to Chapter 1.

In the discussion of memory we noted that some systems require a backup battery power source to enable the processor memory to maintain a program in the event the power source is lost. The batteries may be of several types, including lithium, alkaline, or nickel-cadmium. The lithium is used by many manufacturers because of its excellent operating characteristics. It has a high power density for its size and a long shelf life.

## 15.8 Programming

In Chapter 12 several circuits were shown using relay ladder-type schematic diagrams. When the PLC was introduced, the manufacturers developed a language of commands, instructions, and operations that were presented in a form similar to the relay ladder-type diagram.

A simple electromechanical ladder-type diagram is shown in Figure 15-16. A program written for the circuit as drawn would in effect say: If the push-button switch is operated, then the green light (G) will be energized. If the limit switch remains operated, then the red light (R) remains energized.

However, in drawing the circuit for a PLC, the programmers do not have symbols like those used in electromechanical circuit diagrams (push-buttons, limit switches, pressure switches, etc.). While programming devices will vary with the manufacturer as to the characters they use, in general they use relay contact symbols for all component switches. The circuit from Figure 15-16 will thus appear as shown in Figure 15-17.

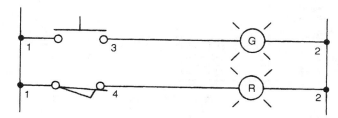

**Figure 15-16**   **A simple electromechanical ladder-type diagram.**

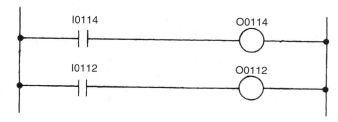

**Figure 15-17**   **Figure 15-16 drawn in programmable form.**

If we were now to take this circuit and show it in a PLC with the inputs, outputs, and bit addresses, it would appear as shown in Figure 15-18.

The program storage or user section of the memory has been given an instruction that if the limit switch is closed, then the red light should be energized. With a 1 in bit 12 of word I01 from the input, there will now be a 1 in bit 04 of word O10 and the red light will be energized.

Another instruction has also been given the program storage: the green light is to be energized when the push-button switch is operated. Therefore, when the push button is operated, bit 14 of word I01 will receive a 1. This action then changes bit 04 of word O10 to a 1 and the green light will be energized.

A program statement thus consists of a condition and an action; for example, "If the limit switch is closed, then energize the red light." In general, then, it can be said that each action is represented by a specific instruction. The instruction tells the processor to do something with the information stored in the data or user table.

Another electromechanical ladder-type circuit diagram is presented in Figure 15-19. The main function of any ladder-type diagram is to present the conditions that are available to control outputs based on inputs. The conditions existing in the circuit shown in Figure 15-19 are as follows:

- The normally closed STOP push-button switch is not operated.
- The normally open START push-button switch is not operated.

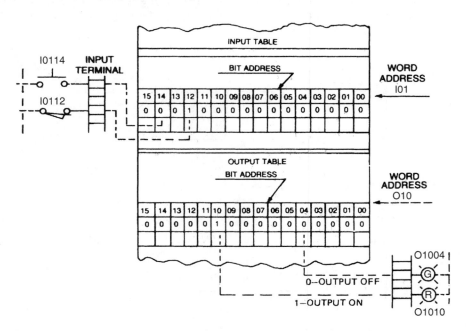

*Figure 15-18    Figure 15-17 combined with inputs, outputs, and bit addresses.*

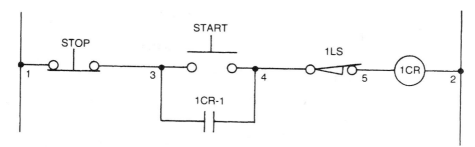

*Figure 15-19* **Typical control circuit.**

- The normally closed limit switch is not operated.
- Relay coil 1CR is not energized.
- Relay contact 1CR-1 is open.

An instruction, then, could be to operate the START push-button switch. The sequence of operation is then:

1. Relay coil 1CR is energized.
    a. Relay contact 1CR-1 closes, interlocking around the START push-button switch.
2. Relay coil 1CR remains energized.

From this sequence it can be stated that for an output to be energized or activated, there must be a complete path for electrical energy from 1 to 2 in the top rung of the ladder, or it can be expressed as logic continuity.

If the circuit shown in Figure 15-19 were programmed for a PLC, it would appear as shown in Figure 15-20. The sketch combines the inputs, outputs, and data tables in the memory.

Each contact and coil is given an address that references it to the data table. This address can be either a connected input or output or an internally stored bit output.

In the electromechanical circuit a contact on relay 1CR was used to interlock around the START push-button switch. In programmable control the holding contact O1004 will have the same address as the output; that is, they are on the same bit (04 of word O10) in the I/O register. The logic equivalent of the holding contact is set to 1 when the output is set to 1.

To start the process in the operation of this circuit as programmed, the processor scans the circuit for a self-check to determine if all the connected inputs and outputs are in the condition shown in the circuit.

The scanner checks the status of all discrete input devices for OFF or ON (0 or 1). It then scans the user program and solves the logic. The next step is to set the discrete output devices ON if conditions indicate that this is the resulting

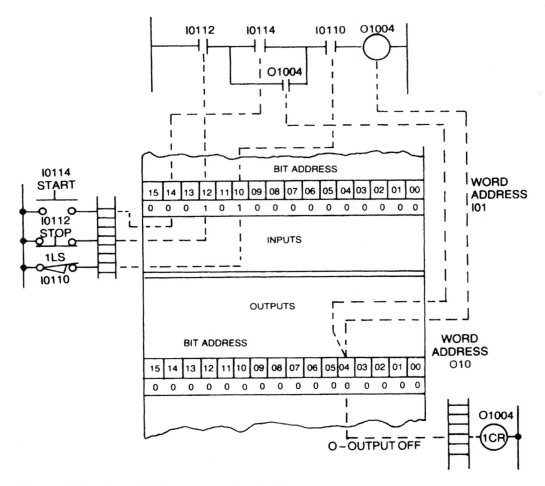

*Figure 15-20*   Figure 15-19 programmed for a PLC.

action required by the program. The process continues: examine inputs, solve logic, determine output status.

The bit status of each input and output device is shown for start conditions in Figure 15-20. As each event occurs through a complete cycle of the machine, the bit status changes to reflect the status of the input and/or the output device.

We can now look at an industrial application involving a plastic injection molding machine with cores in the mold.

A piston and cylinder assembly is shown in Figure 15-21. Fluid power is supplied to the cylinder through a single-solenoid, spring-return operating valve. Piston and cylinder assemblies #2 and #3 (associated with the cores) can be operated with fluid power, pneumatic power, or a mechanical device. Their motion must be detected by limit switches.

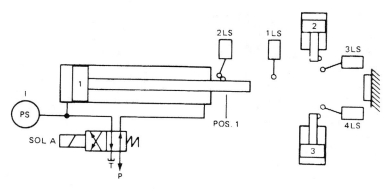

*Figure 15-21*   Piston-cylinder assembly.

Piston #1 must be in position 1 for start conditions. The piston is to move to the right as shown. At the position that #1 piston engages and operates limit switch 1LS, if either #2 or #3 piston has moved to a position operating limit switches 3LS or 4LS, then #1 piston stops and returns to position 1. If limit switches 3LS or 4LS are not operated when #1 piston operates 1LS, the piston continues its travel and builds pressure to a preset level on a work piece. The piston then returns to position 1. Provision must be made to return the piston from any point in its forward travel.

Figure 15-22 shows the electromechanical control circuit for the assembly shown in Figure 15-21. Figure 15-23 shows the same circuit arranged for a PLC. Figure 15-24 shows the connections of the "real world" components as they would be connected into the PC's I/O devices.

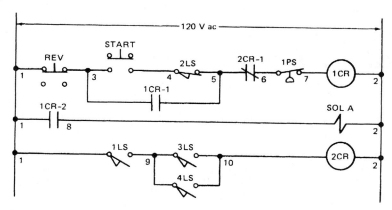

*Figure 15-22*   Electromechanical circuit for Figure 15-21.

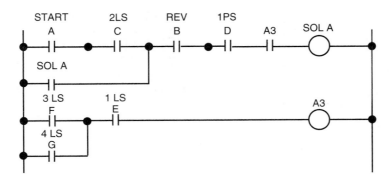

**Figure 15-23**  Circuit in Figure 15-22 arranged for PLC.

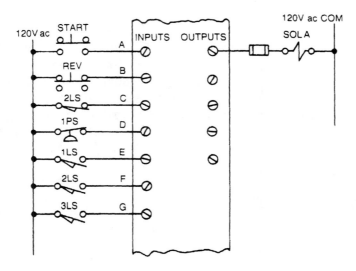

**Figure 15-24**  "Real world" components connected to inputs.

## 15.9  Examine On/Examine Off

One manufacturer of PLCs has used the EXAMINE ON/EXAMINE OFF instructions by the processor. Once again we return to the concept of normally open and normally closed contacts. The EXAMINE ON instruction will pass power when the reference I/O is ON. In a like manner, the EXAMINE OFF instruction will pass power when the reference I/O is OFF. Following the description in Figures 15-18 and 15-20, with an input address on an NO input contact and the input device closed or ON, the processor would then set the appropriate bit in the input register to 1 or ON. The processor will also examine the condition of an input contact for an OFF condition. Here the ladder diagram shows an NC symbol programmed. Since the EXAMINE OFF instruction is looking for an OFF condition, a bit set to 0 is a true condition and power will flow. With a bit set to 1, this is considered an ON condition that makes the EXAMINE OFF contact false, and no power will flow.

# 15.10 Peripheral and Support Devices

By definition, *peripheral* devices are all devices outside the processor, including all hardware. However, by convention peripherals and support devices have come to include only more sophisticated devices such as PCs, printers, programming devices, modems, etc.

Since the number and design of such devices is large and many are complex, only brief descriptions of a few are provided in this text. Every year many changes are made in methods, and new devices are introduced. As a result, it would be a waste of time to describe all of the currently available devices, as many of them will quickly become obsolete.

**15.10.1 Magnetic Tape Drives**   A tape drive provides the ability to save programs for future use. A prerequisite for use is that the tape unit be compatible with the PLC.

The *cassette recorder* provides a means of recording and possibly rerecording the contents of the processor memory. In some cases, manufacturers have incorporated the tape loader into the programming device.

**15.10.2 Disk Drives**   A computer-grade floppy diskette drive is incorporated as a part of the programming terminal. These disks have the advantage of accessing any stored data within a few milliseconds. The disadvantage is that great care must be taken against rough handling, vibration, and impacts. Since they are vulnerable to dust and contamination, they are not acceptable for plant floor use.

The hard disk can store the equivalent of hundreds of diskettes and keep the data protected from environment hazards.

**15.10.3 Printers**   The documenting of a given program for troubleshooting or as a record is important. As with the recorder, the printer must be compatible with the processor, which may include such items as the size of the paper and number of columns to be printed.

An "intelligent printer" has an internal microprocessor to control the various options that the unit may contain.

If the printer is to operate correctly, the communication link between the programming terminal and the printer must be properly installed. A shielded cable is used with the shield grounded at *one end only.*

**15.10.4  Modems**  The term *modem* is derived from MOdulator-DEModulator. It is used to connect the processor to remote field I/O devices and via telephone lines to a computer, printer, tape loader, or another PLC. In the case of connecting remote field I/O devices, it is a great cost saving to avoid wiring all the I/O devices back to the processor.

The modem accepts digital data from either a processor or programming terminal. It is then converted into a frequency-modulated signal, which is then sent over a transmission cable and converted back to a digital signal by another modem.

Telephone modems are used to troubleshoot a remote PLC-based control system. The major difference from the transmission cable used for remote field devices is that the transmission is over commercial telephone lines. Lower transmission frequencies are generally used (300–3300 Hz). This range is the optimum bandwidth of a telephone system.

**15.10.5  PC (Personal Computer—Desktop or Laptop) Data Ports**  The PLC will include a PC and manual data-entry device data ports used to enter machine operating parameters such as time and count set points, positions, temperatures, pressures, speeds, setup conditions, and occasionally process control parameters. The PC is also useful for machine diagnostics and process monitoring. Many of today's systems include electronic storage of these set points, which can allow quick machine setup for process changes (Figure 15-25).

Some PLCs have built-in display units with keypads, while others have display units with touch screens to enter set points, turn switches ON or OFF, read values, graphs, messages, etc.

*Figure 15-25*   **PLC programming system includes PC, PLC programming software and hand-held programmer.** *(Courtesy of Divilbiss Electronics Corporation.)*

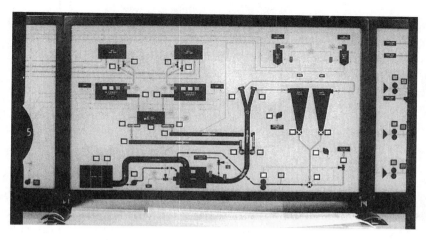

**Figure 15-26** **Semigraphic display.** *(Courtesy of Ronan Engineering Company.)*

With processes becoming more complex, the use of color graphics has become both important and helpful. Many manufacturers have developed color graphics specifically designed for their PLCs. See Figure 15-26 for a graphic display showing a custom process or plant layout with light indicators providing operating status, alarm or pumps, valves, hoppers, tank levels, etc.

Allen-Bradley's Advisor PC color graphic system is a color graphic operator interface that gives plant floor information through color displays that represent the plant operations. The Advisor turns a personal computer into a plant monitoring and control workstation (Figure 15-27).

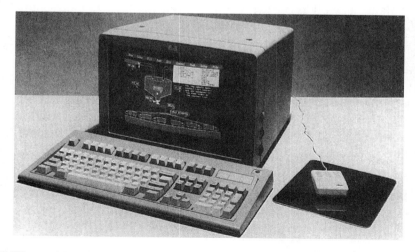

**Figure 15-27** **Advisor PC color-graphic system.** *(Courtesy of Allen-Bradley, a Rockwell International Company.)*

The Advisor PC links plant operators with their programmable controller networks. It reads PLC data tables and dynamically displays alarms and status information in color. It uses a computer keyboard or optional membrane control panel or touch screen to send commands to PLCs to control devices in real time. The unit will operate in on-line mode for live monitoring and control or in off-line mode for system development.

## 15.11  Data Communication Highway

The data highway is communication link that enables PLCs to communicate with other PLCs, various computers throughout a plant, process sensors, and operator consoles. Within a plant, a data highway network consists of twisted-pair cables, coaxial cables, and fiber-optic cables. They carry the digital data between various PLCs and other intelligent microcomputer-based devices within packets of information.

Within a given plant these networks generally extend a maximum of approximately 15,000 feet (cable length). Networks allow information such as control settings and status values to be shared between a wide range of devices. Through a sophisticated system of digital signal transmission and reception, binary data can be organized in a structured format called a *protocol*. A common type of protocol is transmission control protocol and Internet protocol (commonly referred to as TCP/IP). This data format is then electronically moved (or switched) over fiber-optic cable or wire cable. A common system of switching devices and cables is called Ethernet.

Networks can be proprietary (supported by one manufacturer) or conform to industry standards. TCP/IP over Ethernet is an industry standard adopted by manufacturing plants as well as business offices. In the manufacturing plant, there can be one to three types of communication networks:

1. The device network that offers high speed access to plant floor data from a broad range of plant floor devices.
2. The control network that allows intelligent automation devices to share information required for supervisory control, work-cell coordination, operator interface, remote device configuration, programming, and troubleshooting.
3. The information network that gives higher-level computing systems access to plant floor data.

Components of Allen-Bradley's Data Highway Plus are shown in Figure 15-28. It is an industrial local area network designed for factory floor applications and allows connection of up to 64 devices, which may include:

- Programmable logic controllers (PLCs)
- Color graphic systems

**Figure 15-28** Data Highway-Plus communication interface. *(Courtesy of Allen-Bradley, a Rockwell International Company.)*

- Personal computers
- Larger host computers
- Numerical and process controllers

The result is an automation system of computer-controlled devices that can help increase output, improve quality, reduce costs, and speed products to market. A typical Data Highway Plus configuration is shown in Figure 15-29.

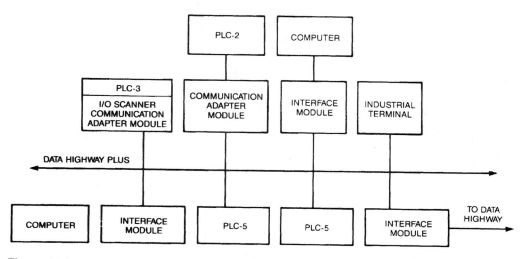

**Figure 15-29** Typical Data Highway-Plus configuration. *(Courtesy of Allen-Bradley, a Rockwell International Company.)*

## 15.12  Converting from Relay Logic to PLC

The programming language adopted by PLC manufacturers uses basic ladder logic symbols. For more complex symbols, such as timers, a generic rectangle is used to store the parameters of the device.

To understand the operation of a control circuit implemented with a PLC, this text shows the conversion of the relay logic circuits in Figures 12-6A, 12-7A, 13-8A, 12-9A, 12-10A, 12-11A, and 12-13A to typical PLC circuits.

In the conversion of a hard-wired, relay-logic circuit to a PLC circuit, one of the critical principles to understand is the state of the input switches. Whether the input switches are normally open or normally closed will affect the state of the logic to be programmed in the PLC. For example, a normally closed STOP switch in a motor control circuit is necessary to interrupt the flow of current to the motor controller. However, in a PLC circuit the switch could be either programmed open or programmed closed. The important condition is the state of the logic in the PLC ladder diagram.

The examples used in the following circuits utilize NO logic. Therefore, you will note that the deenergized state of the PLC logic matches the contacts of the hard-wired circuits.

In industrial applications where the PLC will replace the relay logic circuit, the operator controls and process switches will generally remain as they are wired. Therefore, the PLC logic will need to reflect the normal condition of the operator or process switches.

The following descriptions focus on where the PLC can be applied in completing the circuit requirements.

Figure 12-6A shows the use of a master control switch that will enable or disable the entire circuit. In Figure 15-30, a master control relay MCR-O is applied. Note that the PLC identifies the beginning and the end of the rungs controlled by MCR-O.

Figure 12-7A shows the use of a motorized timer. With separate clutch and motor circuits, motorized timers can be configured in many ways. For PLCs, timers are either on-delay or off-delay.

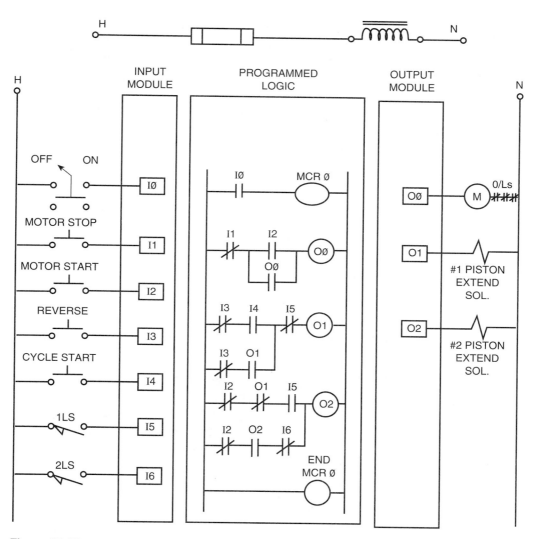

***Figure 15-30*** PLC conversion of Figure 12-6A.

Therefore, in Figure 15-31 an on-delay timer, TON-O, was used to simulate the action of the motorized timer. When using a programmed timer, the programmer must place a value in the Preset Register and the Time Base Register.

Figure 12-8A is a relay only, switching logic circuit. Therefore, Figure 15-32 will reflect the same logical state as the relay circuit (input switches are normally open). One point to recognize in the PLC circuit is that an output point can also be indicated as a relay contact. For example, O0 is also shown in the O0 latching circuit.

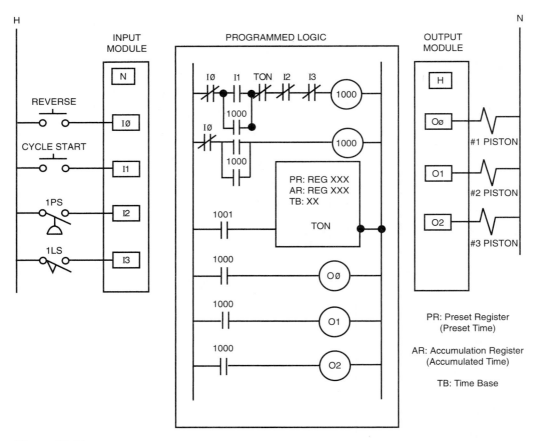

*Figure 15-31*   PLC conversion of Figure 12-7A.

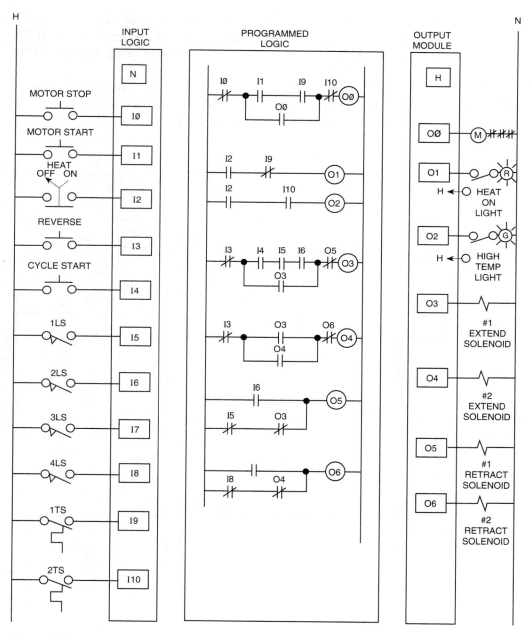

*Figure 15-32* **PLC conversion of Figure 12-8A.**

Figure 12-9A, similar to Figure 12-7A, incorporates a motorized timer. Figure 15-33 employs an on-delay timer. Note the use of an "internal" relay 1000. This designation allows for a relay that is neither an input point nor an output point.

Figure 12-10A incorporates both a motorized timer and a motorized counter, a time delay relay, and a master control device. Figure 15-34 configures the timers using programmed on-delay timers. The counter is also programmable. Like a timer, a counter will require a present value (count).

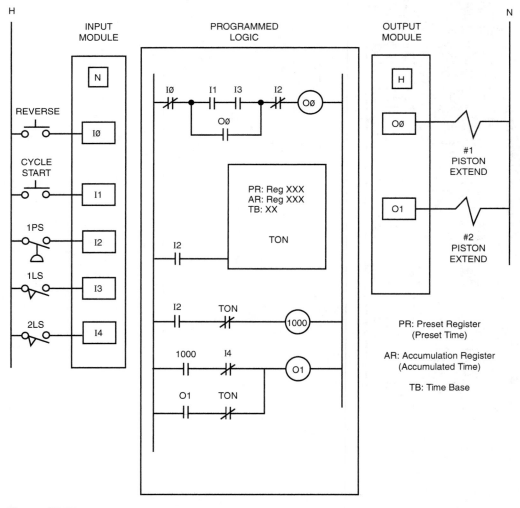

*Figure 15-33*   PLC conversion of Figure 12-9A.

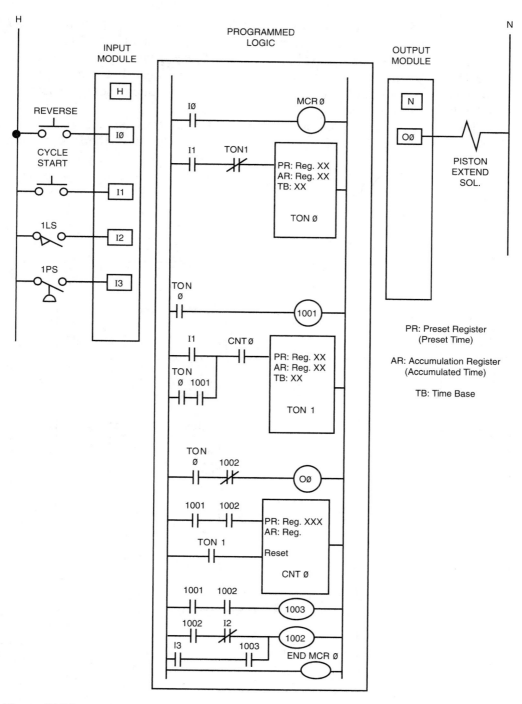

*Figure 15-34*  PLC conversion of Figure 12-10A.

Figure 12-11A utilizes a safety switching logic technique of the START switches wired such that both switches will need to be actuated to extend the cylinder. Also the switches must be released before the next cycle can start. In Figure 15-35 the straightforward application of an on-delay timer and appropriate logic contact configuration will cause the PLC to operate exactly the same as the relay logic circuit.

Figure 12-13A incorporates standard motor control logic. Figure 15-36 shows a technique for controlling the motor circuit external of the PLC circuit.

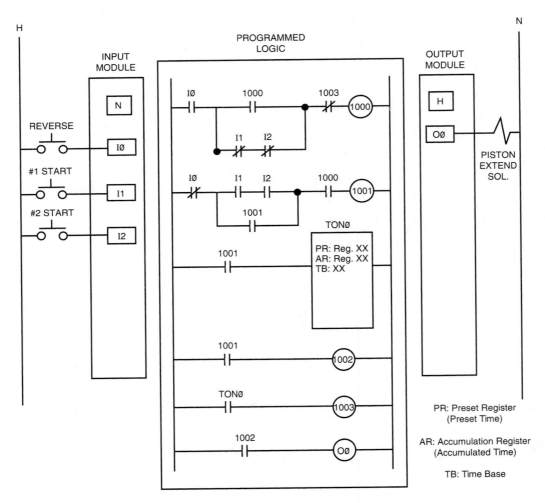

**Figure 15-35**   PLC conversion of Figure 12-11A.

Since the motor control circuit has no logical influence on the logical sequence of the solenoid valves, the entire motor control circuit is external of the PLC. There may be a safety issued involved, especially if the motor must be controlled when the PLC should be disabled for any reason.

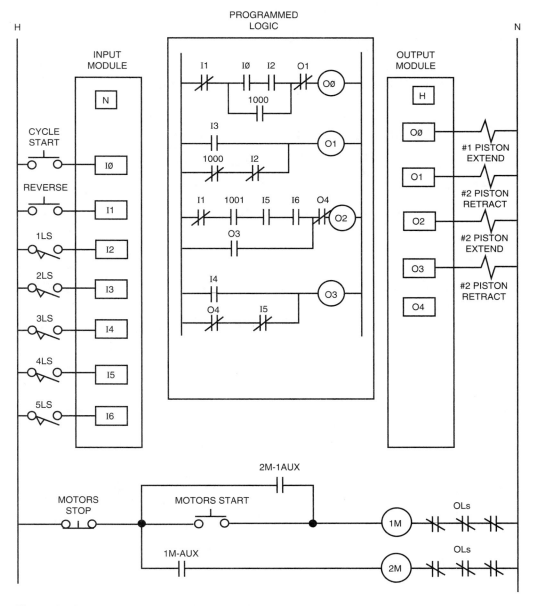

*Figure 15-36*    **PLC conversion of Figure 12-13A.**

## 15.13  PLC Application in Industry

Figure 15-37 shows a programmable logic controller using the total systems approach. Figure 15-38 shows a PLC as used in the plastic injection molding industry.

The HPM Command 90 is a microcomputer control system for injection molding. This microprocessor-based system brings higher levels of product quality and production to injection molding machine process control.

All machine data and process parameters can be accessed during a machine cycle. This system brings advanced levels of control, reporting, and processing, combined with speed and accuracy. An ultrahigh-speed communication bus allows all four of the system's microprocessors to function in parallel.

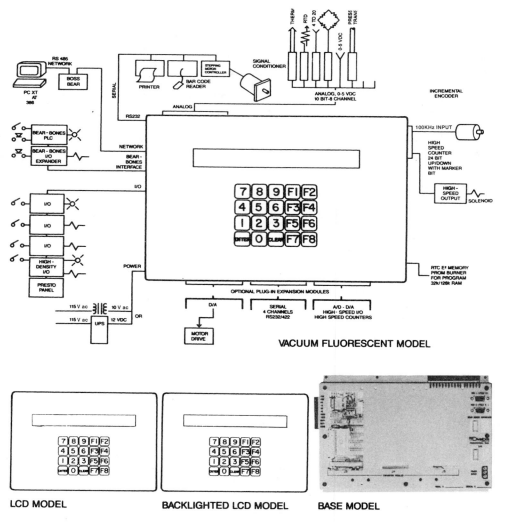

*Figure 15-37*  **Programmable logic controller.** *(Courtesy of Divilbiss Electronics Corporation.)*

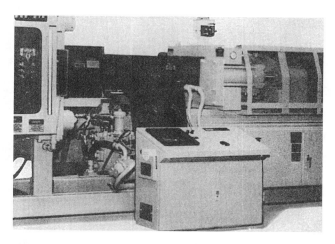

**Figure 15-38**  Industrial application of HPM Command 90. *(Courtesy of HPM Corporation.)*

The heart of this HPM proprietary system is a large, high-resolution, easily read 80-column touch membrane screen with color graphics. This design results in fewer screens to review, yet more complete information on each of the screens for improved efficiency and less operator fatigue.

The Command 90 features oscilloscope-quality graphing capability with real-time graphic display of speeds, pressures, and positions of the clamp and injection. Fast access is provided with a menu button for setup screens, saving setup time.

Figure 15-39 shows the clamp screen, which offers access to all clamp setup data. Figure 15-40 shows the barrel temperatures, providing a graphic

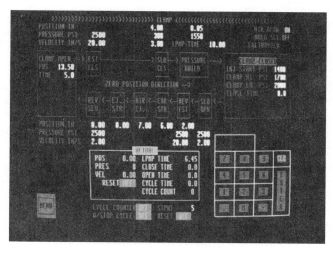

**Figure 15-39**  Clamp screen. *(Courtesy of HPM Corporation.)*

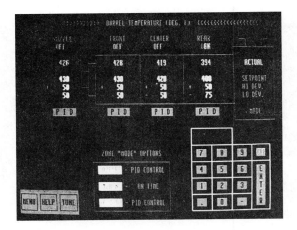

**Figure 15-40**   Graphic representation of the screen showing barrel temperature.   *(Courtesy of HPM Corporation.)*

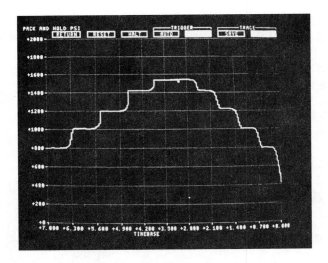

**Figure 15-41**   Graphic representation of hydraulic pressure during pack and hold phase of injection.   *(Courtesy of HPM Corporation.)*

representation of the screw. Figure 15-41 is the pack and hold screen, which provides graphic representation of hydraulic pressures during the pack and hold phase of injection. Figure 15-42 shows a relay ladder diagram (RLD) monitor, which is used by maintenance personnel for troubleshooting and sequence problems (see Chapter 19 on Troubleshooting).

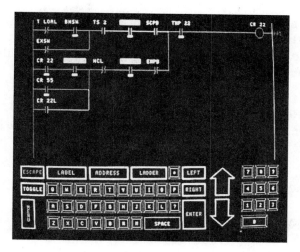

**Figure 15-42**  RLD monitor for troubleshooting and sequence problems. *(Courtesy of HPM Corporation.)*

## Authors' Note

This chapter on programmable logic controllers is a general introduction into a multifaceted subject. Therefore we recommend the following textbook for more detailed coverage of the concepts, programming, and implementation of PLCs: *Introduction to Programmable Logic Controllers*, by Gary Dunning (Delmar/ Thomson Learning, 2000).

## Recommended Web Links

Students are encouraged to view the following Web sites as a supplement to the concepts presented in this textbook. Review and analyze the array of products that are available for electrical control applications. Many of these sites offer technical information that can help in converting the principles to practical applications. To view catalogs, your PC may require Adobe Acrobat Reader software.

### Programmable Controllers

Allen-Bradley
www.ab.com/plclogic
Review: The characteristics and specifications of the different levels of controllers

Cutler-Hammer
www.ch.cutler-hammer.com
Review: The following Web page: Factory Automation (the complete
catalog is available)

Idec
www.idec.com/usa/html/PLCs.html
Review: The characteristics and specifications of the different levels of
controllers

Mitsubishi
www.meau.com/eprise/main/Web_Site_Pages/Public/Products/PLC/P-
PLC-Family
Review: The characteristics and specifications of the different levels of
controllers

Toshiba
www.tic.toshiba.com/productgroups.php?family-plcs
Review: The characteristics and specifications of the different levels of
controllers

GE Fanuc
www.gefanuc.com/products/controllers/index.asp
Review: The characteristics and specifications of the different levels of
controllers

Omron
oeiweb.omron.com/oei/Products-PLC.htm
Review: The characteristics and specifications of the different levels of
controllers

Modicon
www.modicon.com
Review: The characteristics and specifications of the different levels of
controllers shown under the product selection listed as PLC & I/O

Siemens
www.sea.siemens.com/automat/product/plc/auplcov.html
Review: The characteristics and specifications of the different levels of
controllers

The Essentials of Structured PLC
www.controleng.com
Click on "Issue Archive"

1. Click on "2001"
   Click on "February"

under "Product Focus"
Click on "Programmable Logic Controllers"

2. Click on "2001"
   Click on "September"
   under "Departments"
   Click on "Back to Basics"—Article: The Essentials of Structural PLC

## Achievement Review

1. Name and describe the operation and function of four electrical control devices that provide information or input to a control system.

2. Name and describe the operation and function of three electrical control devices that provide decision making or logical sequencing of a control system.

3. Name and describe three electrical control devices that provide work or output from a control system.

4. Given the control circuits developed and explained in prior chapters, using the PLC symbols, convert the following circuits to PLC control: (a) assume all normally open input and (b) assume that all input devices must remain as originally designed.

   | | | |
   |---|---|---|
   | 1. | Figure 7-7 | Pg 110 |
   | 2. | Figure 7-8 | Pg 111 |
   | 3. | Figure 7-9 | Pg 112 |
   | 4. | Figure 8-11D | Pg 162 |
   | 5. | Figure 8-12D | Pg 164 |
   | 6. | Figure 9-23 | Pg 189 |
   | 7. | Figure 9-24 | Pg 190 |
   | 8. | Figure 10-6A | Pg 198 |
   | 9. | Figure 10-6B | Pg 198 |
   | 10. | Figure 10-6C | Pg 199 |
   | 11. | Figure 10-12 | Pg 205 |
   | 12. | Figure 10-13 | Pg 206 |
   | 13. | Figure 11-3 | Pg 211 |

5. List four advantages of a programmable logic controller over hard-wired relay logic circuits.

6. Explain the function of the processor in a PLC. What are the differences in the processor of a small limited function PLC verses a processor for a larger multifunction PLC?

7. Explain how the PLC I/O functions. What are three types of I/O modules?

8. List three different types of devices that can electrically "condition" the input signals to a PLC.

9. Explain how the size of the PLC memory will determine the number of functional I/O points.

10. Name two types of computer memory used in PLCs.

11. Using Figure 15-18 on page 341 as a guide, diagram the input, outputs, and bit addresses of the following circuits in the nonactive state (assume that all input devices must remain as originally designed):

    1. Figure 7-7          Pg 110
    2. Figure 7-8          Pg 111
    3. Figure 7-9          Pg 112
    4. Figure 8-11D        Pg 162
    5. Figure 8-12D        Pg 164

12. Why is the data communication highway important to large manufacturing processes?

13. Identify the different types of input devices to a PLC applied to an injection molding machine (see Figure 15-37, on pg 340).

14. Identify the different types of output devices to a PLC applied to an injection molding machine (see Figure 15-37, on pg 340).

15. In Figure 15-37 on page 340, identify where concepts of data communications are applied.

# Chapter 16  Industrial Data Communications

## Objectives

After studying this chapter, you should be able to:

- Understand the concept of a distributed data factory.
- Describe the architecture, media, and protocol of a local area network.
- Explain the difference between synchronous and asynchronous data transmission.
- Describe interference problems encountered by networks in industrial applications.
- Understand the components of an industrial data communication system.
- Explain the difference between an open network system and a proprietary network.
- Give three reasons why Ethernet is a preferred protocol.
- Name the four layers of an industrial network model.
- Identify five differences between device-level and control-level networks.

## 16.1  Overview

Industry relies on production processes and machines that produce a great deal of data. These data are used to create information such as product quality and production results. Some of the factors that necessitate the conversion of industries into so-called distributed data factories include:

- Production systems are becoming integrated; in other words, a complete manufacturing operation may be composed of several individual processes.
- Processes are dependent on other processes for information to complete their objectives.
- Processes are controlled by any number of individual controllers (PLCs).

- Sensors and actuators are becoming smarter. They produce computer digital signals that can communicate directly with a controller.
- Data are used in both production processes and in business data processing.
- The amount of data created in the factory, and electronically processed into control and management information, has broadened the term *control technology* to *information technology* (IT).

The distributed data factory becomes a complex set of computer-controlled processes. Like building architects, engineers develop the unique design of the interrelated control systems to create an architecture perspective to information technology. This architecture is a road map illustrating how the individual sensors, actuators, displays, and controllers are combined to form a distributed data factory.

Figure 16-1 illustrates a simple production organization that represents a distributed data factory. This facility has many processes, each of which has a need to communicate with other processes for instructions and coordination. This architecture shows the flow of data between the processes.

We are familiar with the control functions of a PLC but this chapter covers the communication tasks the PLC can perform with sensors, actuators, other PLCs, and business computers. Effective communication between machines and other equipment is essential for coordinating the functions of automation systems such as computer-aided manufacturing (CAM), computer-integrated manufacturing (CIM), and control area networking (CAN).

## 16.2 Industrial Information Technology Architecture

Industrial data communications is an important part of the industrial IT architecture. *Data communications* is defined as the transmission of data converted into a digital coded format (1s and 0s) to and from one place to another over an interconnected system. When applied to industrial processes, *industrial data communications* can be defined as the movement of data to and from any place in the factory. Keep in mind that the term *factory* must include the surrounding facilities such as management and business offices, warehouses, and material supply facilities. Also, any process found in construction, transportation, and hospitals can be applied to the data factories concept discussed in this chapter.

For industrial data communications, the architecture of the communication system becomes very important to the success of the manufacturing processes and business objectives by providing data when and where it is needed. Therefore, the concept of digital pulses electronically moving through wires from an originating device to a destination device is an important principle of data communications.

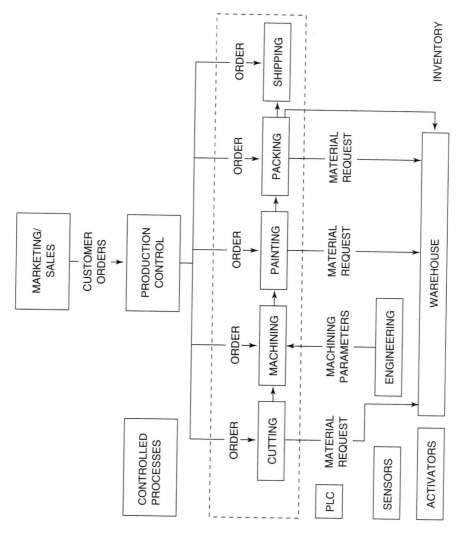

**Figure 16-1  Distributed data factory architecture**

349

An industrial IT architecture must consider the following issues:

- *Connection*—How are we going to connect process and control devices together so data will move when and where needed?
- *Transmission*—Because digital data are electrical signals, what is the best electronics and media (wire or cables) we can use to move data?
- *Access*—Because we have many devices trying to communicate to each other, which device will have priority? How will the devices be recognized?

## 16.3  Data Communication Network Concepts

Data communication networks involve the movement of digital data through transmission cables or wires. The data originate from some source such as a sensor, moves through transmission lines such as twisted-pair cable, is switched or routed to appropriate network segments, and terminates into a controller. The data path would reverse if the data originate at a controller and terminate into an actuator.

Figure 16-2 illustrates the arrangement of a packet of information that is "stamped" with a digital code, which represents source and designation identifiers corresponding to a network protocol. At the end of the packet is an error checking code that follows the data through the network. This code is checked at every point in the network to ensure that data remain the same.

With a set of identifier codes, the movement of the data is switched to and from the process equipment throughout the factory. Figure 16-3 illustrates how a network switch will interpret the data string for the destination and the error checking code. The overall control of the data movement is done by network controllers, such as a switch, that follow the rules of the protocol and, acting on the designation identifiers, switch to specific locations through programmed software at the PLC or process computer level.

The complex assortment of industrial network equipment makes it impossible for this text to discuss each manufacturer's data communication system. However, there are three basic concepts that are important to understand:

1. For a control device to be connected to a network, it must be network compatible. That is, the device must have the physical connection such as the correct connector and it must have the electronics to create and interpret strings of digital data.
2. The string of digital data is structured in a manner that places the device identifier code, the destination code, an error checking code, and the operating value for the device all in the same string.
3. The process of determining the way in which the string fits together, how the communication is started and ended, whose turn it is to send or receive data,

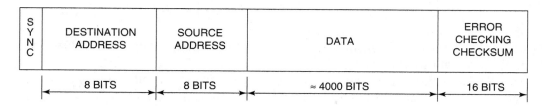

| S Y N C | DESTINATION ADDRESS | SOURCE ADDRESS | DATA | ERROR CHECKING CHECKSUM |
|---|---|---|---|---|
| | 8 BITS | 8 BITS | ≈ 4000 BITS | 16 BITS |

*Figure 16-2* **Information packet.**

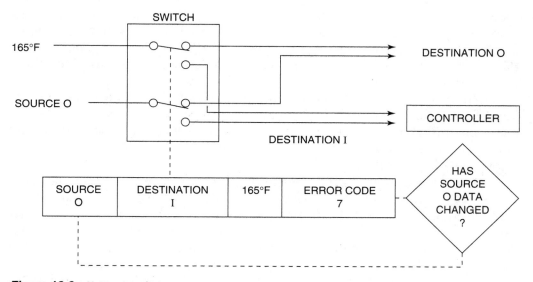

*Figure 16-3* **Network switch.**

and how messages are confirmed is controlled by a *protocol*. The protocol is the traffic cop of the network. Many different protocols exist for industrial data communication applications. Four common protocols are Ethernet, Modbus, ControlNet, and DeviceNet.

## 16.4 Data Transmission

Since data from some industrial devices may be more important than other devices in the process, the amount of time required for the data to move becomes important. Parallel and serial transmissions are two ways data can be physically moved. Also, depending on the timing or sequence of the data movement, synchronous and asynchronous communications can control when data are sent and received. Each of these methods has different speed and equipment issues.

*Parallel transmission* allows for every bit of the data to have its own wire in a cable. Therefore, the entire data string is transmitted at the same time. Parallel transmission is the fasted means of moving data, but it is expensive due to the amount of cabling required.

*Serial transmission* moves the coded data string one bit at a time through one wire. If the data string is 128 bits long, each bit is moved through the wire beginning with the first bit in the string and finishing with the last bit in the string. Compared to parallel transmission, serial is slower but the cost for the cabling is lower. However, in most industrial control applications serial data transmission speeds (rated as bits per second [bps]) are kept to a level that allows controllers to monitor process activities and make appropriate adjustments more frequently. Common serial data transmission standards are RS-232, RS-422, and RS-485. Each of these standards identifies the maximum data rates using specific cables, connectors, and terminators.

Within the different channels for data to move, the issue of coordinating the flow of data is important. *Synchronous data transmission* is the movement of data that is controlled by a clocking device that determines exactly when data should move. This type of transmission makes sure that the source and destination devices are looking for the data all at the same time.

*Asynchronous data transmission* allows for the source data to start and stop when ready. If the destination device is not ready to receive the data, the source keeps sending requests to the destination device until the destination acknowledges the source device. Only after the acknowledgement has taken place does the data move to the destination. The request-acknowledgement process is called *handshaking*.

In process control applications, data such as I/O status, which has time-critical implications, are scheduled to be reviewed during the PLC scan cycle. However, if the data are not so critical, they may be reviewed only where there is a change in the value. In this situation, time-critical applications are synchronous while the not-so-critical data would be asynchronous.

## 16.5 Industrial Data Highway

Industrial data communication implemented through a network takes on the characteristics of a highway. Highways are configured in channels where data move within the channels (serial transmission). The channels can move data in two different directions (bidirectional transmission). The highway has certain rules and regulations for priority and accuracy (protocol). When conditions are right, the flow of data may increase (bit rate).

The type of highway is also interesting to analyze. Secondary channels where data are originating or channels where data are being displayed are called *network segments.* This channel is to get the data to and from its intended source and destination. Primary channels where a large amount of data is switched and controlled are called *network buses* (also called *backbone*).

For industrial data networks, the device level is the secondary channel. Data traffic is slower and the data are programmed as to where they are going. The device level is where I/O devices are connected.

The data are moved to the control level where controllers analyze the state of the devices and compare this data to the programmed control logic, and corrections to the I/O devices are initiated. This level of the network is faster than the device level because the controller is evaluating the data strings of many I/O devices.

Periodically, the controllers send information to computers that evaluate the production data with other data such as warehouse inventory levels (once we produce it, where are we going to store it?), customer orders (how many of these items should we build?), and product transportation needs (when are we going to pick up the product and move it to the warehouse or ship it to the customer?). This level of the network is called the *information level*.

## 16.6 Network Topologies

A *topology* is the pattern of interconnections among nodes. How devices are connected to a network can influence a network's cost and performance. Industrial networks must be rugged and heavy-duty because of environmental conditions such as heat, moisture, and dust.

Networks have three types of topologies: bus, star, and ring. Figure 16-4 provides an example of the connectivity for each of the three topologies. Besides the connection arrangement, there are significant differences in the topologies.

Industrial networks are designed to fulfill the object of data distribution. If the data are distributed locally within the production process, then the bus topology shown in Figure 16-4A is widely used. Most industrial networks are well-suited for bus topology.

Some proprietary networks may use a different topology such as a star. Star has the advantage of higher data rates because of its direct wiring between the I/O devices and the controller. The controller serves as the network controller. Figure 16-4B shows all of the network devices connected to a central hub. Controller 1 serves as the network controller.

In a ring topology, Figure 16-4C, the goal is to achieve the maximum data transfer rate between the devices. In this concept the controllers are designated to be on the ring because of their processing requirements and complex data. Generally, noncontroller devices are not allowed to be on the ring.

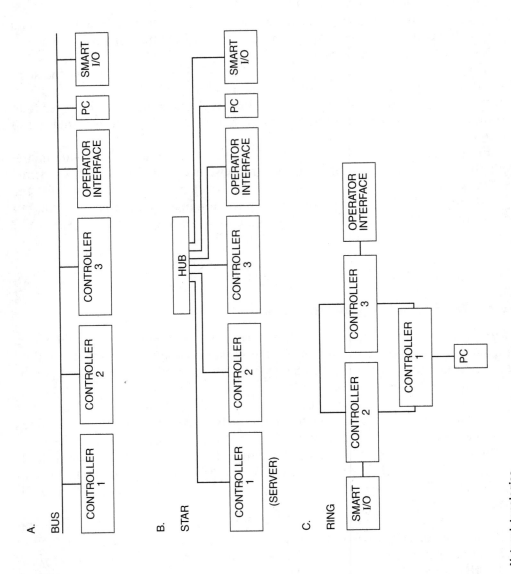

*Figure 16-4* Network topologies.

# 16.7 Industrial Networks

Networks are created to move a variety of data from many different sources to many different destinations. For distributed data applications in industry, data from one source may be needed by several controllers. The output of a controller may be sent to an actuator and as input data to another controller. The only practical solution is to have a network of switched data lines that can be controlled at the appropriate time and place to meet the needs of the process.

A *network* is defined as the interconnection of computers, controllers, and I/O devices to form a path over which to communicate and share data and information. The network circuit is made up of cables, switches, and routers. Each device on the network, called a *node*, has its own unique network address and has the ability to send and receive data on the network as directed by the controller and network software. The assigned unique address is the key to each device being able to recognized as the source or destination of the data. For many factory automation open-system architectures, the address of the device is automatically assigned as units are connected to the network.

The ability for the data to know where they are to go once they are on the network is determined by the switching method used by the network circuit. Generally, the data are formed into data strings that contain the source address, the destination address, the data value in binary coded form, and an error checking code. All of this binary data is formed into a packet that moves together through the network. The network circuit devices recognize the source and destination addresses and makes the correct circuit routes to ensure that the data are moved to the right location.

Figure 16-5 illustrates network architecture for an industrial application. To review this diagram we should look at the two major parts of the network and then focus on the details. First, there are three segments, or parts, to the network: device segment, control segment, and information segment. The purpose of the *device segment* is to move I/O values onto the network. Each I/O device will have its own network address, the source address. When the controller is programmed to identify the I/O device as a control input, the destination code is established for the data string used by the I/O device. The state of the I/O device is the data value. When the I/O is to be recognized by the controller, the network software assembles the data string and the error code is calculated and added to the data string. The *control segment* is the portion of the network that communicates with other PLC controllers and other computer-based devices such as operator terminals and personal computers. Unlike I/O data, controller data is much more complex and the data transfer rates are higher. Therefore, controllers must have network adapter cards to match the appropriate network connection. The *information segment* is considered to be the interface to the computers that are not

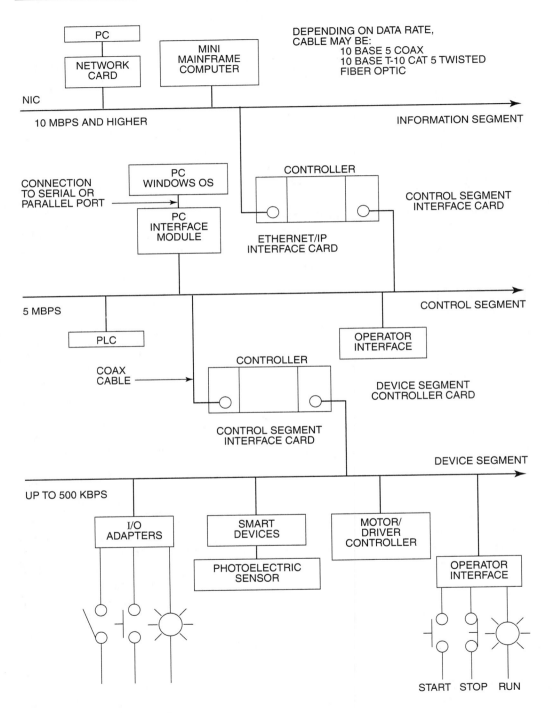

*Figure 16-5*    **Network architecture.**

process controllers. These computers can be various computers located in the distributed data factory as shown previously in Figure 16-1.

Second, each segment has a defined data transmission rate. This rate is determined by the type of cable used to connect the devices, the network protocol, and the specifications of the network interface card in the controller. In Figure 16-5, the data rate increases from the device segment to the information segment because the data string becomes more complex due to the variety of information passing through those network segments.

Finally, industrial networks use a network concept called *peer-to-peer network*. In this architecture, each controller is an equal participant on the network; there is no master network controller. In business networks such as client/server networks a master network controller is called a *server*.

## 16.8  Ethernet and the Information Highway

The Ethernet protocol is considered to be a stable and reliable protocol for common local area networks (LANs). Many business offices have adopted Ethernet as their LAN protocol. Therefore, when industrial controllers are sending information to a computer used in the business office, the network segment from the controller to the business computer uses Ethernet. Ethernet uses a bus or star topology.

For companies having several factories and the need to control the data between all factories, the network becomes a wide area network (WAN). Transmission control protocol/Internet protocol (TCP/IP) is widely used. This protocol is the standard for the Internet. Data utilizing this protocol can be transported to a server connected to the Internet.

## 16.9  Transmission Media

*Media* is the type of substance used to carry the electrical signals that represent the digital data. While telecommunications systems use exotic medial such as satellite and microwave, industrial applications typically use cable. Industrial data communication media may use three types of transmission media: twisted-pair (TP) wires, coaxial cables, and fiber-optic cables. Some applications may use unshielded twisted-pair wires. A common data media is eight wires (four twisted pairs) terminating in an RJ45 data jack. In some cases fiber-optic cable is used for high transmission rate network segments such as the network backbone.

As examples of media usage, the Ethernet standard can be implemented in five different ways:

| | |
|---|---|
| 10Base5 | Standard "thick" coaxial cable |
| 10Base2 | "Thin" coaxial cable |

| 10BaseT | Unshielded twisted pair (UTP) |
| 100BaseT | Unshielded twisted pair (UTP) |
| 10BaseFL or FOIRL | Fiber optics |

## 16.10  Data Transfer Rate

The unit of measure for network data transfer rate is bits per second (bps or b/s). Even though the rated transfer rate of a network is very high, the typical bit rate is much lower. For example, at the device level the transfer rate may be less than 1 million bps. At the control level the transfer rate may be 5 million bps (5 megabits per second or 5Mbps). At the information level, the transfer rate may be rated at 10Mbps to 100Mbps.

## 16.11  Interference

The biggest problem of industrial data communications is external electrical interference. When exposed to interference, network controllers, like all sensitive computer circuits, can lose data and create transmission errors. Voltage drops, power surges, and power outages are the most common problems. Fiber-optic cable is not affected by electromagnetic interference. Also, with the advances in microwave technology, more industrial applications are using wireless technology.

Cable routes should be planned to avoid fluorescent light fittings and power cables (exceptions can be made in the case of fiber-optic cable). They should not be run in the same conduit as power, or the same cable tray of a trunking system. Crossing power cables is allowed but it must be at right angles, and some form of bridge should be used.

## 16.12  Open Systems Versus Proprietary Systems

The electronics, which control data networks, are a mix of computer systems, some of which follow industry standards (called *open systems*) while others are *proprietary* (unique) *systems* designed by a specific company. Proprietary systems are constructed of data equipment following the same proprietary software and hardware. The benefit of a factory data communication network architecture using proprietary network equipment is that all of the computer systems within the factory can be connected to the same network and the transmission of information is controlled through specific software. However, proprietary systems are only available from a specific manufacturer.

The benefit of open-system architecture is that the data communication equipment can be purchased from several different manufacturers. However, incompatibility between equipment is a problem the technologist must solve and technicians must maintain.

## 16.13  Network Layers

When you carefully examine how data move from a series of bits coming from a sensor to a numerical display on an operator interface panel, you begin to realize that data are constantly being pushed, tested, and transformed as they move through a network. The concept of layering of information breaks down the complex process of transporting data into pieces that can be managed by single network devices.

Here is an example of network layering:

| | | |
|---|---|---|
| Layer 7 | Application | Defining the meaning of data and determining where data originates and terminates |
| Layer 6 | Presentation | Building the data string (handled by controller software) |
| Layer 5 | Session | Opening and closing communication paths (handled by controller software) |
| Layer 4 | Transport | Error checking (handled by controller software) |
| Layer 3 | Network | Determining data paths in the network (handled by controller software) |
| Layer 2 | Data Link | Managing data transmission, source, destination, and checksum (handled by controller software) |
| Layer 1 | Physical | Defining voltage levels and signal connections (handled by interface cards, repeaters, and hubs) |
| Layer 0 | Transmission | Defining physical data transport such as the type of media to use |

From this example you begin to see that industrial data communication occurs at four critical points: Layer 7, where data originates; Layers 6, 5, 4, 3, and 2, where the controller software adjusts the data string; Layer 1, where the network electronics control the voltage and signal levels; and Layer 0, where the data are actually conducted through wires. Therefore, for the technician, troubleshooting a network is reduced to four areas: (1) the sensor or actuator; (2) the controller program and operating system; (3) the network interface cards and switching electronics; and (4) the network wires and connectors.

## 16.14  Typical Network Systems

Figure 16-6 describes three network systems produced by Allen-Bradley.

| | DeviceNet Network | ControlNet Network | EtherNet/IP Network |
|---|---|---|---|
| Function | Connects low-level devices directly to plant-floor controllers without interfacing them through I/O modules | Supports transmission of time-critical data between PLC processors and I/O devices | Plant management system tie-in (i.e., material handling); configuration, data collection, and control on a single high-speed network; time-critical applications with no established schedule |
| Typical devices networked | Sensors, motor starters, drivers, PCs, push-buttons, low-end man–machine interface (MMI), bar code readers, PLC processors, and others | PLC processors, I/O chassis, man–machine interface (MMI), PCs, drives, robots | Mainframe computers, PLC processors, robots, human–machine interface (HMI), I/O and I/O adapters |
| Data transmission | Small packets; data sent as needed | Medium-size packets; data transmissions are deterministic and repeatable | Large packets; data sent regularly |
| Number of nodes (max) | 64 | 99 | No limit |
| Data transfer rate | 500, 250, or 125kpbs | 5Mbps | 10Mbps to 100Mbps |
| Architecture | Open system | Open system | Open system |
| Example functions | Control, configure, and collect data; networking sensors and actuators to a PLC or a PC to reduce field wiring and increase diagnostics | Control, configure, and collect data; PLC processor controlling remote I/O chassis, peer-to-peer messaging with other controllers using redundant media connections for time-critical applications | Control, configure, and collect data; using a single PC for data acquisition from many PLC processors, or using a single PC to program up/download multiple PLC processors for non-time-critical messaging between controllers |

**Figure 16-6** Types of Allen-Bradley network systems.

# Recommended Web Links

Students are encouraged to view the following Web sites as a supplement to the concepts presented in this textbook. Review and analyze the array of products that are available for electrical control applications. Many of these sites offer technical information that can help in converting the principles presented to practical applications. To view catalogs, your PC may require Adobe Acrobat Reader software.

General Electric
www.geindustrial.com/cwc/products?famid=10
Review: The information provided about protocol, I/O, and network requirements

Allen-Bradley
www.ab.com/networks
Review: The information provided about protocol, I/O, and network requirements

Modicon
www.modicon.com
Review. The information provided about protocol, I/O, and network requirements shown under the product selection listed as "Networks"

## Device Networking

Review: Networking Fundamentals
www.controleng.com

click on "Issue Archive"
click on "2001"
click on "May"

Look under "Back to Basics"—Article: Device Networking 101: Your First Installation

# Achievement Review

1. Describe how a distributed data factory is different from a factory that does not use computers to control its processes.

2. What is the purpose of taking an architecture approach to documenting a production process?

3. What are the three issues that are addressed by the industrial IT architecture?

4. What is the central goal of an industrial data communication architecture?

5. Describe at least three principles of data communications.

6. What is the purpose of an error checking code in the packet of information flowing through a network?

7. What are the three basic concepts all manufacturers require of their data communication systems?

8. Describe the difference between parallel and serial data transmission wiring methods.

9. Describe the difference between synchronous and asynchronous data.

10. Provide a general description of an industrial data highway.

11. Contrast and compare the three network topologies.

12. Give three reasons why Ethernet is a preferred network protocol.

13. Name the four layers of an industrial network model. Also, list typical devices responsible for the operation of each layer

# Quality Control

## Objectives

After studying this chapter, you should be able to:

- Define quality.

- Explain why quality is important.

- Describe the role of control systems in quality control.

- Design a simple alarm circuit.

- Understand the importance of control parameters and standards.

- Describe the purpose of a normal distribution chart.

- Describe the purpose of a frequency distribution chart.

- Explain the purpose of a data acquisition system.

- Provide several reasons why quality control systems can be, or become inaccurate.

## 17.1  Defining Quality and Quality Control

The issue of product quality and cost effectiveness is critical to surviving in business. Competition between companies throughout the world for limited markets have driven the need to produce products of the highest quality and at the lowest price.

*Quality* is a term that can mean many things to different people. A consumer may define quality as long lasting, maintenance-free, good appearance, and low cost compared to the benefit received. However, for the manufacturer the concern is how to meet the expectations of their customers, within a price that is acceptable by the customer, keeping in mind that the manufacturer's customer may be another manufacturer or an assembler.

As readers of this book, the question you are looking to have answered is: How are electrical controls used to maintain the quality of a product? The answer is: through the use of controls, operated by qualified personnel.

The objective of quality control systems is to utilize technology in a manner that will allow process operators to maintain the standards defined as quality by the company. Most companies strive for "zero defects." However, the nature of the company's manufacturing process (production rate, tolerance standards, level of technology being utilized, etc.) will determine whether "zero defects" is a justifiable goal.

## 17.2 Electrical and Electronic Circuits Used in Quality Control

Circuits are used to monitor the real functional tolerances required for engineering specifications. The data obtained from the monitoring process are evaluated with statistical analysis. The result of the analysis will inform quality control personnel of the production of defective parts beyond an acceptable limit. Variations in product tolerances can be attributed to: (1) changing conditions within the fabricating equipment, (2) quality of the raw materials being used, and (3) operator error.

Today, all manufacturing companies employ one of two types of sample monitoring processes: (1) where only a few parts out of the production of many are evaluated or (2) continuous quality monitoring process where every part is evaluated. As process control techniques, utilizing computer technology improves, the goal of achieving continuous (100%) quality monitoring can be economically achieved.

## 17.3 Quality Achieved Through Machine and Process Monitoring

In processes in which there is a reliance on an operator to control the actuators that influence the quality of products, alarm circuits are provided. These circuits use process sensors and visual or audible alarm devices.

Figures 17-1A and 17-1B illustrate the switching of pilot lights at designated temperature points to provide an operator with a visual representation of the process's current temperature. This temperature indicator circuit relies on the operator to be fully trained on the significance of the lights and about any actions needed when certain lights are on or off.

Figure 17-2 is an example of alarm circuit that can provide both a visual and an audible indication of the status of the float switch. Alarm horns are important in situations where the operator can be some distance from the process actuators.

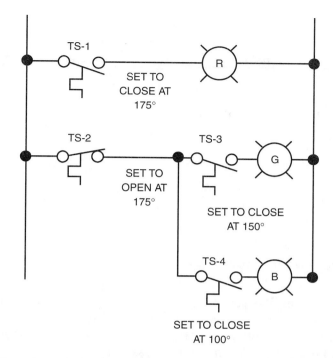

***Figure 17-1A*** **Temperature indicator circuit utilizing red (R), green (G), and blue (B) pilot lights.**

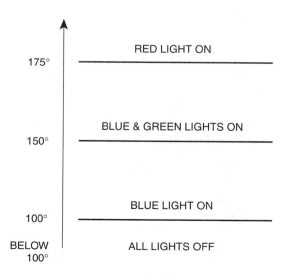

***Figure 17-1B*** **Temperature indicator pilot-light sequence.**

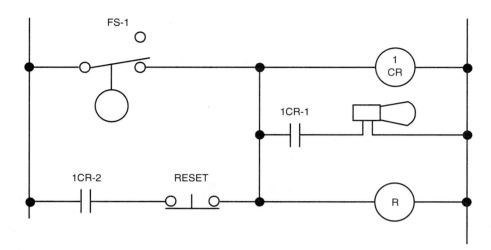

**Figure 17-2**   Fluid level alarm circuit.

Figure 17-3 is an illustration of an annunciator panel that can centralize all of the critical measurement points at one location. The operator can easily see the status of the measurement points. Process switches and analog sensors can be connected to the panels. Technicians must calibrate the analog sensors to the alarm points.

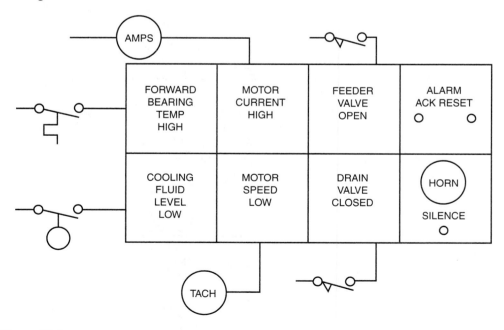

**Figure 17-3**   Annunciator alarm panel.

## 17.4 Process Tolerance (Standards)

*Tolerance* is the measurement that defines the range of values acceptable in meeting the specifications of quality. In machining processes, tolerances can be measured one part at a time or continuously. Each method requires a different application of sensors and indicator circuits.

For example, tolerances can be measured using depth gauges and calipers with digital display. These gauges can use linear displacement transducers or linear potentiometers as measurement sensors.

In order to classify the data obtained by measurement devices, people who are responsible for maintaining or assuring quality need to present the data in a format that will easily identify problems and show how well or how badly the process is operating. One method of graphing the measurements is with a *normal distribution*. Figure 17-4 illustrates how the measurements are within a "pass" range and within a defined tolerance. Also, the curve shows the range of measurements located outside the tolerance range. Distribution curves use statistical calculation of mean and standard deviation.

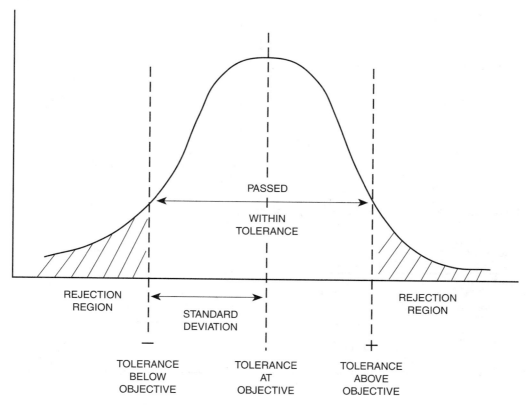

*Figure 17-4* Normal distribution graph.

There are four ways to interpret measurement readings:

1. *Mean* is the true average of the sample data readings.
2. *Median* is the middle value of all the sample data readings, in ascending order.
3. *Mode* is the value that occurs the most frequently in the sample data.
4. *Standard deviation* is the value that determines the spread between the sample data.

Other graphic techniques are:

- *Frequency distribution*. Figure 17-5 illustrates a bar graph that shows the measured values and the frequency (number of times this value was measured).
- *Control chart*. These charts are applied to continuous measurement tolerances. Figure 17-6 shows this very important graph that identifies the

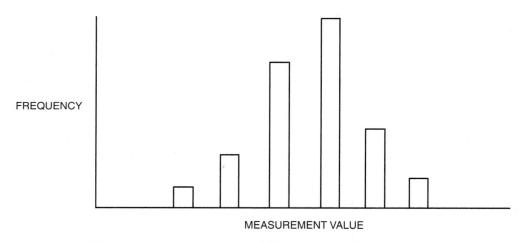

*Figure 17-5*   **Frequency distribution bar graph.**

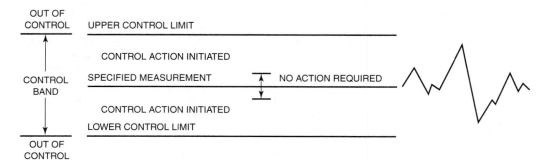

*Figure 17-6*   **Control chart with limits.**

points in the process measurements where a control action is required. Quality is maintained when measurements are within the "specified measurement" region of the graph.

## 17.5 Information Systems

In order to control according to process standards, data must be accumulated to verify that the operating process is indeed outside the limits of the standards. For many industries, the use of a data acquisition system providing data to a personal computer has become an inexpensive yet accurate method for providing information on the quality of a process.

**17.6.1 Data Acquisition Systems**   A data acquisitions system is generally a self-contained circuit that allows for the direct connection of process sensors. The output is a multiplexed digital signal. Figure 17-7 illustrates the general block diagram of a data acquisition system. Many manufacturers have developed circuits that plug directly into a PC card slot.

The manufacturer's specifications usually define the type of sensors that can be connected to the input channels, the sampling rate of that channel, and the bit resolution of the analog to digital converter. The acquisition system has the ability to receive the measurement parameters from an operator using a personal computer. Generally, manufacturers of acquisition systems will develop PC software to allow the control of the acquisition system.

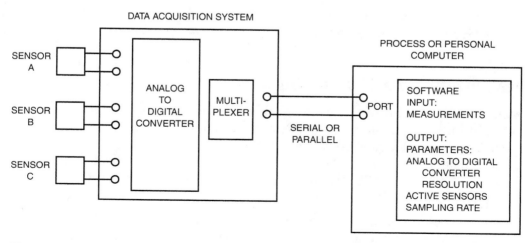

**Figure 17-7**   **General block diagram of process measurement and control system**

**17.6.2 Personal Computer Software**   As the personal computer receives the data from the acquisition system in sequence with the PC's clock, the manufacturer's software will store, print, or display the data. The software is generally adequate for basic quality analysis. However, for advanced statistical analysis, the data files will need to be analyzed by other software such as a statistical analysis software or an electronic spreadsheet. Graphics software can be used to display the data in many graphic forms, including 3-D in multiple colors.

## 17.7  Maintaining Quality

Maintaining quality is the ability to correct any process or machine errors quickly. Therefore, quality is related to the type of control being used. The application of closed-loop control (see Chapter 6) to manufacturing systems is used extensively to keep parts within quality tolerances because tolerances ensure better fit with other parts resulting in less maintenance and longer product life.

Sensors are the eyes of the control system. Without working sensors, the ability of a closed-loop system to correct errors is lost. Therefore, good maintenance must be practiced. The following is a list of commonly found problems that cause quality control systems to become inaccurate:

- Sensors not calibrated properly
- Worn-out parts (i.e., leaking pressure lines)
- Sensors not properly bonded to the process
- Loss of power or intermittent power
- Voltage differential between ground points (faulty grounds)
- Incorrect closed-loop parameters
- Operators not being observant—reliance on open-loop control
- General lack of maintenance

## Recommended Web Links

Students are encouraged to view the following Web sites as a supplement to the concepts presented in this textbook. Review and analyze the array of products that are available for electrical control applications. Many of these sites offer technical information that can help in converting the principles to practical applications. To view catalogs, your PC may require Adobe Acrobat Reader software.

NDC
www.korins.com/m/ndc/products/products.html
Review: Sensors, controllers, and analyzers applied to industrial quality control

Mettler Toledo
www.mt.com/home/products/en/ind/sqc/sqc.asp
Review: The principles and applications of quality control

InfinityQS
www.infinityqs.com
Review: The applications of and articles (pdf) about quality control

# Achievement Review

1. Interview a working member of your family, a neighbor, or a supervisor, and develop reasons why quality is important to their company.

2. With three or four other students in your group, provide a list of criteria that define the quality of a product. You can use common products such as stereos, automobiles, computers, etc.

3. Develop a list of four sensors that can be used to monitor the quality of a product. Each sensor should relate to the criteria that defines the quality.

4. Design an alarm circuit that will sound a horn and activate an alarm light when the fluid level in a tank is too low.

5. Explain the role of the analog to digital converter in a data acquisition system.

6. Develop a list of five reasons why quality control systems can be, or become, inaccurate. From these reasons, develop a way to correct the problem.

7. What is the objective of quality control circuits?

8. What mathematical method of analysis is used to determine the level of quality being processed?

9. State two causes for variations in product tolerances.

10. Is "zero defects" a realistic goal? Why?

# Chapter 18 Safety

## Objectives

After studying this chapter, you should be able to:

- Explain the problems associated with unsafe worker areas.
- Describe three ways workers can be protected from moving machinery utilizing electrical controls.
- Name two organizations that publish electrical safety literature.
- Describe two advantages for utilizing a programmable controller to monitor industrial safety.
- Diagram a motor control circuit with multiple stop switches.
- Explain why machine safety and protection should be a concern of a company.
- Describe three types of sensors used in machine safety applications.
- Identify the sequence of steps within a control program to recover from a machine malfunction.

## 18.1 Worker Safety

Worker safety is a major concern for businesses, particularly those engaged in the production of industrial machines and manufactured products. Besides personal injury to workers, the direct and indirect costs of accidents on the job can be a large financial loss for a company. Therefore, it is very important for companies to incorporate monitoring and shielding devices within their machines that will guard workers from mechanisms that could cause bodily harm. Approximately one-third of all general industry citations handed out by the Occupational Safety and Health Administration (OSHA) point to the lack of appropriate machine guarding.

The first level of operator safety is to stop the motion of any moving mechanism when any portion of the worker's body enters into a designated area near the mechanism where there is the possibility of injury. This level of safety can be achieved with a switched device located in the stop leg of the motor control circuit (as shown in Figure 18-1).

Devices used for safety circuits include:

- Ultrasonic switches
- Photoelectric switches
- Limit switches
- Pressure switches
- Proximity switches
- Push-buttons with oversized activators

The type of switched device will depend on the manner of activation. For example, when operators place their hands in close proximity to a moving mechanism, they could:

- Break the light beam of a photoelectric switch
- Move a gate that activates a limit switch
- Move from a foot pad that activates a pressure switch

There are two methods for utilizing electrical controls in safety circuits:

1. Circuits that will activate a guard to prohibit the operator from reaching into the danger zone at any time (machine stopped or running).
2. Circuits that will immediately stop the machine when any portion of the operator's body is in the danger zone.

There are two types of safety control techniques:

1. Multiple stop devices in series
2. Redundant stop devices in case of failure of one device

Safety circuits are usually designed specifically for the shape and geometry of the machine, where the operator is relation to the machine, the speed of the mechanism, and other factors that can sometimes only be determined after watching how the operator moves and works. Therefore, safety is a continuous process constantly being refined and improved.

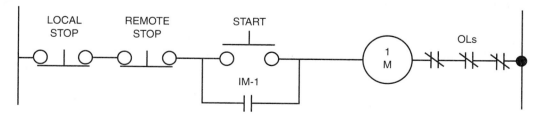

***Figure 18-1*** **Motor control circuit with safety switch.**

Safety should also be considered at points physically away from the actual mechanism but from which the movement of the mechanism can be affected, for example, the circuit breakers that control the electrical power to the machine and the push-button controls for the machine. When maintenance personnel are working within the danger area of the machine, the power should always be disconnected and the control device locked with a key. This key to unlock and reactivate the power should usually be held by a person responsible for the safety of the workers in the area.

Under the law, employers are obligated to provide their employees with a safe place to work and rules for them to perform their work safely. The federal government enacted the Occupational Safety and Health Act, which created OSHA as an agency within the Department of Commerce. This group is responsible for monitoring the safety of workers and prosecuting any company violating the policies of the Agency. Besides OSHA, many other agencies are involved in worker safety, including the following:

- Insurance companies
- Industry-specific agencies
- Local and state government agencies

A person employed in a position requiring work with electrical switching, control devices, and circuits should review the *National Electrical Safety Code®*, the National Electrical Code®, American Society of Testing and Materials, and other key publications. Other organizations that review and support electrical safety are The American National Safety Institute (ANSI), The Institute of Electrical and Electronics Engineers (IEEE), and American Society of Safety Engineers (ASSE).

Safety should be a topic that is reviewed frequently by management, and discussed and practiced constantly by workers and technicians.

## 18.2 Machine Safety

Machines that have moving parts that extend beyond their base are capable of damaging other machines. For example, in complex manufacturing work cell designs, many independent machines move in and out along controlled paths that intersect the paths of other machines. If any one of the machines stopped or malfunctioned, others could crash or become entangled with the defective one. This type of industrial accident would not involve an employee, but damage to expensive equipment could stop a company's entire production line. Articulated machines such as robots and coordinated machines such as machining centers need protection from themselves and other machines. For example, in machining cells, robot arms frequently pass through the line of travel of adjacent machines.

Machine safety in a coordinated and programmed manufacturing system involves the use of detectors and programmed safety sequences. Examples of common detectors are:

- Zero speed switches on motors for detecting stopped motors
- Overtorque switches for detecting jammed parts
- Proximity switches to detect overtravel
- Flow switches to detect loss of lubricating oil
- Pressure switches to detect jammed hydraulic actuators
- Temperature switches to detect overheating

Safety sequences are difficult to implement because each machine is designed to work independently. Therefore, when they are placed in a coordinated work process with other machines, a master controller is needed. This master controller is not only responsible for the correct sequencing of each machine in the process pattern, but also the recognition of a problem, the determination of the severity of the problem, and the correct action to be taken to ensure machine safety. The corrective action is in the form of a programmed sequence of moves given to each machine in the process. The sequence would involve:

1. Stop all machines.
2. Evaluate the current positions of each machine.
3. Begin to retract each machine to a home position.
4. Notify the operator of the detected problem.
5. Identify the defective machine and the source of the problem.
6. Wait for a restart command from the operator.

# 18.3  Diagnostic Systems

Diagnostic systems are used for the detection of faults with a manufacturing process and appropriately notifying the machine operator. In this system, the operator has the responsibility of identifying the problem and making any corrective actions. For example, a diagnostic system could detect a pump malfunction through the detection of loss of pressure in a tank.

More advanced diagnostic systems will store the number of times the fault event has taken place and what were the conditions when the event occurred. From this type of information, maintenance personnel can better predict when fault events may take place and take corrective action before any damage to the process can occur or poor quality material is produced.

# 18.4  Machine Safety Circuit

As an illustration of machine safety, Figure 18-2 shows a circuit designed to monitor the pressure across hydraulic fluid filters to detect and warn operators of

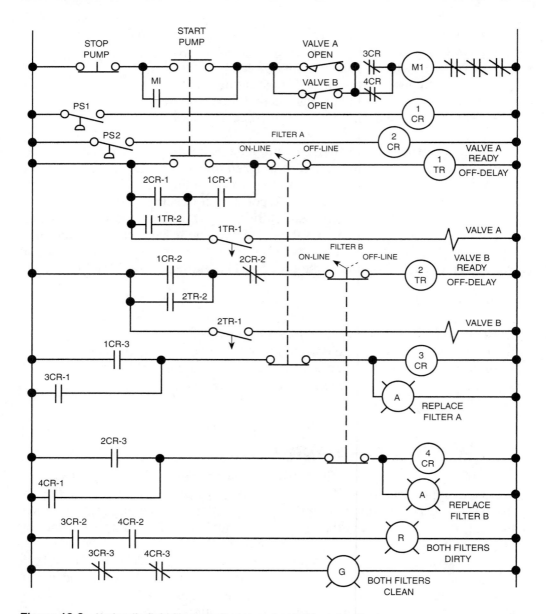

*Figure 18-2* **Hydraulic fluid filter monitoring and switching circuit.**

clogged filters, automatically switch to a clean filter, and shut down the electrical motor driving the hydraulic pump. The circuit operates as follows.

Valves A and B are solenoid valves with a position switch. Pressure switches PS1 and PS2 are differential pressure switches that activate when the pressure across the filters increases due to a clogged filter. When the pump motor is started, a secondary contact will actuate 1TR, which latches through the instantaneous contact 1TR-2 and 1CR-1. Then 1TD-1 closes immediately and stays closed for a short period of time—just enough time to switch filters without shutting down the motor.

When a filter gets dirty, the differential pressure across the switch goes up and associated relay contacts deactivate one solenoid valve while activating the other. The operator is alarmed. By taking the filter off-line, the opposite filter carries the load of the fluid.

Should both filters become clogged, valve A and valve B close, thereby alarming the operator and shutting down the pump motor. Additional contacts of 3CR and 4CR will ensure complete shutdown.

## 18.5 Programmable Controllers in Safety

Programmable controllers have been used to monitor safety. As an intelligent device acting as a "Safety Engineer," the controller can be programmed to monitor machine as well as human safety. However, the controller must have the appropriate detection devices strategically located to detect a safety violation or hazardous condition. The benefits of the safety controller are:

- Immediate response to a change in conditions
- A stored "preplanned action" to the response
- Notification to appropriate personnel
- Maintenance of an on-going log of safety violations
- Evaluation of conditions prior to a problem to identify the actual fault that created the problem

Figure 18-3 illustrates the principle of disabling the power to a programmable controller's input, output, and processor modules. The power can be disabled by the operator or by a PLC fault. Alarm horns are effective in gaining an operator's attention.

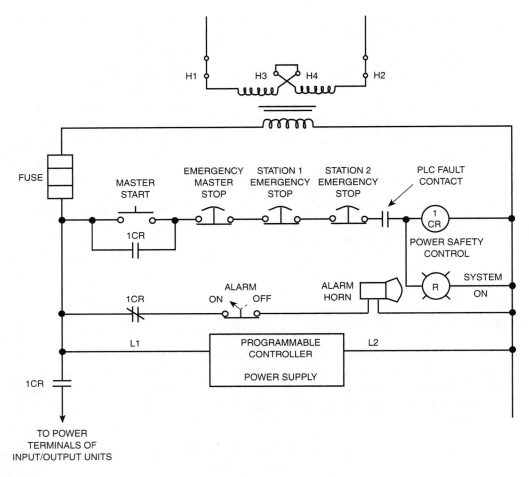

*Figure 18-3*   PLC power control circuit.

# 18.6 Other Safety Conditions

Many industries employ processes in operating environments conducive to explosion and fire, and that are very wet. Therefore, electrical sensors and actuators must be utilized that have been designed and certified to operate within these environments. Only devices meeting or exceeding environmental specifications should be used; for example:

- Intrinsically safe devices in explosive environments
- Electrically safe devices used underwater
- Ground fault sensors
- Lightning arrestors

# Recommended Web Links

Students are encouraged to view the following Web sites as a supplement to the concepts presented in this textbook. Review and analyze the array of products that are available for electrical control applications. Many of these sites offer technical information that can help in converting the principles to practical applications. To view catalogs, your PC may require Adobe Acrobat Reader software.

Honeywell
content.honeywell.com/sensing/prodinfo/safetysensors/
Review: Sensors and actuators applied to machine/human safety

Pinnacle Systems, Inc.
www.pinnaclesystems.com
Review: Sensors and actuators applied to machine/human safety

# Achievement Review

1. With a group of three or four other students, travel through your school and observe any areas where student safety can be at risk.

2. Design a motor control safety circuit with two operator stop stitches, a bearing overheating thermal switch, and zero speed switch.

3. Go to your local public library and obtain a copy of the *National Electrical Code®*. Describe four points where electrical safety is emphasized.

4. Should or should not programmable controllers provide all safety detection and alarm?

5. Make an appointment with a manager of a local manufacturing company and ask him/her why safety is important to the company and what is done to ensure safety.

6. Describe three types of sensors used in machine safety applications and provide examples of how they could be used.

7. Go to your school's machine shop and prepare a "safety manual" for people who will be using the machines.

8. What is the advantage of storing the number of times a fault event has taken place and what the conditions were when the event occurred?

9. Provide three advantages for a "safety controller."

10. Besides OSHA, what other agencies are involved in matters related to worker safety?

# Chapter 19 Troubleshooting

## Objectives

After studying this chapter, you should be able to:

- List five areas that should be considered when starting a troubleshooting job.
- Describe two methods that can be used in troubleshooting a job.
- Explain a procedure for checking fuses.
- Discuss why a loose connection in an electrical power circuit can be a major problem.
- Explain why contacts on electrical operating equipment should not be filed.
- List several problems resulting from low line voltage.
- Discuss the merits of good housekeeping.
- Show how an open in a common line can result in a faulty circuit operation.
- Explain one method for checking an electrical control circuit.
- List four steps of the procedure to locate problems resulting from momentary faults.
- Explain the relationship among voltage, current, and resistance or impedance expressed in Ohm's law.
- List several problems that can occur in an electric motor.
- Explain some of the problems in a PC that can be checked with a meter.

## 19.1 Safety First*

Testing, care, and maintenance of electrical equipment should be performed by trained and experienced electricians who are thoroughly knowledgeable about electrical systems. High voltages and currents are present that may cause injury or extensive equipment damage.

*Reproduced with permission from the Fluke Corporation.

All possible precautions that must be taken cannot be anticipated in testing all the different equipment. Always be certain that all power has been turned off, locked out, and tagged in any situation in which you must actually come in contact with the circuit or equipment. Be sure the equipment cannot be turned on by anyone but you.

Use only well-designed and well-maintained equipment to test, repair, and maintain electrical systems and equipment. Use appropriate safety equipment, such as safety glasses, insulating gloves, flash suits, hard hats, insulating mats, etc., when working on electrical circuits.

Make sure that multimeters used for working on power circuits contain adequate protection on all inputs, including fuse protection on *all* current measurement input jacks.

## 19.2  Analyzing the Problem

Effective troubleshooting starts with an analysis of the problem. Too frequently, troubleshooting is approached in a hit-or-miss fashion. This approach generally creates more expense and wastes time.

To analyze any problem, it helps to break it down into types of sections to limit the size of the job. In troubleshooting, the following causal areas might be considered:

- Electrical, electronic
- Mechanical
- Fluid power
- Pneumatic
- Personnel

In many cases, the problem may be a combination of two or more of these areas.

The following three examples illustrate how breaking down a problem into types of causes can simplify the troubleshooting procedure.

1. From advance information supplied by a machine operator or from early examination by an electrician, what first appears to be an electrical problem may turn out to be a mechanical one.
2. The failure of a limit switch to function properly may be caused by problems in the electrical contacts or the mechanical operator. The result of a cycle failing to complete is the same, however, regardless of which of these two items caused the problem.
3. As long as people operate machines, problems will arise that do not respond to the usual form of troubleshooting. Such problems may be intentional or unintentional. They may stem from misunderstanding, lack of cooperation, or lack of knowledge of the machine.

Whatever the cause of a problem, it will be recognized quickly by the troubleshooter who takes the approach outlined here. The problem should be handled carefully and diplomatically so that the machine can be returned quickly to its intended job.

Problems can be further separated into physical location or type of operation. For example, in a large machine in which the cycle of operation moves from one section of the machine to another, the trouble may be localized in one section. Localizing the area of the problem may immediately eliminate 75% of the total machine as a possible trouble source. A practical application here could be trouble developing in loading or unloading equipment on a press. If the proper clearance signal has been given to the loader or unloader from the press control, trouble developing after this cycle starts can generally be localized in the loading or unloading control.

Success in troubleshooting is the ability to segregate the problem area from other unrelated circuitry.

Before getting into the mechanics of troubleshooting, let us examine some of the problem spots.

## 19.3  Major Trouble Spots

It would be impractical, if not impossible, to list all potential trouble spots. However, the following areas contribute to a large percentage of troubles.

**19.3.1  Fuses**    Checking fuses is generally a good place to start when a problem has occurred. Too often this is overlooked. The details for checking fuses in a complete power circuit is covered in Section 19.6.

The replacement of an open (defective) fuse can be an important safety factor. As discussed in Chapter 2, there are three different types of fuses. Within each type there are different voltage and current ratings. Too often just any type of fuse is used as a replacement. Unless changes have been made in the machine circuit and components, the replacement fuse should be exactly the same type, voltage, and current rating as the fuse removed.

The policing of fuses can be a problem. However, the replacement policy given here must be rigidly maintained if safety to personnel and the machine is to be realized.

One extremely important case involves a machine connected to a power source that has a high short-circuit current available. In such cases, it may be advisable to use current-limiting fuses with high interrupting capacity.

**19.3.2  Loose Connections**    There may be hundreds of connections on a machine. Each of these spots may be a source of trouble. Many advancements have been made in terminal block and component connectors to improve this condition. The use of stranded conductors in place of solid conductors has, in general, improved the connection problem.

The problem starts when the machine is built and continues throughout the life of the machine. It may be of greater importance in power circuits, as the current handled is of greater magnitude. A loose connection in a power circuit can generate local heat. This heat spreads to other parts of the same component, other components, or conductors. An example of where direct trouble can arise is thermally sensitive elements. These can be overload relays or thermally operated circuit breakers.

For the correction of loose connections, the best advice is to follow a good program of preventive maintenance in which connections are periodically checked and tightened.

**19.3.3 Faulty Contacts**   Potential problems with faulty contacts exist in such components as motor starters, contactors, relays, push-buttons, and switches.

A problem that appears quite often and one of the most difficult to locate is the normally closed contact. Observation indicates that the contact is closed but does not reveal if it is conducting current.

Any contact that has had an overload through it should be checked for welding.

Such conditions as weak contact pressure and dirt or an oxide film on contact will prevent it from conducting. Many times contacts can be cleaned by drawing a piece of rough paper between the contacts. **Caution:** Use only a fine abrasive to clean contacts. Do not file contacts. Most contacts have a silver plate over the copper. If this plating is destroyed by filing, the contact will have a short life. If contacts are worn or pitted so badly that a fine abrasive will not clean them, it is better to change the contacts.

Another problem that may occur with a double-pole, double-break contact is cross-firing, that is, one contact of the double break travels across to the opposite contact, but the other remains in its original position. If both the NO and NC contacts are being used in the circuit, a malfunction of control may occur.

**19.3.4 Incorrect Wire Markers**   The problem of incorrect wire markers usually appears on the builder's assembly floor or in reassembly in the user's plant. The error can be difficult to locate, as a cable may have many conductors running some distance to various parts of the machine.

One common problem is the transposition of numbers. For example, a conductor may have a 69 marked on one end and a 96 on the other end. Another problem that may occur is in connecting conductors into a terminal block. With a long block and many conductors, it is a common error to connect a conductor either one block above or below the proper position.

**19.3.5 Combination Problems**   Reference has been made to combination problems, but their importance should be emphasized. The following are typical types of combination problems:

- Electrical-mechanical
- Electrical-pressure (fluid power or pneumatic)
- Electrical-temperature

The greatest problem is that the observed or reported trouble is not always indicative of which aspect of the combination is at fault. It may be both.

It is usually faster to check the electrical circuit first. However, both systems must be checked as both may contribute to the problem.

As an example, very few solenoid coils burn out due to a defect in the coil. Probably over 90% of all solenoid trouble on valves develops from a faulty mechanical or pressure condition that prevents the solenoid plunger from seating properly, thus drawing excessive current. The result is an overload or a burned-out solenoid coil.

**19.3.6 Low Voltage**    If no immediate indication of trouble is apparent, one of the first checks to make is the line and control voltage. Due to inadequate power supply or conductor size, low voltage can be a problem.

The problem generally shows up more on starting or energizing a component, such as a motor starter or solenoid. However, it can cause trouble at other spots in the cycle.

A common practice in small shops is the addition of more machines without properly checking the power supply (line transformers) or the line conductors. The source and line become so heavily loaded that when they are called on for a normal temporary machine overload, the voltage drops off rapidly. This drop may result in magnetic devices such as starters and relays dropping off the line (opening their contacts) through undervoltage or overload protective devices.

Heat is one result of low voltage that may not be noticed immediately in the functioning of a machine. As the voltage drops, the current to a given load increases, producing heat in the coils of the components (motor starters, relays, solenoids), which not only shortens the life of the components but may cause malfunctioning. For example, where there are moving metal parts with close tolerances, heat can cause these parts to expand to a point of sticking. In cases in which electrical heating is used, the heat is reduced by the square of the voltage. For example, if the voltage is dropped to one-half of the heating element's rated voltage, the heat output will be reduced to one-fourth.

**19.3.7  Grounds**

**19.3.7.1 Typical Locations.**    There are many locations on a machine where a grounded condition can occur. However, following are a few spots in which grounds occur most often.

- *Connection points in solenoid valves, limit switches, and pressure switches.* Due to the design of many components, the space allowed for conductor entrance and connection is limited. As a result, a part of a bare conductor may be against the side of an uninsulated component case. Where bolt and nut connections are made, the insulating tape may not be wrapped securely, or it may be of such quality that age destroys its insulating properties. This problem may occur in the field (user's plant), where due to the urgency for a quick change, sufficient care is not taken when handling the wiring and connections on a replaced unit.

- *Pulling conductors.* In pulling conductors through conduit where there are several bends and 90-degree fittings, or into pull boxes and cabinets, the conductor insulation may be scraped or cut. If care is not taken to eliminate sharp edges or burrs on freshly cut or machined parts, cuts and abrasions occur. Avoidance of scrapes and cuts is one good reason for the use of the insulation required for machine tool wiring.

- *Loose strands.* The use of stranded wire has greatly reduced many problems in machine wiring. However, care must be taken when placing a stranded conductor into a connector. All strands must be used. One or two strands unconnected can touch the case or a normally grounded conductor, creating an unwanted ground. Even if the ground condition does not appear, the current-carrying capacity of the conductor is reduced.

**19.3.7.2 Means of Detection.** For the best operation of an electrical system, some means of detecting the presence of grounds should be available. There are two methods shown here. Each has merits.

Using the grounded method shown in Figure 19-1, the circuit is deenergized by the opening of the fuse. This condition means that the machine will be down until the ground is located and removed. In the small shop this inconvenience is usually not too serious. The ungrounded method shown in Figure 19-2 is usually used in large production shops. This method has the advantage that production can continue with one ground until there is time to locate the ground and remove it.

In the first method, all coils are tied solidly to a common line, and this line is grounded. The opposite side of the control power source is protected by a fuse or circuit breaker.

If a ground condition should appear in the circuit shown in Figure 19-1 on circuit point 8 between contact 1CR-2 and solenoid A, the load (coil) is bypassed because there is a direct path to ground on both sides (relay 1 CR energized—1CR-2 closed). This direct path to ground puts a direct short on the control power source, opening the protective device and removing the control power from the circuit.

When the ground is located and removed, the fuse can be replaced or the circuit breaker reset. Thus, the control circuit is again ready for operation. Note that the ground condition must be removed first.

In Figure 19-2, the common side is not grounded. A set of two ground detector lights (standard 120-V indicating lights) are connected in series across the power source. A solid ground is then placed between the two lights.

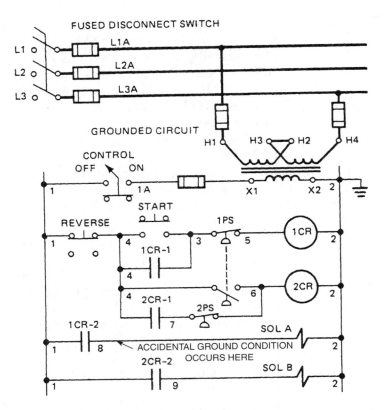

**Figure 19-1** Control circuit for potential gound at circuit point 8. Conductors carrying load current at line voltage are denoted by heavy lines.

With the system showing no grounds, both lights will glow at half brilliance, since two 120-V bulbs are connected in series across 120 V (60 V on each).

If a ground appears in the system, one of the lights will go out. The other will go to full brilliance. The determining factor of how these two lights perform depends on which side of the load the ground appears. For example, if a ground should appear on circuit line 16 between the fuse and solenoid A, lamp A will go out when contact 3CR-1 closes. Lamp B will glow at full brilliance.

An advantage of this method is that in some cases a ground can be present but not cause any immediate trouble. A circuit that keeps ground faults from disabling a machine gives maintenance time to check for the ground without immediately taking the machine out of production.

The important point is that some system should be used to detect the presence of a ground.

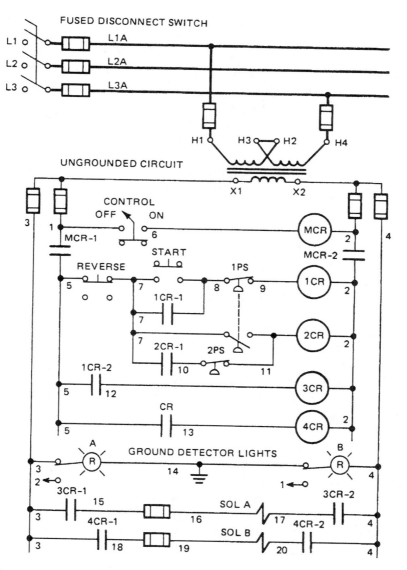

**Figure 19-2** Typical control circuit for ground fault detection and solenoid protection. Conductors carrying load current at line voltage are denoted by heavy lines.

### 19.3.7.3 Wiring and Grounding*.

Incorrect wiring and grounding can result in major problems for companies with computers and computer controlled equipment. These problems often show up as unexpected shutdowns, burned-out equipment, scrambled data, and equipment lockups.

*Courtesy of the Fluke Corporation.

The Electric Power Research Institute estimates that 80% of all power quality problems are the result of poor grounding, wiring, or load placement on the user's circuits. By making a few simple checks with your digital multimeter (DMM) you may be able to eliminate some common power quality problems caused by wiring and grounding.

It is recommended that all circuits that serve critical loads should be on dedicated circuits and neutrals. Check wiring for correct bonding, tied neutrals and grounds, overloaded neutrals, and shared neutrals to critical loads and to ensure that the neutral is not used for equipment grounding.

A visual check of wire condition may be all that is necessary to eliminate many problems associated with poor equipment performance.

**19.3.8 Poor Housekeeping**   Poor housekeeping leads to more work for the troubleshooter. There is an overall economy in having a clean machine and a well-organized and well-executed preventive maintenance program.

Dust, dirt, and grease should be removed periodically from electrical parts. Their presence causes mechanical failure and forms paths between points of different potential, causing a short circuit.

Moving mechanical parts should be checked, particularly in large motor starters. Such items as loose pins and bolts and wearing parts are sources of trouble.

Overheated parts generally indicate trouble. Without proper instruments, it is difficult to determine the temperature of a part or how high a temperature can be sustained. Certainly any signs of smoke or baking of insulation are cause for immediate concern.

Manufacturers of components have done considerable design work to prevent dust, dirt, and fluids from entering.

When it is necessary to remove a cover or open a door for troubleshooting, immediately replace it after the trouble is corrected.

Many users have gone to great lengths to develop and rigidly enforce a good electrical maintenance program. Records are kept of each reported trouble and the work is done to correct the problem. These records are compiled periodically and are available to the supervisor. Such records not only lead to faster troubleshooting in the future, but also give the supervisor an indication of why the production in a given department may be down.

**19.3.9 Trouble Patterns**   As troubleshooting work progresses over a period of time with a particular machine or group of machines, a pattern of trouble may develop. For example, it may be necessary to increase the production rate with a particular machine. Certain areas of this machine may not have been originally designed to handle an increased rate of operation. Unless the machine is redesigned and rebuilt, a pattern of trouble may develop.

Another example is a machine that is relocated to another section of the user's plant where the environment may be different. A change in the atmosphere and a presence of dust, dirt, or metal chips may create a pattern of trouble for a machine if it is not designed to operate under these conditions.

**19.3.10 Opens in a Common Line**   Many circuits have two or more common connection points for multiple connections. For example, from the elementary diagram shown in Figure 19-3, line 1 has two points of connection. Line 2 has five points of connection. Line 4 has six points of connection. As it is not good wiring practice to place more than two conductors under any one terminal, connections are generally jumpered on the panel components or brought back to the terminal block. It is not unusual for a condition to arise in which a jumper is omitted. In Figure 19-3, a jumper from circuit connection #4 is omitted from relay 2CR to circuit connection #4 on either relay 1CR or to connection point #4 on the terminal block. Also, the common coil connection on relay coil 4CR is not jumpered to relay coils 1CR, 2CR, or 3CR, or brought back to the terminal block. The schematic circuit is shown as it would appear with these two jumpers missing to illustrate the effect on the condition of the circuit and the resulting faulty operation.

In this rather simple circuit, it does not appear that this error would be committed. However, there are cases in which there may be many of these common connections. In larger and more complicated circuit diagrams, the probability of jumpers being missed is much greater.

**19.3.11 Wiring the Wrong Contact**   Many components such as relays, limit switches, and pressure and temperature switches have NO and NC contacts available for use.

A wiring error is often made, particularly when only one of the two available contacts is used. The error consists of wiring the wrong side of the contact; that is, the NO contact may be put into the circuit where the NC should be used. The reverse of this situation may also be true.

Unless the person who does the wiring is completely familiar with the component and double-checks the work, this error will occur frequently.

The routine checks discussed in the following section will reveal this error quickly.

**19.3.12 Momentary Faults**   The momentary fault is one of the most difficult problems to troubleshoot. With this type of fault, the machine or control can be under close observation for hours with no failures. However, the fault may occur at any time, and if the observer's attention is even briefly diverted, any direct evidence that might have been seen is lost.

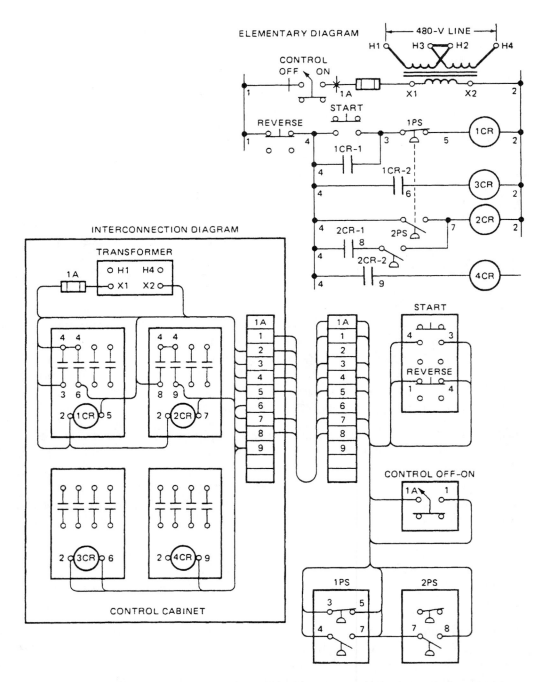

**Figure 19-3** Typical control circuit and its equivalent control panel, operator, and sensor wiring scheme. Conductors carrying load current at line voltage are denoted by heavy lines.

There is no direct solution to this problem. The best approach is a well-organized analysis. The following steps might be helpful:

1. Attempt to localize within the total cycle. If the fault always occurs at the same place in the cycle, generally only the control associated with that part of the cycle is involved. If the fault occurs at random spots through the cycle, then the spots to examine are those that are common through the entire cycle. Examples are a drum or selector switch used to isolate an entire section of control.

2. Examine for loose connections, particularly in the area of the fault. Attention should be paid to areas where mechanical action may have damaged conductors, pulled them loose from connectors, or cut or broken them. The complete break of a stranded conductor within the insulation is rare.

3. Localize attention to components. Many times casual observation will not disclose the trouble. In these cases the complete replacement of the component(s) in question is the quickest and best solution. Here, the plug-in components have a distinct advantage in returning a machine to operating condition in a minimum of time.

4. Examine the circuit for unusual conditions. This type of trouble rarely occurs in the user's plant. However, there are cases in which, either through an oversight on the part of the circuit designer or by a change of operating conditions on the machine, a circuit change is indicated as a solution to the problem.

## 19.4 Equipment for Troubleshooting

Some of the most important tools for a troubleshooter include:

- A working knowledge of Ohm's law.
- Meters that can be used to measure/indicate the important units involved in Ohm's law.
- Meters that can be used to measure/indicate temperature.
- Safe and applicable test probes.

Figure 19-4 outlines the relationship among important areas of electrical work: voltage, current, resistance or impedance, and power.

Control circuits, relay or PLC, should have visual indicators to identify logic and actuator status. They may be LEDs, incandescent lamps, or neon lamps. Their operation supplies important information to the troubleshooter. However, there is still important information needed when a system or machine fails to operate properly. This information may include voltage (ac or dc), current in a given line, and resistance/impedance in a given circuit.

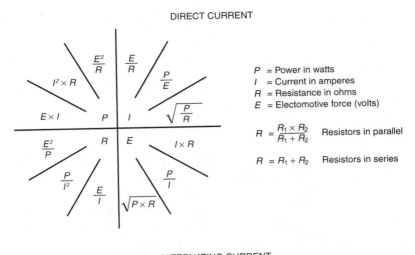

DIRECT CURRENT

$\frac{E^2}{R}$  |  $\frac{E}{R}$

$I^2 \times R$    $\frac{P}{E}$

$E \times I$    $P$  |  $I$    $\sqrt{\frac{P}{R}}$

$\frac{E^2}{P}$    $R$  |  $E$    $I \times R$

$\frac{P}{I^2}$    $\frac{P}{I}$

$\frac{E}{I}$    $\sqrt{P \times R}$

$P$ = Power in watts
$I$ = Current in amperes
$R$ = Resistance in ohms
$E$ = Electomotive force (volts)

$R = \frac{R_1 \times R_2}{R_1 + R_2}$    Resistors in parallel

$R = R_1 + R_2$    Resistors in series

ALTERNATING CURRENT

$\frac{E^2}{Z}$  |  $\frac{E}{Z}$

$I^2 \times R$    $\frac{P}{E}$

$E \times I$    $P$  |  $I$    $\sqrt{\frac{P}{Z}}$

$\frac{E^2}{P}$    $Z$  |  $E$    $I \times Z$

$\frac{P}{I^2}$    $\frac{P}{I}$

$\frac{E}{I}$    $\sqrt{P \times Z}$

$f = \frac{1}{2\pi LC}$ OR $\frac{1}{2\pi C X_C}$

$X_L = 2\pi f L$

$X_C = \frac{1}{2\pi f C}$

$L = \frac{X_L}{2\pi f}$

$C = \frac{1}{2\pi f X_C}$

$Z = \sqrt{R^2 + X^2}$ OR $\sqrt{R^2 + (X_L - X_C)^2}$

$f$ = Frequency in hertz
$C$ = Capacitance in farads
$L$ = Inductance in henrys
$Z$ = Impedance in ohms
$Z$ = $R$ when $X_L = X_C$
$Z_L$ = Inductive reactance in ohms
$Z_C$ = Capacitive reactance in ohms

**Figure 19-4**  Ohm's law.

For example, an open circuit will show infinite resistance, and a closed circuit will show zero resistance. Figures 19-5, 19-6, and 19-7 show some of the latest instruments available for troubleshooting. Figures 19-8 and 19-9 show examples of well-designed test probes.

**19.4.1 Multimeters\***  One of the most basic tasks of a DMM is measuring voltage. A typical dc voltage source is a battery (like the one used in your car); ac voltage is usually created by a generator. The wall outlets in your home are

\*Reproduced with permission from the Fluke Corporation.

**Figure 19-5** Multimeter. *(Reproduced with permission from the Fluke Corp.)*

**Figure 19-6** Clamp-on ac/dc current probe. *(Reproduced with permission from the Fluke Corp.)*

common sources for ac voltage. Some devices convert ac to dc; for example, electronic equipment such as TVs, stereos, VCRs, and computers that you plug into an ac wall outlet use devices called *rectifiers* to convert the ac voltage to a dc voltage. This dc voltage powers the electronic circuits in these devices.

Testing for proper supply voltage is usually the first thing done when troubleshooting a circuit. If there is no voltage present, or if it is too high or too low, the voltage problem should be corrected before investigating further.

The waveforms associated with ac voltages are either sinusoidal (sine waves) or nonsinusoidal (sawtooth, square, ripple, etc.). DMMs display the *rms* (root-mean-square) value of these voltage waveforms. The rms value is the effective or equivalent dc value of the ac voltage.

Most meters, called *average responding* meters, give accurate rms readings if the ac voltage signal is a pure sine wave. Averaging meters are not capable of measuring nonsinusoidal signals accurately. Special DMMs called *true-rms* DMMs will accurately measure the correct rms value, regardless of the waveform, and should be used for nonsinusoidal signals.

**Figure 19-7**   **Digital thermometer.**
*(Reproduced with permission from the Fluke Corp.)*

**Figure 19-8**   **Industrial test probes.**
*(Reproduced with permission from the Fluke Corp.)*

**Figure 19-9**   **Banana-plug test probes.**   *(Reproduced with permission from the Fluke Corp.)*

A DMM's ability to measure ac voltage can be limited by the frequency of the signal. Most DMMs can accurately measure ac voltages with frequencies from 50 Hz to 500 Hz, while others can measure ac voltages with frequencies from 20 Hz to 100 kHz. DMM accuracy specifications for ac voltage and ac current should state the frequency range of a signal the meter can accurately measure.

Voltage measurements determine:

- Source voltage
- Voltage drop
- Voltage unbalance

**19.4.2 Voltage Measurements\*** Voltage measurements must sometimes be made in cramped quarters. You must give full attention to touching the test probes to the correct test points and keeping your hands clear of high voltages and moving machinery. A slip while holding the test probes and trying to look at the meter display can result in injury to you or damage to the equipment (Figure 19-10).

It is critical to ensure that the proper voltage is supplied to all electrical equipment on the circuit. Additional loads (new machinery, new outlets, etc.) that have been added to existing circuits may cause excessive voltage drop and energy losses. The *National Electric Code®* states that for reasonable efficiency of operation, voltage drop in branch circuit conductors should be limited to no more than 3% at the farthest outlet (Figure 19-11).

In order to conserve energy and reduce equipment costs, motors are typically matched closely to load requirements and have very little reserve power. Therefore, with motors operating close to full load, the chance of motor burnout and energy waste due to unbalanced voltage is increased. Voltage imbalance results in unbalanced currents in the stator windings and can be caused by single-phase loads that have been tapped onto three-phase supply circuits (Figure 19-12).

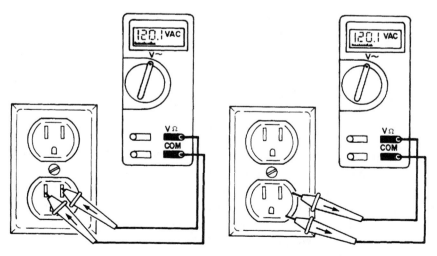

**Figure 19-10** **Measuring voltage.** *(Reproduced with permission from the Fluke Corp.)*

\*Reproduced with permission from the Fluke Corp.

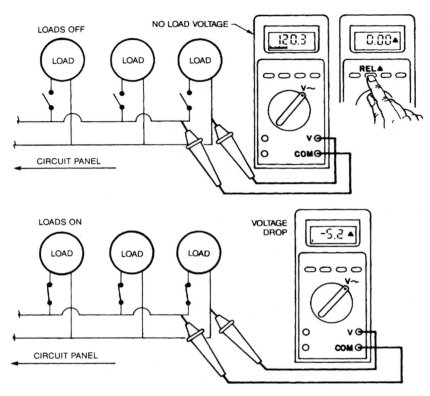

***Figure 19-11*** **Measuring voltage drop in branch circuits.** *(Reproduced with permission from the Fluke Corp.)*

To calculate voltage imbalance:

$$\text{Percent voltage unbalance} = \frac{\text{Maximum deviation from the average voltage}}{\text{Average voltage}}$$

**19.4.3 Current Measurements\*** Current tests should be made to ensure that the continuous load rating noted on the motor's nameplate is not exceeded and that all three-phase currents are balanced. If the measured load current exceeds the nameplate rating or the current is unbalanced, the life of the motor will be reduced due to high operating temperature. Unbalanced current may be caused by voltage imbalance between phases, a shorted motor winding, or a high resistance connection (Figure 19-13).

*Reproduced with permission from the Fluke Corp.

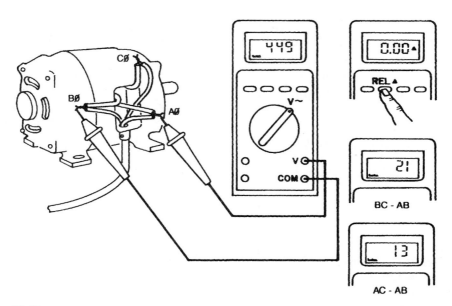

**Figure 19-12** Checking for voltage imbalance with relative mode. *(Reproduced with permission from the Fluke Corp.)*

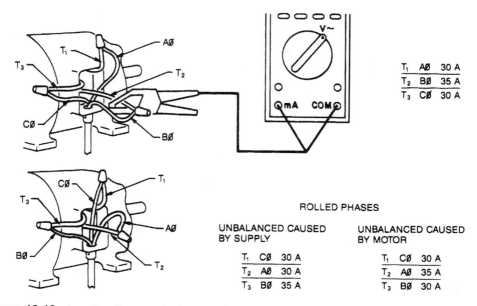

**Figure 19-13** Locating the source of current imbalance—motor or supply. *(Reproduced with permission from the Fluke Corp.)*

Starting current for an induction motor is typically six to ten times the full load current of the motor. Overvoltage of 10% or more of rated voltage or excessive motor loading can cause higher-than-normal inrush current. Excessive inrush current can cause low-voltage on feeder circuits and result in untimely opening of low-voltage protection equipment, dropout of sensitive protective devices, and reduced motor torque on other motors served by the feeder (Figure 19-14).

**19.4.4 Contact Bounce**   Contact bounce from vibration cannot be detected by a traditional VOM needle or ordinary digital multimeter display. Analog or ordinary digital displays will not respond fast enough to allow you to see the brief bounce.

The Fluke Corp. manufactures a fast-responding analog bar graph meter. When used in conjunction with the resistance mode, it is great for detecting contact bounce from vibration (Figure 19-15). When the contact bounce opens, its resistance value changes momentarily from zero to infinity and back. The analog bar graph is ten times faster than the digital reading and will display at least one segment the moment the contact opens. Remember to open and lock out the circuit before making resistance measurements across the contacts.

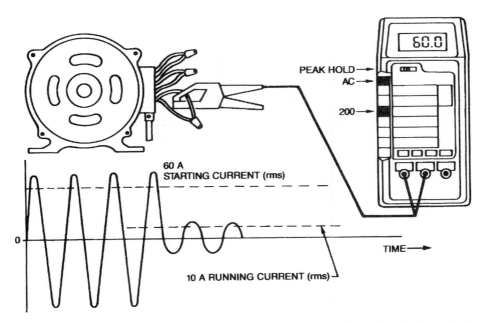

**Figure 19-14**   **Using peak hold to measure motor starting current.** *(Reproduced with permission from the Fluke Corp.)*

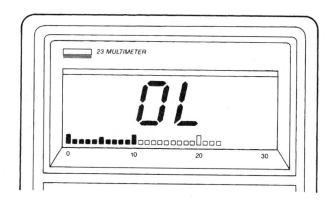

***Figure 19-15*** **Determining contact bounce from vibration.** *(Reproduced with permission from the Fluke Corp.)*

**19.4.5 Temperature Measurements\*** Overheated parts in a circuit are a positive indication of problems. Measure the temperature of the component if you suspect overheating. Compare the temperature reading to those of similar components that are known to be operating correctly and consult the manufacturer's specifications.

Fluke has accessories to convert the Fluke DMM into a handy thermometer. They can be used with a K-type thermocouple for measuring the following temperatures:

- Motors and bearings (see Figure 19-16)
- Thermal overload relays and circuit breakers
- Ambient temperatures for correcting conductor ampacities
- Operating circuit components (Figure 19-17)

**19.4.6 Safety Checklist for Meters\*** The following safety precautions should be practiced when working with meters:

- Use a meter that meets accepted safety standards.
- Use a meter with fused current inputs and be sure to check the fuses before making current measurements.
- Inspect test leads for physical damage before making a measurement.
- Use the meter to check continuity of the test leads.
- Only use test leads that have shrouded connectors and finger guards.
- Only use meters with recessed input jacks.
- Select the proper function and range for your measurement.

\*Reproduced with permission from the Fluke Corporation.

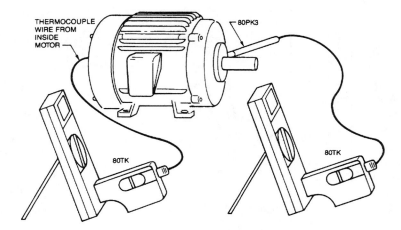

**Figure 19-16** **Measuring motor operating and bearing temperatures.** *(Reproduced with permission from the Fluke Corp.)*

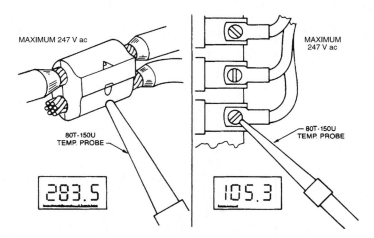

**Figure 19-17** **Measuring the temperature of live circuit components.** *(Reproduced with permission from the Fluke Corp.)*

- Be certain the meter is in good operating condition.
- Follow all equipment safety procedures.
- Always disconnect the "hot" (red) test lead first.
- Do not work alone.
- Use a meter that has overload protection on the ohms function.
- When measuring current without a current clamp, turn the power off before connecting into the circuit.
- Be aware of high-current and high-voltage situations and use the appropriate equipment, such as high-voltage probes and high-current clamps.

## 19.5 Motors

Properly troubleshooting a motor will depend on the type and rating characteristics of the motor. Human senses, such as sight, sound, feel, and smell should also be incorporated into the troubleshooting techniques.

**19.5.1 Single-Phase Motors** There are several types of single-phase motors that require a centrifugal switch to open the start circuit after the motor is up to speed. If the motor hums but will not start, try rotating the shaft by hand. If it now starts, then the centrifugal switch is not operating properly or the start winding is open. Check the winding with an ohmmeter.

**19.5.2 Polyphase Motors** If a polyphase motor does not start, the applied voltage may be too low (more than 10% less than rated voltage). Low voltage causes low starting torque. It is also possible that one phase is open. This condition is called *single phasing*. Check each phase with an ohmmeter. This condition may be recognized by excessive noise level and rapid heat buildup.

If the motor is overloaded, heat buildup may result. Motors will generally operate for short periods on overload but not continuously. Check each phase with a tong ammeter and compare with the nameplate rating.

**19.5.3 Overheating of Bearings** Bearings can overheat due to lack oil, dirty oil, or oil not reaching the shaft. There is also the problem of misalignment of shaft and bearing.

**19.5.4 Noise** Improper balance can cause excessive vibration. Be sure that the motor is properly mounted and coupled to the load. In some cases an uneven air gap due to worn bearings may be the problem, or dirt in the air gap may be the problem. Compressed air can be used to blow the dirt away.

**19.5.5 Speed** If a rotor of a squirrel-cage induction motor has an open circuit or high resistance, the speed will drop. If the speed in a universal motor is high, the problem may be a shorted field coil.

**19.5.6 Component Coils (Relays, Contactors, and Full-Voltage Starters)** Figure 19-18 shows a flow chart for use in troubleshooting component coils. In many cases, shortcuts can be made and success realized. However, whatever system is used, a systematic approach should be followed. Too many times a "hunt-and-try" method only leads to wasted time and possible damage to equipment.

## 19.6 Troubleshooting a Complete Control Circuit

Figure 19-19 shows a power and control circuit in which two 480-V, three-phase, 60-Hz motors with magnetic full-voltage starters are used. The control voltage is taken from the secondary of an isolated secondary transformer and is 120 volts.

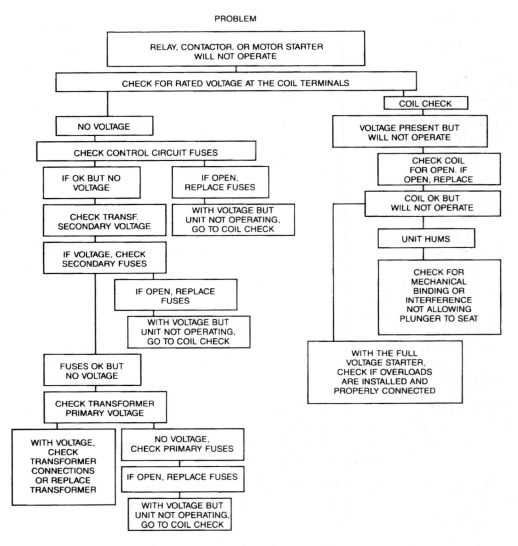

*Figure 19-18*   Use of flow chart in troubleshooting.

Depending on the desired machine speed, one or two motors can be used. Heating elements are used in the machine operating fluid to bring the fluid temperature up to 70°F as required by the machine. The machine must be in a given position for start condition and reverses by pressure.

The approach to effective troubleshooting, as noted earlier, is to first segregate the section where the trouble is observed to be occurring. In the circuit shown in Figure 19-19, there are three sections indicated: motors control, heat control,

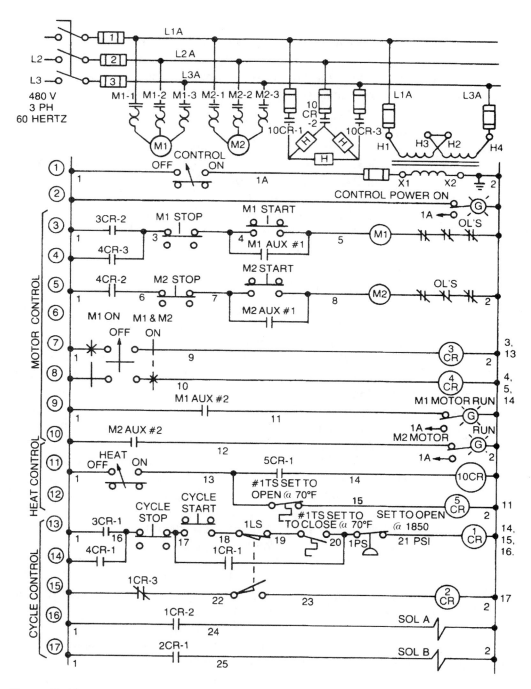

**Figure 19-19** Control circuit for troubleshooting practice.

and cycle control. In a particular problem you have observed that the motors and heat control are okay. The problem is that solenoid A is not energizing at the proper time. You should then be concerned only with the cycle-control section.

However, let us first take a look at this entire circuit. Always keep in mind the possible causes of trouble as outlined earlier in the chapter.

**1. Motors and Motor Control**

a. Using the ac voltmeter (600-V range), check the voltage from L1–L2, L2–L3, and L3–L1. With approximately 480 V showing on all three checks, the plant line voltage is established.

b. Close the line fused disconnect switch. Check the voltage from L1A–L2A, L2A–L3A, and L3A–L1A. If any of the three checks shows no voltage, check the fuses in the disconnect switch. For example:

  • No voltage from L1A–L2A
  • No voltage from L2A–L3A
  • Voltage from L1A–L3A

  You know that fuse #2 is open. Open the disconnect switch and replace the fuse. Note that the same general checking procedure can be used to determine which of the three fuses is open. If there is no voltage at any of the three checks, then two of the fuses or possibly all fuses are open.

c. Turn the control OFF/ON selector switch to ON. The control power on light should be illuminated. If not, using the 150-V scale on the ac voltmeter, check the voltage from terminals X1–X2 on the transformer secondary. This voltage should read approximately 120. If there is voltage here, check from X2–1A. With no voltage here, the control fuse is probably open. Open the line disconnect switch and replace the control fuse. Reclose the disconnect switch and recheck voltage from X1–X2. With no voltage here, the selector switch may be wired incorrectly. Again, open the disconnect switch and check the wiring on the selector switch. Always remember in all checking that there is a possibility of an open conductor at a component terminal or at a terminal block. With voltage established at X1–X2 and the control power light not illuminated, depress the push-to-test pilot light to determine if the bulb is burned out. When you originally checked for voltage at X1–X2 you found no voltage. The fuses in the transformer primary should be checked (Figure 19-20). When the open fuse is located, open the fused disconnect switch and replace the fuse or fuses.

d. Operate the M1–M2 motor starter selector switch to M1 ON. Relay 3CR should be energized, closing relay contact 3CR-2. If not, using the 150-V range on the ac voltmeter, check from circuit number 2 to circuit number 9. With no voltage here, either the selector switch is wired incorrectly or there is an open conductor. Turn the control OFF/ON selector to OFF and check the wiring on the selector switch and for an open conductor.

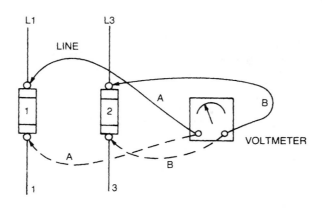

Put lead A on L1; B on L2 — Should read line voltage.
　　　　　　　　　　　　　　If not, the problem is in the supply.
Put lead A on L1; B on 3  — With line voltage reading, the #2 fuse is good.
　　　　　　　　　　　　　　With no reading, #2 fuse is open.
Put lead A on 1; B on L2  — With line voltage reading, #1 fuse is good.
　　　　　　　　　　　　　　With no reading, the #1 fuse is open.

*Figure 19-20*　Checking fuses.

e. Operate the M1 motor start push-button switch. M1 motor starter coil should be energized. If not, start checking the circuit line from circuit number 2 to circuit number 1, again using the 150-V range ac voltmeter. From this point on, there can be two methods used.

  • *Series checking.* With this method, one prod is placed on point 2 and the other progressively moves from circuit number 5 to circuit number 4 to circuit number 3, and finally to circuit number 1. At the point where voltage is indicated, the open circuit is in the first component to the right. For example, with the motor start push-button switch held operated, there is voltage at 3 but not at 4; the open, then, is in the M1 stop push-button switch.

  • *Half-split checking.* With this method, one prod is placed on point 2 and the other directly to 4. Thus, in the problem indicated above, only one additional move with the second prod is required to locate the problem. While in this particular case it does not seem to have an advantage, there are cases in which there are many components in series. In these cases, considerable time can be saved with the half-split method.

f. Turn the M1-OFF-M1 and M2 selector switch to M1 and M2. Relay 4CR should be energized, closing relay contacts 4CR-2 and 4CR-3. Operate the M1 motor start push-button switch and the M2 motor start push-button switch. M1 motor starter coil and M2 motor starter coil should be energized. With one or both of the motor starter coils not energized, follow the same general procedure in checking as indicated for M1 only.

Apart from the possibility of an open motor starter coil, about the only other problem that may come up in the motor starter circuits is the possibility that one of the overload relay contacts in series with the motor starter coil is defective and open or a jumper is missing on the starter.

**2. Heat control.** Operate the heat OFF/ON selector switch to ON. With the temperature of the machine operating fluid above 70°F, neither relay coil 5CR nor contactor coil 10CR will be energized.

With the temperature below 70°F, temperature switch NC contact will be closed, allowing relay coil 5CR to be energized. Relay contact 5CR-1 closes, energizing contactor coil 10CR. Contactor contacts 10CR-1, 10CR-2, and 10CR-3 close, energizing the heating elements.

If the contactor 10CR is not energized:

a. Check to see if relay coil 5CR is energized.
b. Check the voltage from 2–15. If voltage here, the coil is open or there is an open conductor.
c. No voltage at 2–15 but voltage 2–13. Check the temperature switch. It may be defective or incorrectly wired.
d. No voltage at 2–13, the heat OFF/ON switch may be incorrectly wired or there may be an open conductor.

**3. Cycle Control.** The same general pattern is followed in checking the cycle control as was followed with the motor and heat control. For example, with the M1 motor starter control energized, relay 3CR is energized, closing the NO 3CR-2 contact. Operate the cycle start push-button switch.

a. Control relay 1CR does not energize.

Using the series checking method explained, place one voltmeter prod on circuit point 2. Then, using the other prod, proceed consecutively through points 21, 20, 19, 18, 17, and 16. At the point that voltage is detected, the open circuit is in the component immediately to the right of this point. Remember that the cycle start push-button switch must be held operated during these checks.

An example of a problem would be voltage was noted at point 18 but not at 19. There can be three possibilities:

a. Operating cam not holding the switch in the operated condition.
b. Limit switch incorrectly wired.
c. Open conductor.

Using the half-split method, you would probably start with the second probe at point 18. In this case it would have required only one move to locate the problem, thus reducing the time to troubleshoot.

The problem of open conductors has been mentioned several times. Remember that push-button switches, selector switches, and pilot lights are generally located on an operator's panel. Relays, contactors, and motor starters are located in a control cabinet. Limit switches, pressure switches, and temperature switches can be mounted remotely on the machine. These placements result in several terminal block connections. Be sure these connections are kept tight.

## 19.7 Troubleshooting the Programmable Logic Controller

It is difficult to offer a complete and comprehensive approach to troubleshooting the PLC since each PLC manufacturer has taken a slightly different approach to the design of the unit. These differences mean that a diagnostic chart that would apply to one unit would not necessarily apply to another.

There is one area that will generally apply to all units: the peripheral devices. This hardware at least will be the same in most cases. It will consist of components supplying input information for motion, pressure, and temperature, and components controlling all outputs, such as solenoids, relays, motor starters, visual indicators, and alarm systems. The area between the inputs and outputs is the responsibility of the central processing unit. This area is seldom approached by the troubleshooter, apart from replacement of units within the processor.

In most cases the PLC manufacturer will supply visual indication of all incoming and outgoing signals. However, there are several areas in which the use of a meter to check volts as well as an ohmmeter to check continuity is useful. Starting with the power supply and continuing through all the hardware, information can be obtained even from the absence of visual indication.

Position indicators that are mechanically attached to a machine or process line may be out of alignment with the operator. Temperature sensors, such as thermocouples, may not be seated properly or may be open. Pressure sensors may be defective and not responding correctly to a given pressure. All these components can be checked for correct voltage with a meter. Continuity can be checked with an ohmmeter. Problems with output components, such as actuator solenoids, motors, and indicating lamps can be resolved by checking the output signal voltage and the continuity of a suspected circuit.

One manufacturer of PCs has provided diagnostic monitoring I/O modules. They provide the troubleshooter with information to quickly identify the possible source of a problem. In practice, the module actually "learns" the operation of a process. When the operator is satisfied that the process is correct, the monitoring module is used to observe the PC and the system hardware being controlled.

In Chapter 15 we described the HPM COMMAND 90. One of its features is the *power flow* screen, which can be of tremendous help to the maintenance personnel when troubleshooting machine sequence problems. Referring to Figure 19-21, the screen is a security level 4, accessed from the main menu.

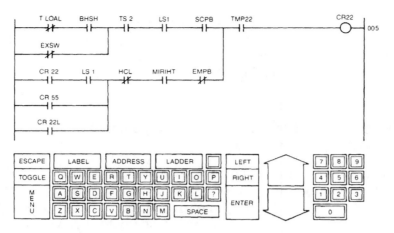

**Figure 19-21**    **Power flow screen.**   *(Courtesy of HPM Corporation.)*

Once in power flow, a single ladder from the RLD will be displayed along with a keyboard and some special function keys. Power flow is indicated from left to right, the color red indicating passing power. A white bar will be displayed under a contact when the reference address controlling that contact is on.

Timers and counters will be shown as a green box symbol. Within the box will be a label: "R" (run) input, "E" (enable) input, or "O" (output). The set point and actual values are also shown.

The right-most element in the ladder is the coil. It is blue if no path was found to turn it on. A red coil indicates one or more paths were found to turn it on, and all references to the coil will also be on. The green number to the right of the coil indicates the ladder number currently being displayed.

Two large arrows between the alpha and numeric keypads permit scrolling up or down one ladder at a time. When a touch on the arrow is recognized, it will turn white.

## 19.8  Electronic Troubleshooting Hints*

The electrician does not necessarily need to be an electronics technician to locate and repair many of the problems associated with solid-state equipment. The following hints will help you locate common problems associated with solid-state equipment.

---

*Reproduced with permission from the Fluke Corporation.

1. Actual test voltages and other measurements should be recorded when the equipment is in good working order. They can then be compared to measurements during troubleshooting. Having records to refer to will speed up troubleshooting.
2. Look for obvious problems by visual inspection (physical damage, overheated terminations, broken wires, corrosion, etc.).
3. Measure supply voltage and current. Is it within the limits of the equipment?
4. Check all circuit breakers and fuses. If circuit breakers or fuses are open, they should not be closed until the fault has been cleared. Otherwise, if the circuit is closed on a fault, operator injury or equipment damage could result.
5. High temperature is the leading cause of failures of electronic equipment. Equipment should be placed in a well-ventilated area so that heat can escape.
6. Transient ac voltage peaks or voltage surges in control circuits may last only milliseconds but can cause problems in the control circuit. A DMM (similar to that shown in Figure 19-6) with peak-hold function can be used to determine whether a transient ac peak problem exists.
7. Do not change control settings unless you are familiar with the equipment. If the settings and adjustments are incorrect, resetting them can be a time-consuming process. If you cannot fix the equipment or you are unsure of proper and safe troubleshooting procedure, call the manufacturer.

## Recommended Web Links

Students are encouraged to view the following Web sites as a supplement to the concepts presented in this textbook. Review and analyze the array of products that are available for electrical control applications. Many of these sites offer technical information that can help in converting the principles to practical applications. To view catalogs, your PC may require Adobe Acrobat Reader software.

Fluke Corporation
http://www.fluke.com/us.asp
Review: Solutions and Products (also see Calibrators)

BK Precision
http://www.bkprecision.com/
Review: Application Notes and Products

Tektronix
http://www.tektronix.com
Review: High-Precision Instruments

Test Equipment Depot
http://www.testequipmentdepot.com/newcatalog.htm
Review: Test Instruments

### Troubleshooting Papers

Siemans Energy and Automation, Inc.
http://www.sea.siemens.com/motorsbu/product/White%20Papers/
Troubleshooting.pdf
Review: "Troubleshooting Induction Motors" by William R. Finley

### Maintenance and Troubleshooting of Electrical Motors

Reliance Electric
http://www.reliance.com/prodserv/motgen/h7000.htm
Review: The complete three chapter tutorial.

### Troubleshooting Allen-Bradley Products

Allen-Bradley
www.ab.com
Enter the word "troubleshooting" in the Search box. Troubleshooting
guides are provided for individual products.

## Achievement Review

1. Explain two different methods of troubleshooting an electrical circuit.

2. Draw a sketch that explains how to check fuses.

3. What types of contacts may cause trouble and yet appear to be in good condition?

4. Why should you never file a contact?

5. What is a *combination problem* as applied to machine control?

6. Explain two methods of detecting grounds in an electrical circuit.

7. Explain how an open in a common line can cause trouble.

8. List a few locations where you may be able to find grounds that have been created in wiring a machine.

9. What are some of the electrical problems with machines that could have been avoided by good housekeeping practices?

10. What condition may lead to a low-voltage problem in an industrial shop?

11. What is the value of a good preventive maintenance program?

12. In the circuit shown in Figure 19-19, the electrician has incorrectly wired the NO and NC limit switch contacts. NO contact is wired to 22–23 and the NC contact is wired to 18–19. With this change, can a cycle be started? If not, why not?

13. In the circuit shown in Figure 19-19, the NC contact on relay 1CR is found in which circuit line?

    a. 14
    b. 15
    c. 16

14. In the circuit shown in Figure 19-19, with the M1-OFF-M1 and M2 selector switch set to OFF, which motors can be started?

    a. M1
    b. M2
    c. Neither

15. Using Ohm's law, calculate the current in line A–B of the following diagram:

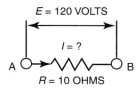

16. Using an ohmmeter, how can you tell if a circuit is open or closed?

17. What is the purpose of the centrifugal switch in some single-phase motors?

18. What can cause noise to develop in a motor?

19. What is recommended for all circuits that serve critical loads?

# Chapter 20 Designing Control Systems for Easy Maintenance

## Objectives

After studying this chapter, you should be able to:

- List several important points in the assembly of a machine using electrical control.
- Show how to cross-reference relay contacts with circuit line numbers.
- List several items that can be of help in making better drawings for electrical control.
- Draw a typical control panel showing where components such as relays, motor starters, and disconnecting means should be placed.
- List the advantages and disadvantages of returning all circuit connections to the panel.
- Discuss several concerns with the use of mechanical limit switches.
- Explain the advantages of having all electrical components accessible.
- Name several places where indicating lights can help the troubleshooter.

## 20.1 Design Considerations

In its final form, a design generally involves some compromise. Considerations such as safety, cost, manufacturing, assembly, and operating conditions enter into the final machine. All of these considerations, and others, show that careful design planning is a prerequisite to ease of maintenance.

Maintenance problems are considered in the design through analyzing items such as:

- The economics of building a competitive product
- The environment in which the machine is to be used
- The problem of installing electrical components after other mechanical and fluid power work has been completed

- The possible presence or development of vibration, shock, or other mechanical motion that may affect the electrical components

For the best chance of success in dealing with maintenance problems, the designer should:

- Present adequate wiring diagrams, layouts, and instructions
- Display evidence of experience in the electrical control field
- Avoid sloppy workmanship

In the assembly of a machine, several points are important:

- Circuit wiring should be exactly as called for on the electrical prints. In case of an error on the prints, corrections should be made before the machine is checked out.
- All conductors and terminals should be properly marked with numbers corresponding to the wiring diagram.
- A sufficient number of terminal block checkpoints should be provided.
- All terminal connections should be double-checked to be sure the conductors are tight in the connectors.
- Care should be taken in pulling conductors through conduit and fittings so that no conductor insulation is cut or scraped.
- Conduit fill, as set up in electrical standards, should not be exceeded.
- In long or difficult-access conduits, three or four spare conductors should be pulled.

In the field installation, a greatly improved overall job can be obtained by close cooperation between the user and the builder. The user must know:

- Total power requirements with the amount of possible low-power factor load (induction motor)
- Foundation and assembly with relative location of separate control enclosure, if used
- Location of main power connection
- Location of terminal disconnect points on the machine (where machine is disassembled for shipment)

The builder must know:

- User's power voltage, phase, and frequency
- Any special conditions or limitations on power supply
- Unusual ambient temperature condition
- Unusual atmospheric conditions
- User's specification of manufacturer or components (if any)
- User's specification on motor starters, if not full voltage
- A complete and accurate sequence of functions the machine is to follow,

which can be difficult to obtain due to:

   a. Inexperience, particularly in the case of a new process
   b. General lack of knowledge
   c. Problems of communication, generally best solved by a cooperative effort on the part of the user and builder

In all cases, a well-studied, careful analysis and procedure should be followed in selecting, applying, and installing the electrical control. This approach always pays off in reduced maintenance.

In supplying some practical help as a guide to designing for easy maintenance, there are two major areas to examine:

   1. Diagrams and layouts
   2. Locating, assembly, and installing components

## 20.2 Diagrams and Layouts

There are aids to making original elementary diagrams and layouts more useful and to cutting maintenance and troubleshooting time. The numbering of elementary circuit lines and cross-referencing the relays and their contacts are two such aids. As shown in Figure 20-1, normally closed contacts are so indicated by a bar under the line number. Note that the line numbers are enclosed in a geometric figure to prevent mistaking them for circuit numbers.

All contacts and the conductors connected to them should be properly numbered. Numbering should carry throughout the entire electrical system. Continuous numbers may involve going through one or more terminal blocks. The incoming and outgoing conductors as well as the terminal blocks should carry the proper electrical circuit numbers. If at all possible, connections to all electrical components should be taken back to one common checkpoint.

All electrical elements on a machine should be correctly identified with the same marking as shown on the wiring diagram. For example, if a given solenoid is marked "solenoid A1" on the drawing, the actual solenoid on the machine should carry the same marking, "solenoid A1."

A drawing should be made showing the relative location of each electrical component on the machine. The drawing need not be a scale drawing, but it should be reasonably accurate in showing the location of parts relative to each other and in relative size. For example, if solenoid A1 is located on the left-hand end of the machine base, it should be so shown on the machine electrical layout. Figure 20-2 illustrates these points.

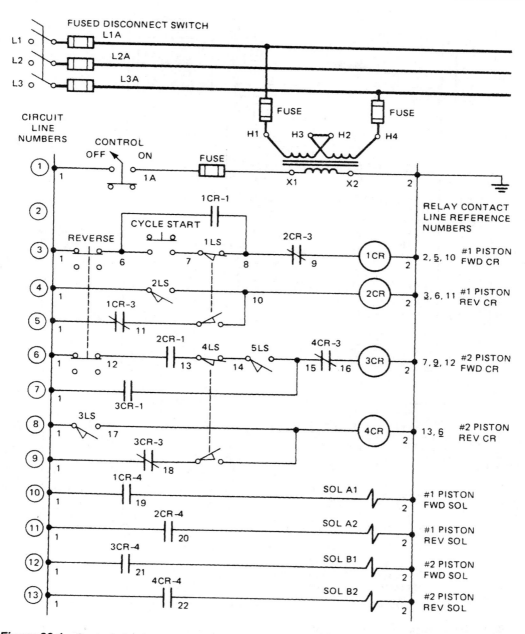

**Figure 20-1** Control circuit showing proper wire number identification of its various components. Conductors carrying load current at line voltage are denoted by heavy lines. For complete circuit and description, see Figures 12-13A and 12-13B.

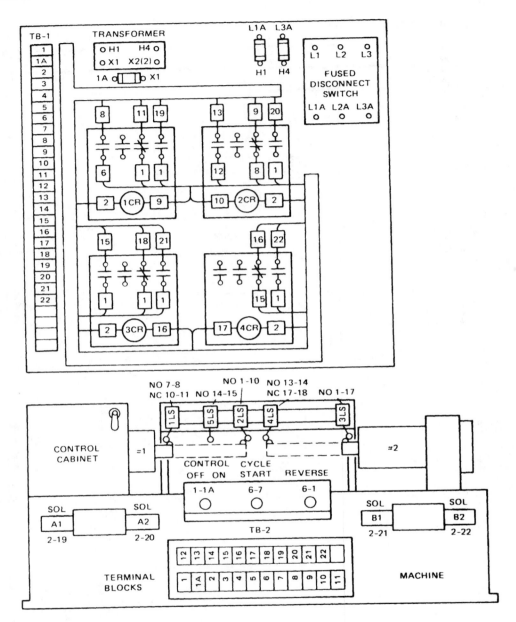

**Figure 20-2**  Machine and panel layout.

Figure 20-3 illustrates the advantages of displaying the identification and approximate location of various control components on a machine. In this specific case all inputs and outputs are indicated.

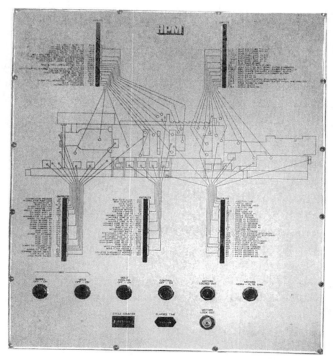

**Figure 20-3**   **Display of approximate location of input and output control units.** *(Courtesy of HPM Corporation.)*

It is important to list components on the sheet with the elementary diagram and/or layout. They should be cross-indexed in some manner with the components as they appear in the circuit diagram or layout.

The component should be described so that the user can obtain a replacement if necessary. An example is shown in Figure 20-4. Only basic information is used to describe the components and to cross-reference them to the circuit.

Within any organization, maintenance or troubleshooting can be assisted by standardization of circuit numbers and component numbers of designation. For example, in a motor starter circuit, the numbers 1, 2, 3, and 4 can always be used in the same location. After some experience, the maintenance worker or troubleshooter can remember that the coil is always 4,1; the interlock and START button is 3,4; and the STOP button is always 2,3.

It is understandable that a given arrangement cannot always be used completely. However, with a group of machines that are similar in design, this pattern of standard circuit numbers can be followed to a large extent.

It follows that components can carry a similar standardization. For example, solenoid S1 may be a clamp forward solenoid. A limit switch 1LS may be a clamp forward stop.

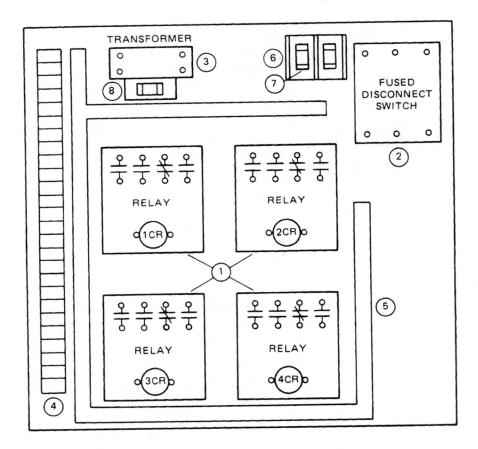

(1) 600-V MAX, CONTROL RELAY — FOUR-POLE, 120-V, 60-Hz COIL, OPEN TYPE

(2) REMOTE OPERATED FUSED DISCONNECT SWITCH, THREE POLE, 60 A, 600 V

(3) CONTROL CIRCUIT TRANSFORMER, OPEN TYPE, WITH BUILT-IN FUSE BLOCK; 350 VA, 240–480-V PRIMARY; 120-V SECONDARY

(4) CHANNEL MOUNTING-TYPE TERMINAL BLOCK, 300 V, TUBULAR SCREW TYPE WITH PRESSURE PLATE, NO. 14 WIRE

(5) 1" WIDE X 2" HIGH PANEL WIRING CHANNEL

(6) TWO-POLE FUSE BLOCK, 30 A, 600 V

(7) 10-A, 600-V FERRULE-TYPE FUSE

(8) 15-A, 250-V FERRULE-TYPE FUSE

**Figure 20-4**  Elementary diagram and description of panel component.

In a group of similar machines, it helps the maintenance personnel if the same designations for specific electrical components can be maintained on all the machines. The worker can then become acquainted with the functions of specific components, such as solenoids and limit switches, regardless of the machine involved.

Another help is to advance circuit numbers for similar components in the tens place. For example, if more than one motor starter is used, the #1 motor starter would carry circuit numbers 1, 2, 3, and 4. The #2 motor starter would carry circuit numbers 21, 22, 23, and 24. The #3 motor starter would carry circuits 31, 32, 33, and 34. The numbers should mean more than just numbers. For example, with the proper numbers in the ones and tens places, the troubleshooter can immediately spot the circuit numbers of 32 and 33 as the STOP button on the #3 starter.

In an elementary circuit diagram carrying more than one solenoid, each solenoid should carry a reasonable description of its function. This description will help maintenance or service personnel quickly locate a possible source of trouble without referring to the fluid power circuit. The description will aid in quickly locating the solenoid on the machine. This suggestion can also apply to the relays, in which a specific relay controls the initial energizing of a solenoid. For example, in Figure 20-1, a circuit is shown with several solenoids, not only indicating their function but also picturing their relative location.

An item that is almost a must with complex circuits is some form of sequence of operations. This information may be written out in detail, listing component by component each step in the operation of the circuit. However, a shorter method that gives nearly the same information is the sequence bar chart. (Several examples are shown in Chapter 12.)

The size of the drawing is important, first of all because of the problem of storing them. If every size and shape were allowed, the task of systematic and protective filing of drawings could be burdensome. Page sizes of 8 1/2″ × 11″ or 9″ × 12″ and multiples thereof are generally accepted. The maximum size is usually 36″ × 48″.

The drawing size can also be a problem for the troubleshooter or maintenance personnel. If the drawing is too large, it is unwieldy to handle at the machine. If it is too small, it is hard to read the schematic.

In the mechanics of drawing the elementary diagram, there are a few suggestions to "clean up" the drawing and make it easier to follow:

- Evenly space circuit lines. Approximately one-half inch apart is satisfactory in most cases.
- Space component symbols on the line so they are clear and so that space is left for the circuit numbers.

- It is not necessary to carry all circuit lines to their completion. For example, consider the line connection on a push-to-test pilot light. An arrow leading from the line connection on the light and indicating the proper circuit number termination of the arrow is sufficient (see Figure 12-8A).
- When the contacts of limit switches or selector switches are separated in a circuit drawing, there may be a problem for the maintenance personnel reading the drawing. There are three things that can be done to help when the circuit is drawn:

  a. Arrange the operating switches in the circuit so that the contacts are grouped in close proximity to each other (see Figure 12-13A).
  b. If the NO and NC limit switch contacts can be located in the circuit drawing so there are no other lines or contacts between, the two symbols can be joined by a broken line. Do not use the broken line and cross several other circuit lines. This usage only tends to confuse the reader. Figure 20-1 shows an example of the proper use of the broken line connection.
  c. A cross-reference can be made between limit switch contact symbols by using an arrow and line number to "tie" the two contacts (Figure 20-5). This arrangement can be used for widely spaced limit switch contact symbols on the same page, or where two contacts of the same limit switch appear on different pages of the circuit diagram.

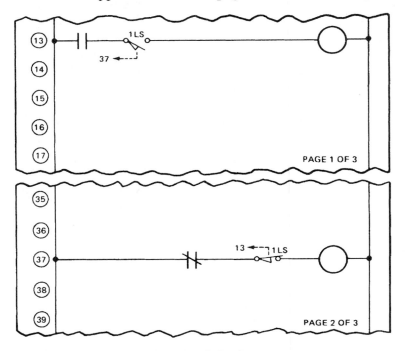

*Figure 20-5*   Limit switch designation in a control circuit.

To increase their utility, many machines are designed with optional features. These may not always be required by the user on the initial job. However, to make it easy to add the option at a later date in the field, provision can be made in the circuit to show its function. The conductors are brought to a terminal block and given circuit numbers. The opening in the circuit that is to receive the optional feature is then jumpered. Appropriate notes instruct the user on adding the option. The auxiliary circuit may be shown as a broken line if desired.

Auxiliary circuits are also helpful when the user has equipment to work in conjunction with the machine, for example, a safety device, material feeder, or material removal equipment.

The circuit in Figure 20-6 shows an example of an auxiliary circuit.

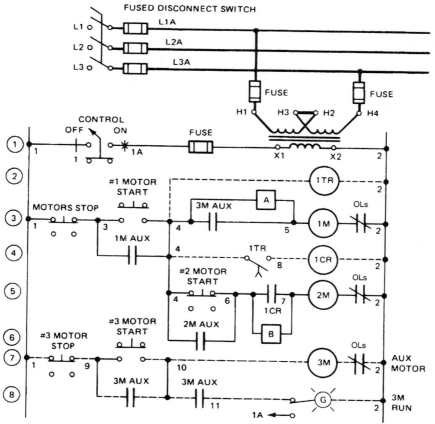

3M AUX PILOT MOTOR. OPTIONAL CIRCUITS SHOWN IN BROKEN LINE. TO REQUIRE 3M MOTOR STARTS BEFORE MAIN MOTORS 1M & 2M START, REMOVE JUMPER A. TO REQUIRE TIME DELAY BETWEEN 1M & 2M MOTORS STARTING, REMOVE JUMPER B.

**Figure 20-6** Control circuit showing auxiliary circuits. Conductors carrying load current at line voltage are denoted by heavy lines.

## 20.3 Locating, Assembling, and Installing Components

After the control circuit is designed, one of the first areas to examine is the control panel (Figure 20-7). While more actual physical work is probably done on the machine, the control cabinet is generally the key to locating trouble.

Maintenance work can be reduced by starting with a systematic and standard arrangement of components. Figure 20-7 shows the layout of a typical control panel with explanatory notes on requirements for good design.

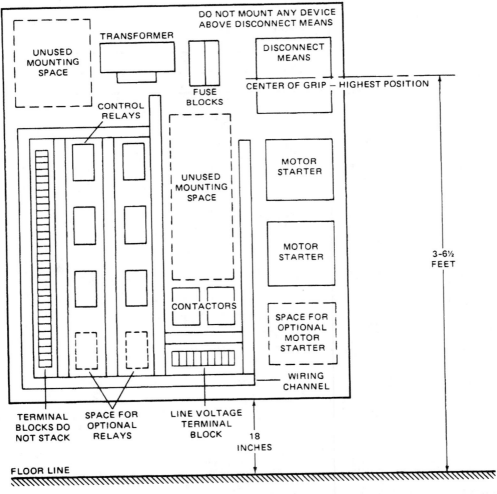

*Figure 20-7*   Panel layout examples. Panel should be removable through enclosure door opening. All line voltage devices (above 120 V) should be mounted above or to the side of control voltage devices. All control devices normally panel mounted should be in one enclosure.

The wiring duct and plastic ties for cabling are universally used. Both save time and expense in the original wiring of the panel and machine. The greatest saving for maintenance is in checking a circuit. It is time consuming to search out and check a particular conductor that may be at the center of a group of 40 or 50 conductors, all tightly laced. After locating the conductor and changing or correcting the trouble, the cable is usually altered.

Wiring duct is available in various widths and heights. It can be obtained with slots or holes in the sides or open slots to the top. Relay manufacturers use a top strip only, fitted between the relay and supported by center posts. This arrangement further reduces component space requirements.

Where wiring duct is not applicable, plastic ties can be used. They can be installed either by a tool or by hand. If it is necessary to get into the cable, the plastic ties can be quickly cut and new ones applied later in a short time.

Figure 20-8 shows various types and sizes of wiring duct. Figure 20-9 shows the use of wiring duct and plastic ties.

The use of terminal blocks for electrical control on a machine is important. They can be of tremendous aid in checking circuits, thus reducing maintenance time. They are valuable when machines must be disassembled for shipment and reassembled in the user's plant. In such an instance, the terminal block on each part containing electrical equipment is a must.

Figure 20-10 shows a variety of terminal blocks available for industrial use. Figure 20-11 shows the use of terminal blocks on a machine.

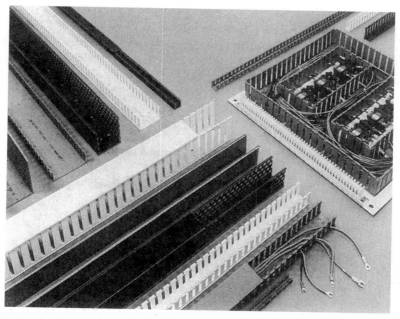

**Figure 20-8** **Wiring duct.** *(Courtesy of Panduit Corp., Electrical Products Group.)*

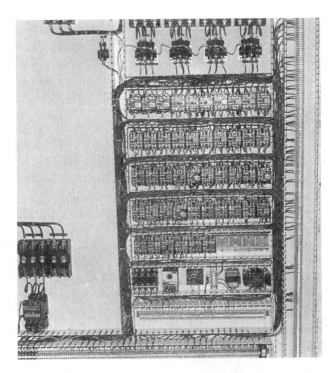

***Figure 20-9*** **Use of wiring duct and plastic ties.** *(Courtesy of HPM Corporation.)*

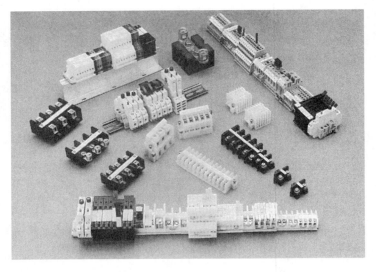

***Figure 20-10*** **Examples of terminal blocks.** *(Courtesy of Allen-Bradley, a Rockwell International Company.)*

**Figure 20-11** Use of terminal blocks. *(Courtesy of HPM Corporation.)*

Provide a central point for checking by bringing all connections between components back to the control panel. In some cases this design may add expense in the initial wiring of the machine. It also increases the amount of conductor used and the size of conduit. Additional terminal blocks and more wiring time are required. However, these disadvantages must be carefully weighed against the maintenance time saved in the user's plant. A much faster job of troubleshooting can be done if all connections are available at one point.

For example, look at the circuit in Figure 20-12. Limit switches 1LS, 2LS, and 3LS and solenoid A are probably on the machine. Normally it would be necessary only to bring points 1, 2, and 3 back to the control panel because the source of power and relay 1CR are available here. To provide checkpoints in the control panel for the entire circuit, points 4, 5, and 6 could also be brought back to a terminal strip in the control cabinet.

Many designers find it economical to use either fuses or overload heating elements in the solenoid circuit. They find that it requires less time and is less expensive to replace a fuse or reset an overload than to replace a solenoid coil. In this case the solenoid coil connection would be brought back to the control panel to pick up this protection. An example of this arrangement is similar to that

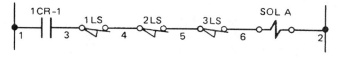

**Figure 20-12** Circuit without fuse protection.

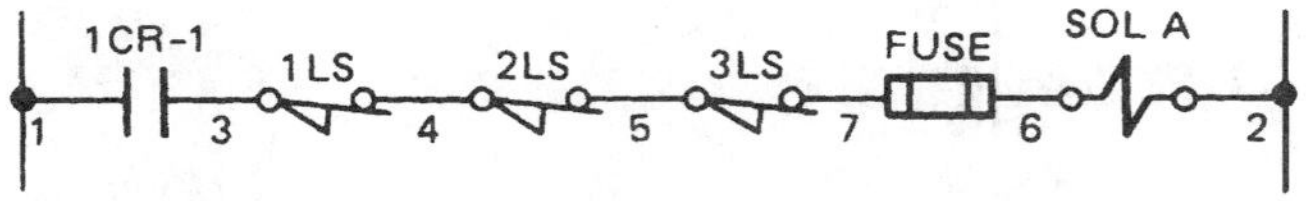

*Figure 20-13*  Circuit with fuse protection.

shown in Figure 20-12, except that a protective device is added as shown in Figure 20-13. The fuse selected in this case should have a short time lag to override the momentary inrush current but should open on sustained overload.

**20.3.1 Limit Switch Problems**  Experience has shown that a high percentage of limit switch problems results from their application, not from basic design of the unit.

Much of the material in this section is drawn from experience and covers mechanical, vane, and proximity switches.

## 1. Mechanical Switches

   a. **Environment.** Limit switches are usually located on the machine and are thus subjected to the working area environment. Flying liquids, metal chips and debris, corrosive fumes, excessive temperatures, and shock or vibration are some of the problems.

   Generally, a temperature limit of approximately 175°F should not be exceeded. In cases of exposure to higher temperatures, thermally insulated barriers can be of some help.

   If heavy shock or vibration is present in the machine, the electrical circuit may be interrupted by the contacts opening. Under these conditions, a low control mass and high contact pressure are important. The switch should be installed so that the direction of the shock forces is in a different plane from the plane in which the contacts operate.

   Very few of the housings and/or integral parts of a limit switch will stand up for long in the presence of corrosive fumes. It is helpful to seal the switch at the conduit end, or use a small, light-duty, hermetically sealed switch with a relay.

   The problem of flying chips, liquids, or other debris can sometimes be handled by relocation, change of switch operators, or addition of mechanical guards. If the problem is liquids entering the switch through the conduit, change the design of the conduit, or use a multiconductor synthetic rubber cable with a compression seal.

   b. **Actuation.** One of the most common problems is the design of the actuating cam. Such problems as impact forces, direction of travel, speed of the moving part, frequency of repeat operation, and overtravel are always present.

For example, the operating frequency should not exceed 2 times per second. The impact speed should not exceed 400 feet per minute. The minimum designed overtravel should be approximately one-third of that required to actuate the switch. The maximum designed overtravel should not exceed 10 degrees. The angle of the cam engaging the switch actuator should not produce forces that tend to deflect the actuator.

Figure 20-14 shows examples of various cam designs as they are applied to limit switch operation. Solutions to many of the previously described problems are illustrated.

Excessive impact from improperly designed actuating systems is the leading cause of premature failure of the electromechanical limit switch. At slow speed, impact is rarely troublesome, but as speed increases, impact applied to the switch becomes critical. In today's higher speed machines, therefore, it is important to give proper consideration to correctly designed actuating systems. The following recommendations are designed to assist you in obtaining greater life from limit switches.

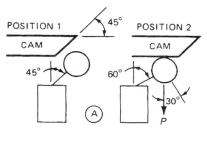

Switch levers should be positioned as nearly parallel with leading edges of cams as possible. Cam A is satisfactory for speeds up to 50 rpm; cam B for speeds up to 200 rpm (nonuniform acceleration of switch lever); cam C for speeds to 400 rpm (uniform or other controlled acceleration).

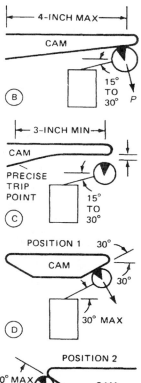

Cam D is designed with a trailing edge so it can override the switch lever and then return. Cam actuates switch on return also. Lever angle is very important in applications of this type.

The black sector in the roller indicates recommended design limits of angle of pressure P. Pressure applied by actuating mechanism to switch operating lever should approximate direction of lever rotation with a variation not to exceed 30° angle of pressure. Changes drastically with rotation of lever. Cam must be designed for proper pressure angles of all positions of lever travel.

**Figure 20-14**   **Cam designs for limit switch operation.** *(Courtesy of R. B. Denison Mfg. Co.)*

Figure 20-15 shows an actual application of limit switches on a machine. A conductor termination enclosure is also shown. This enclosure carries terminal blocks for the various electrical components located in that area.

c. **Contacts.** The electrical contacts on limit switches are very similar to contacts on other electrical components. They will not hold up under excessive overloads. Even at normal loads, excessive speed of operation tends to create local heating in the contact, which can be damaging. Excessive voltage and high inductive loads will also contribute to contact failure.

A few of the factors that affect sensing and therefore the output switching action are:

- The speed and direction of the material that is to be sensed
- The size, shape, and material that is to be sensed
- Sensitivity adjustment required to properly use the type and amount of energy received by the sensing head

In general, the use of the proximity switch requires greater overall knowledge of electrical work. Due to its relatively complex design as compared to the mechanical or vane switches, more complex maintenance problems can be expected. Maintenance problems generally lie with the amplifier and/or switching unit. Unless there is a complete familiarity with these units, it is better practice to change the complete unit and allow the manufacturer to make the repairs. Plug-in elements make this arrangement relatively simple.

*Figure 20-15*   Industrial application of a limit switch with circuit conductor terminal enclosure.
*(Courtesy of HPM Corporation.)*

Opposite polarities should not be connected to the contacts of a limit switch unless the switch is designed for such service. Likewise, power from different sources should not be connected to the contacts of a single switch. Normally, if good circuit design is followed, applying proper polarity does not become a problem.

In many cases the use of double-pole switches can replace two single-pole switches. This arrangement may eliminate a relay that would otherwise be required to provide the extra circuit.

**2. Vane Switches.** Many of the problems noted for mechanical limit switches can be avoided with the vane limit switch. For example, all integral parts are encapsulated within a cast-aluminum enclosure. This construction prevents the entrance of dust, dirt, coolants, oil, and lubricants into the switch mechanism. **Caution:** Misuse can lead to maintenance problems.

The following are factors to consider when using the vane switch.

    **a. Excessive contact load.** The contacts on the vane switch are rated at 0.75 A make and 0.2 A continuous at 115 V, 60 Hz. This rating is considerably below the contact rating of the heavy-duty mechanical limit switch. The normal procedure is to use a relay in conjunction with the vane switch to increase its output capacity.

    **b. Vane design.** The operation of the vane switch is accomplished by passing an operating vane through the vane slot. The proper size and shape of the vane and the direction and accuracy of the vane entrance into the slot become design problems for each specific application to give maximum operating efficiency.

Figure 20-16 shows various designs of vane switch operators.

**3. Proximity switches.** Most of the information noted for the vane switch can be applied to the proximity switch. In particular, the proximity switch has the advantage of operating in adverse environments.

Under some conditions, the requirements for a separate output unit for switching action may be a disadvantage. One example is supplying space for the amplifier and switching units where many switches are required.

**20.3.2 Accessibility of Components**  Concerns about accessibility should include the components in the control cabinets, push-button station, and on the machine. Any maintenance problem can be made easier by accessibility.

Problems with accessibility may arise from inadequate planning in the original design state or from lack of cooperation between mechanical and electrical assembly and installation.

For the maintenance worker, the first problem is to locate the trouble. Then the faulty component and/or conductors must be repaired or replaced.

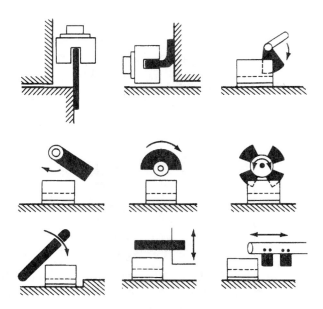

**Figure 20-16**   **Various designs of cams for use with vane limit switches. Arrows indicate direction of cam travel.** *(Courtesy of General Electric Co.)*

The second problem is time. Downtime is directly proportional to the time it takes to locate the trouble and repair or replace faulty parts. Downtime is a loss both in production and labor. Accessibility of components is important for easy maintenance in order to minimize downtime.

For the components in the three groups mentioned (the control cabinets, push-button station, and machine), a few suggestions follow.

### 1. Control Cabinets
    a. Keep the components within an easy and safe range of working height: approximately 2 feet to $5\frac{1}{2}$ feet above the floor.
    b. Design the control panels so they can be removed through the cabinet door opening.
    c. Do not stack terminal blocks on top of one another.
    d. Space components and terminal blocks so that the terminals are accessible for adding or removing conductors.
    e. Use relays with features such as building block assembly for time delay, mechanical latch, and additional standard relay contacts. The changing of NO and NC contacts or the universal contact arrangement is useful.

### 2. Push-Button Stations
    a. Space the individual units so that the wiring to the contact block is accessible.
    b. Where multiple units are used, it may be desirable to hinge the mounting plate to the enclosure so the units are accessible for checking.

### 3. The Machine

    a. Motors should be located so the mounting means are accessible and the motor can be easily removed. The conduit box should be clear so the cover can be removed.

    b. In the installation of pressure switches, shut-off valves should be provided in the fluid power lines for the removal of the switch unit.

    c. Temperature switches installed in a reservoir for temperature indication or control should be easily removed without draining the reservoir.

    d. Solenoids on valves should be clear of connecting tubes, pipes, conduits, or other mechanical devices. The electrical connections to the solenoids should be accessible for disconnecting. Their connection should be of a bolt-and-nut type and insulated with approved tape. The bolts or screws securing the solenoid to the valve should be accessible for complete removal of the solenoid from the valve.

        Except for the oil-immersed solenoids now used on many valves, solenoid valves should not be installed immersed in an oil tank.

    e. Planning for conduit and/or wireways and pull boxes within the bed or main structure of a machine always pays off. It clears, from the outside of the machine, a multitude of conduits that might otherwise interfere with component installation. Also, it provides for access to the conduits through pull boxes for easier adding or pulling of conductors within the machine.

**20.3.3 Numbering and Identification of Components**    The use of the layout drawing is of help in locating the area for components installed on a machine. A further help is to identify these units on the machine with a permanent name plate. The nameplate should carry the identification of the unit exactly as noted on the layout and elementary diagram.

    If at all possible, the nameplate should not be attached directly to the component. It should be attached to one side and on a permanent base, because if it is on the component being removed, the nameplate may be lost or may not be changed to the new unit installed.

**20.3.4 Use of Indicating Lights**    It is possible to use many indicating lights on a machine, but this condition is generally the exception. More often, the problem is the absence of indicating lights that give valuable information.

    The number of lights is usually in direct proportion to the size and complexity of the control system. For example, a machine with ten motors may have ten motor run indicating lights, as compared to one light with one motor.

    When a machine goes through multiple sequences, it is valuable to know that each preceding sequence has been completed and the next sequence started.

    The machine may have several types of operations available, set up through selector or drum switches. In such cases a light should indicate which type of operation is being used.

Safety and indicating lights are closely tied. For example, with multiple operators, each operator should have a visual indication of the action that the other operators have taken. Other examples of valuable indicating lights are oil temperature, over or under safe operating conditions; safety guards that may have been removed; excessive loading; overtravel; and malfunction of a component in a critical part of a cycle.

**20.3.5  Use of Plug-in Components**   Components such as relays, limit switches, solenoids, timers, and pyrometers are available in convenient plug-in designs.

The use of a plug-in component provides many advantages. However, it does not solve all problems. The following points should be considered:

- Does experience indicate a high rate of failure in a particular application? Can a different component give a better life? Can operating conditions be improved, or is replacement the best solution?
- Will downtime for changing the component seriously affect production?
- Does the nature of the component (complexity of circuit) indicate a change rather than repair on the machine?
- Can repair work be accomplished at a bench easier and faster than on a machine?
- Is the machine a prototype or custom design that may require many changes before a final design is reached?
- Is the level of maintenance experience at a given user's shop such that disconnecting and reconnecting circuits on a component may become a problem?
- Are changes being considered by the designer or user for an additional or different type of control after a short time or use?

# Recommended Web Links

Students are encouraged to view the following Web sites as a supplement to the concepts presented in this textbook. Review and analyze the array of products that are available for electrical control applications. Many of these sites offer technical information that can help in converting the principles to practical applications. To view catalogs, your PC may require Adobe Acrobat Reader software.

## Industrial Control Panels

The Panel Shop
http://www.thepanelshop.com/
Review: Projects

Bold Systems
http://www.boldsystems.com/Industrial.html
Review: Custom Enclosures and Consoles

AEC Corp.
http://www.aec-corp.com/appelec/index.htm
Review: Industrial Control Panels

ESL Electrical Systems
http://www.eslsys.com/
Review: Control Panels

Gerlach, Inc.
http://www.gerlachinc.com/service/service.htm
Review: The seven photos linked at bottom of page

## Enclosures:

Hoffman
http://www.hoffmanonline.com/
Review: Markets—Industrial

Adalet
http://www.adalet.com/
Note: Adalet specializes in explosion-proof enclosures

Enclosure Manufacturers
http://www.enclosuremanufacturers.com/
Review: The major enclosure manufacturers and scan their product lines

## Control Panel Accessories

Panduit
http://www.panduiteg.com/default.asp
Review: The various components used in the construction of a control panel

## Terminal Blocks

Allen-Bradley
http://www.ab.com/industrialcontrols/products/t_blocks/
Review: The different features and electrical characteristics

Extensive Review of Terminal Blocks
www.controleng.com

click on "Issue Archive"
click on "2001"
click on "March"
Look under "Product Focus—Terminal Blocks"

## Wire

Okonite
http://www.okonite.com/

USA Wire and Cable
http://www.usawire-cable.com/
Review: The electrical characteristics of the different types of wire and cable.

## Assembly Tools

Ideal Industries
http://www.idealindustries.com/tools_and_totes/
Review: The different hand operated and automatic tools used in control circuit construction and maintenance.

## Control Panel Layout Software

Hoffman PanelDraw
http://www.hoffmanonline.com/PanelDraw/index.htm

AutoDesk—Building Electrical
http://www.autodesk.com
Review: Products & Solutions, Building Electrical, Product Information

Microsoft Visio
http://www.microsoft.com/office/visio/default.asp
Review: The feature or functions of this software.

## Photos of PLC Panels

EDS-ISI
http://www.eds-isi.com/isi_plc.htm
Note the different layouts and components.

## Achievement Review

1. What is the importance of the use of circuit numbers? Why would a pattern of numbers as used on circuits between various machines in a plant help troubleshoot problems? Explain your answer.

2. List a few items that would help to clean up an electrical circuit diagram and make it easier to read.

3. Sketch a typical control panel showing the location of components, terminal blocks, wiring duct, and the clear unused space. Use your own judgment as to what would be good design.

4. List the advantages and disadvantages of bringing all conductors from limit switches, solenoids, and oil-tight units back to a control panel for checking purposes.

5. In using mechanically operated limit switches, what are the limitations on impact speed, overtravel, and contact rating?

6. Are there contact limitations when using vane limit switches? Would you recommend the use of a vane switch in a dusty atmosphere?

7. List five suggestions to follow when mounting and installing electrical components in a control cabinet or push-button station.

8. What are some advantages of indicating lights?

9. What factors would you consider in deciding on the use of plug-in components?

10. The generally accepted size of drawings is _____ and multiples thereof.
    a. $8\frac{1}{2}'' \times 11''$
    b. $10'' \times 20''$
    c. $9'' \times 12''$

11. Why is the use of terminal blocks for electrical control on a machine important?

    a. To aid in checking circuits.
    b. They are always required when pilot lights are used.
    c. For changing voltage levels.

12. When mounting components on a panel in a control cabinet, they should be located within a range of _____ feet above the floor.

    a. 1–3
    b. 2–5½
    c. 1–6

# Appendix A
# Summary of Electrical Symbols

## Common Electrical Symbols Used in This Textbook

Conductors Connected

Conductors Not Connected

Battery

Thermocouple

Relay Coil

NO Relay Contact

NC Relay Contact

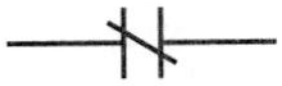

Time-Delay Relay Coil

Instantaneous Contact, NO

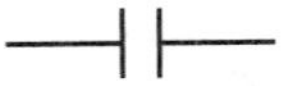

Instantaneous Contact, NC

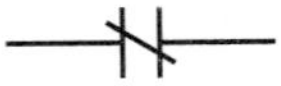

No Timing Contact Delay After Energizing

NC Timing Contact Delay After Energizing

NO Timing Contact Delay After Deenergizing

NC Timing Contact Delay After Deenergizing

Two-Circuit Push-Button Switch

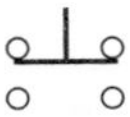

Voltmeter

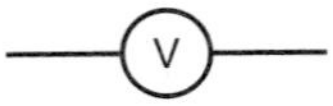

Ammeter

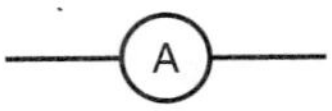

Solenoid Coil

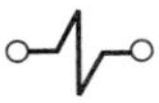

Timer
Motor

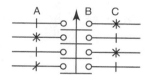

Timer
Clutch

Contacts
(operational sequence must be
shown above contact: open 0, closed ×)

Selector or Drum Switch
(—*— indicates contact closed)

Dual-Primary, Single-Secondary
Control Transformer

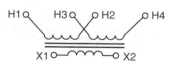

Full-Voltage Magnetic
Motor Starter

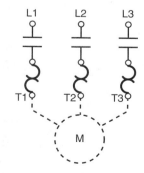

Limit Switch, NO

Limit Switch, NO, Held Closed

Limit Switch, NC

Limit Switch, NC, Held Open

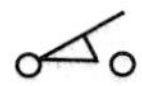

Pressure Switch, NO

Pressure Switch, NC

Temperature Switch, NO

Temperature Switch, NC

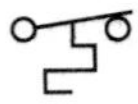

Proximity Limit Switch, NO

Proximity Limit Switch, NC

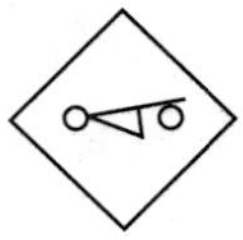

Ferrule-Type Fuse

Contactor Coil

Contacts

Heating Element

Thermal Overload Relay

Ground Connection

Three-Pole Fused Disconnect
Switch, Ganged with Handle

Three-Pole Circuit Interrupter,
Ganged with Handle

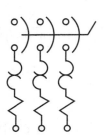

Three-Pole, Thermal-Magnetic Circuit
Breaker, Ganged with Handle

## Graphic Symbols for Solid-State Logic Diagrams (Distinctive Shape)

Three-input AND

Three-input OR

Two-input EXCLUSIVE-OR

Three-input NAND

Three-input NOR

NOT or INVERTER

# Appendix B
# Units of Measurement

**Metric System of Measurement**
Principal unit for length: meter
Principal unit for capacity: liter
Principal unit for weight: gram

**Prefixes Used for Subdivisions and Multiples**

pico-(p) = $10^{-12}$  deka-(da) = $10^{1}$
nano-(n) = $10^{-9}$  hecto-(h) = $10^{2}$
micro-($\mu$) = $10^{-6}$  kilo-(k) = $10^{3}$
milli-(m) = $10^{-3}$  mega-(M) = $10^{6}$
centi-(c) = $10^{-2}$  giga-(G) = $10^{9}$
deci-(d) = $10^{-1}$  tera-(T) = $10^{12}$

## Measures of Length
10 millimeters (mm) = 1 centimeter (cm)
10 centimeters (cm) = 1 decimeter (dm)
10 decimeters (dm) = 1 meter (m)
1000 meters (m) = 1 kilometer (km)

## Square Measure
100 square millimeters ($mm^2$) = 1 square centimeter ($cm^2$)
100 square centimeters ($cm^2$) = 1 square decimeter ($dm^2$)
100 square decimeters ($dm^2$) = 1 square meter ($m^2$)

## Cubic Measure
1000 cubic millimeters ($mm^3$) = 1 cubic centimeter ($cm^3$)
1000 cubic centimeters ($cm^3$) = 1 cubic decimeter ($dm^3$)
1000 cubic decimeters ($dm^3$) = 1 cubic meter ($m^3$)

## Dry and Liquid Measure
10 milliliters (mL) = 1 centiliter (cL)
10 centiliters (cL) = 1 deciliter (dL)
10 deciliters (dL) = 1 liter (L)
100 liters (L) = 1 hectoliter (hL)

1 liter = 1 $dL^3$ = the volume of 1 kilogram of pure
water at a temperature of 4°C (35.6°F).

## Measure of Weight
10 milligrams (mg) = 1 centigram (cg)
10 centigrams (cg) = 1 decigram (dg)
10 decigrams (dg) = 1 gram (g)
10 grams (g) = 1 dekagram (dag)
10 dekagrams (dag) = 1 hectogram (hg)
10 hectograms (hg) = 1 kilogram (kg)
1000 kilograms (kg) = 1 (metric) ton (t)

# Metric and U.S. Customary Conversions

## Linear Measure

1 kilometer (km)   = 0.6214 mile
1 meter (m)        = 3.2808 feet
1 centimeter (cm)  = 0.3937 inch
1 millimeter (mm) = 0.03937 inch

1 mile (mi) = 1.609 kilometers (km)
1 yard (yd) = 0.9144 meter (m)
1 foot (ft)  = 0.3048 meter (m)
1 inch (in)  = 2.54 centimeters (cm)

## Square Measure
1 square kilometer ($km^2$) = 0.3861 square mile = 247.1 acres
1 hectare ($ha^2$) = 2.471 acres = 107.640 square feet
1 acre ($a^2$) = 0.0247 acre = 1076.4 square feet
1 square meter ($m^2$) = 10.764 square feet = 1.196 square yards
1 square centimeter ($cm^2$) = 0.155 square inch
1 square millimeter ($mm^2$) = 0.00155 square inch

1 square mile ($mi^2$) = 2.5899 square kilometers
1 acre (acre) = 0.4047 hectare = 40.47 acres
1 square yard ($yd^2$) = 0.836 square meter
1 square foot ($ft^2$) = 0.0929 square meter = 929 square centimeters
1 square inch ($in^2$) = 6.452 square centimeters = 645.2 square millimeters

## Cubic Measure

1 cubic meter ($m^3$) = 35.314 cubic feet = 1.308 cubic yards
1 cubic meter ($m^3$) = 264.2 U.S. gallons
1 cubic centimeter ($cm^3$) = 0.061 cubic inch
1 liter (L) (cubic decimeter) = 0.0353 cubic foot = 61.023 cubic inches
1 liter (L) = 0.2642 U.S. gallon = 1.0567 U.S. quarts

1 cubic yard ($yd^3$) = 0.7645 cubic meter
1 cubic foot ($ft^3$) = 0.02832 cubic meter = 28.317 liters
1 cubic inch ($in^3$) = 16.38716 cubic centimeters
1 U.S. gallon (gal) = 3.785 liters
1 U.S. quart (qt) = 0.946 liter

## Weight

1 metric ton (t) = 0.9842 ton (of 2240 pounds) = 2204.6 pounds
1 kilogram (kg) = 2.2046 pounds = 35.274 ounces avoirdupois
1 gram (g) = 0.03215 ounce troy = 0.3527 ounce avoirdupois
1 gram (g) = 15.432 grains

1 ton (t) (of 2240 pounds) = 1.016 metric ton = 1016 kilograms
1 pound (lb) = 0.4536 kilogram = 453.6 grams
1 ounce (oz) avoirdupois = 28.35 grams
1 ounce (oz) troy = 31.103 grams
1 grain (gr) = 0.0648 gram

1 kilogram per square millimeter ($kg/mm^2$) = 1422.32 pounds per square inch
1 kilogram per square centimeter ($kg/cm^2$) = 14.223 pounds per square inch

1 kilogram-meter (kg·m) = 7.233 foot-pounds (ft-lb)
1 pound per square inch (psi) = 0.0703 kilogram per square centimeter
1 kilocalorie (kcal) = 3.968 British thermal units (Btu)

# Heat

### Thermometer Scales

On the Fahrenheit (F) thermometer, the freezing point of water is marked at 32 degrees (°) on the scale, and the boiling point of water at atmospheric pressure is marked at 212°. The distance between these points is divided into 180°.

On the Celsius (C) thermometer, the freezing point of water is marked at 0° on the scale, and the boiling point of water is marked at 100°.

The following formulas are used for converting temperatures:

$$\text{Degrees Fahrenheit} = \frac{9 \times \text{degrees Celsius}}{5} + 32$$

$$\text{Degrees Celsius} = \frac{5 \times (\text{degrees Fahrenheit} - 32)}{9}$$

## Units Derived from the Base Electrical Unit (Ampere, A)

| Measurement | Derived Unit of Measurement | Formula |
|---|---|---|
| Electrical potential | volt (V) | V = W/A |
| Electrical resistance | ohm (Ω) | Ω = V/A |
| Electrical capacitance | farad (F) | F = A·(s/V) |
| Quantity of electricity | coulomb (C) | C = A·s |
| Electrical inductance | henry (H) | H = Wb/A |
| Magnetic flux | weber (Wb) | Wb = V·s |

*Note*: s = seconds

# Appendix C
# Rules of Thumb for Electric Motors

Here are some simple ways to take the mystery out of determining several important factors concerning electric motors.

## C.1 Horsepower Versus Amperes

For three-phase motors we find that approximately:

1 hp - 575 V requires 1 A of current
1 hp - 460 V requires 1.25 A of current
1 hp - 230 V requires 2.50 A of current
1 hp - 2300 V requires 0.25 A of current

The key to these relationships is 1 hp - 1 A on 575 V. All others then become a direct ratio of this figure, e.g., 575/460 times 1.0 equals 1.25 A for 1 hp required on 460 V. Now, for example, with these facts at hand, you can say that a 150-hp motor on 230 V will require approximately 375 A. This fact is obtained as follows: 1 hp on 230 V requires 2.5 A per horsepower, so you simply multiply 2.5 times 150 hp.

## C.2 Horsepower Revolutions per Minute—Torque

*Torque* is simply a twisting force that causes rotation around a fixed point. For example, torque is something we all experience every time we pass through a revolving door. *Horsepower* is what is required when we pass through the door because we are exercising torque at a certain rate of revolutions per minute (rpm). The faster we go through the revolving door, the more horsepower is required. In this case, the torque remains the same. In relating this concept to motors, we use this rule of thumb:

> 1 hp at 1800 rpm delivers 3 ft.-lb. of torque
>
> 1 hp at 900 rpm delivers 6 ft.-lb. of torque

Using this simple rule, we can see that a 10-hp motor at 1800 rpm delivers 30 ft.-lb. of torque, a 20-hp motor at 1800 rpm delivers 60 ft.-lb. of torque, and a 1-hp motor at 1200 rpm delivers 4.5 ft.-lb. of torque. Here is how to figure torque at a different operating speed. Multiply the torque at 1800 rpm by the ratio 1800/1200. Torque, then, is the inverse ratio of the speed. In other words, for the same horsepower: SPEED DOWN—TORQUE UP.

A quick estimate for the torque of a 125-hp, 600-rpm motor can be figured by the following procedure: 125 hp times 3 ft.-lb. equal 375 ft.-lb. of torque for a 125-hp motor at 1800 rpm. Now, to convert to 600 rpm, multiply 375 times 1800/600 rpm or 1125 ft.-lb.

This rule will enable you to quickly determine the torque a motor is capable of delivering, down to 10 rpm. Below this speed, other factors must be taken into consideration.

## C.3 Shaft Size—Horsepower—Revolutions per Minute

Remember the number 1150. It stands for the fact that a 1-inch diameter shaft can transmit 1 hp at 50 rpm. As the shaft speed goes up, so does the horsepower, and by the same ratio. Therefore, if you double the speed, you double the horsepower capacity of the shaft. However, when you double the shaft diameter, the capacity of the shaft to transmit horsepower is increased 8 times. Thus, whatever the shaft size is in inches, cube it and multiply the resulting figure by the proper speed ratio, and the horsepower-transmitting ability of the shaft is determined.

To express this relationship in a formula, use the following:

$$\text{Shaft horsepower} = \frac{(\text{Shaft diameter in inches}) \times \text{rpm}}{50}$$

However, it is advisable to be conservative, so modify the results by 75%.

# Appendix D
# Electrical Formulas

1. Synchronous speed, frequency, and number of poles of ac motors and generators

$$\text{rpm} = \frac{120 \times f}{P} \qquad\qquad f = \frac{P \times \text{rpm}}{120} \qquad\qquad P = \frac{120 \times f}{\text{rpm}}$$

where:  rpm = revolutions per minute
   $f$ = frequency, in cycles per second
   $P$ = number of poles

## 2. Power in ac and dc circuits

### Alternating Current

| To Find | Three Phase | Two Phase *(Four Wire) | Single Phase | Direct Current |
|---|---|---|---|---|
| Amperes when hp is known | $I = \dfrac{746 \times hp}{1.73 \times E \times Eff \times PF}$ | $I = \dfrac{746 \times hp}{2 \times E \times Eff \times PF}$ | $I = \dfrac{746 \times hp}{E \times Eff \times PF}$ | $I = \dfrac{746 \times hp}{E \times Eff}$ |
| Amperes when kW is known | $I = \dfrac{1000 \times kW}{1.73 \times E \times PF}$ | $I = \dfrac{1000 \times kW}{2 \times E \times PF}$ | $I = \dfrac{1000 \times kW}{E \times PF}$ | $I = \dfrac{1000 \times kW}{E}$ |
| Amperes when kVA is known | $I = \dfrac{1000 \times kVa}{1.73 \times E}$ | $I = \dfrac{1000 \times kVa}{2 \times E}$ | $I = \dfrac{1000 \times kVa}{E}$ | — |
| Kilowatts input | $kW = \dfrac{1.73 \times E \times I \times PF}{1000}$ | $kW = \dfrac{2 \times E \times I \times PF}{1000}$ | $kW = \dfrac{E \times I \times PF}{1000}$ | $kW = \dfrac{E \times I}{1000}$ |
| Kilovolt-amperes | $kVA = \dfrac{1.73 \times E \times I}{1000}$ | $kVA = \dfrac{2 \times E \times I}{1000}$ | $kVA = \dfrac{E \times I}{1000}$ | — |
| Horsepower output | $hp = \dfrac{1.73 \times E \times I \times Eff \times PF}{746}$ | $hp = \dfrac{2 \times E \times I \times Eff \times PF}{746}$ | $hp = \dfrac{E \times I \times Eff \times PF}{746}$ | $hp = \dfrac{E \times I \times Eff}{746}$ |

I   = Amperes  
E   = Volts  
Eff = Efficiency, in decimals  
PF = Power factor, in decimals  

kW  = Kilowatts  
kVA = Kilovolt-amperes  
hp  = Horsepower output  

*For two-phase, three-wire balanced circuits, the amperes in common conductor = 1.41 times that in either of the other two.

# Appendix E
# Use of Electrical Codes and Standards

## E.1  Major Goals

All electrical codes and standards are written with at least two major goals in mind: maximum safety in operation and long, trouble-free operating life. These goals are best accomplished through good basic engineering design.

It is hard to write a single comprehensive code or standard because of the problems involved in applying electrical control. There are many different types of machines and user requirements. From the small machine shop to the large production-line manufacturer, conditions and requirements differ. Even within a given user's plant, conditions change from section to section and from time to time within the same section. Also, the needs or requirements of today will not necessarily be those of tomorrow.

During the past years, many groups have coordinated their efforts in writing codes and standards. In most cases they are written from the experience and knowledge of those groups. Therefore, codes and standards reflect the concepts of good design as held by the group that writes them. Differences, additions, or omissions among groups do not necessarily rank one group above another. Rather, it is an indication that they are written for a specific application in a certain locality or field. This specificity, of course, presents a problem to the manufacturer who supplies electrically controlled machines to every group.

One of the manufacturing problems for the builder is economics. (This statement is not meant to imply that good design is in direct proportion to cost. Often a poor design costs more.) The problems vary from one builder to another and arise in the fields of engineering, marketing, and production.

Problems are usually resolved by close cooperation between the builder and user on each specific machine or group of machines. Both the builder and user must be concerned with the need for codes and standards. From a practical sense they must weigh the relative merits of economy against complete adherence to safety requirements. Safety must be the top consideration, however.

The builder and user should be acquainted with the NFPA 79, *Electrical Standard for Industrial Machinery*, which applies to metal working machine tools and plastics machinery. The NFPA electrical standard and others are reviewed, revised, and reissued on a regular basis. In addition, codes of local municipalities and states should be followed where applicable.

The codes and standards referred to in this text are those of application. They cover, for example, how the circuit is to be presented, and how, where, and what components are to be installed on a machine or in a control cabinet.

## E.2 Recent Development of the NFPA 79 Standard

In June 1981 the Joint Industrial Council (JIC) Board of Directors acknowledged the dated state of the electrical and electronic standards and requested that the NFPA 79 incorporate the material and topics covered by the JIC electrical (EMP-1-67, EGP-1-67) and electronic (EL-1-71) standards with the intention that the JIC standards would eventually be declared superceded. The NFPA Standards Council approved the request with the stipulation that the material and topics incorporated from the JIC standards be limited to areas related to electrical shock and fire hazards. The 1985 edition reflected the incorporation of the appropriate material from the JIC electrical (EMP-1-67, EGP-1-67) standards not previously covered. The 1991 edition includes additional references to international standards. The 1994 edition of NFPA 79 was reformatted in accordance with the ANSI style manual (8th edition)

The next publication of NFPA 79 is scheduled for release in 2002/2003. In this release, the writing committee attempts to harmonize the NFPA 79 standard with HS1738, the SAE Electrical Standard for Industrial Machines and with EN60204, the European general requirements for electrical equipment of industrial machines. The ANSI style format is maintained in the release; however, the clauses are rearranged to match the European standard.

## E.3 Tables

The following 14 tables are to be used only for general information and reference. In specific industrial applications, the current and applicable codes and standards should be consulted. The tables that follow are reproduced from the 1994 NFPA 79 *Electrical Standard for Industrial Machinery* through the courtesy of the National Fire Protection Association Inc.

**Table 1 — Fuse and circuit breaker selection: motor, motor branch circuit, and motor controller**

| Application | Maximum setting or rating[1] (fuse and circuit breaker) | | |
|---|---|---|---|
| | Type[2] | | |
| Fuse class with time delay[3] | AC-2 | AC-3 | AC-4 |
| RK-5[4] | 150 | 175 | 175 |
| RK-1 | 150 | 175 | 175 |
| J | 150 | 175 | 225 |
| CC | 150 | 300 | 300 |
| Instantaneous trip C/B[5] | 700 | 700 | 700 |
| Inverse trip C/B[6] | 150 | 250 | 250 |

NOTES (to Table 1):

1) For motors with locked-rotor code letters a-e, see ANSI/NFPA 70, Table 430-152.

2) Types:
— AC-2: Slip-ring motors starting, switching off, or all light-starting duty motors.
— AC-3: Squirrel-cage motors; starting, switching off while running, occasional inching, jogging, or plugging but not to exceed 5 operations per minute or 10 operations per 10 minutes. All wye-delta and two-step autotransformer starting, or all medium-starting duty motors.
— AC-4: Squirrel-cage motors; starting, plugging, inching, jogging, or all heavy-starting duty motors.

3) Where the rating of a time-delay fuse (other than CC type) specified by the table is not sufficient for the starting of the motor, it shall be permitted to be increased but shall in no case be permitted to exceed 225 percent.

4) Class RK-5 fuses shall be used only with NEMA rated motor controllers.

5) Magnetic only circuit breakers are limited to single motor applications. These instantaneous trip circuit breakers shall only be used if they are adjustable, if part of a combination controller having motor-running and also short-circuit and ground-fault protection in each conductor, and if the combination is especially identified for use, and it is installed per any instructions included in its listing or labeling. Circuit breakers with adjustable trip settings shall be set at the controller manufacturer's recommendation, but not greater than 1300 percent of the motor full-load current.

6) Where the rating of an inverse time circuit breaker specified in the table is not sufficient for the starting current of the motor, it shall be permitted to be increased but in no case exceed:
— 400 percent for full-load currents of 100 amperes or less, or
— 300 percent for full-load currents greater than 100 amperes.
NOTE: IEC 947-4 defines the terms Type 1 and Type 2 coordinated protection as follows:
— Type 1 Protection: Under short-circuit conditions the contactor or starter may not be suitable for further use without repair or replacement.
— Type 2 Protection: Under short-circuit conditions the contactor or starter shall be suitable for further use.
The maximum allowable values in Table 1 do not guarantee Type 2 protection. Type 2 protection is recommended for use in applications where enhanced performance and reliability are required.

**Table 2 — Relationship between conductor size and maximum
rating or short-circuit protective device for power circuits**

| Conductor size AWG | Max. rating non-time-delay fuse or inverse time circuit breaker | Time delay or dual element fuse |
|---|---|---|
| 14 | 60 | 30 |
| 12 | 80 | 40 |
| 10 | 100 | 50 |
| 8 | 150 | 80 |
| 6 | 200 | 100 |
| 4 | 250 | 125 |
| 3 | 300 | 150 |
| 2 | 350 | 175 |
| 1 | 400 | 200 |
| 0 | 500 | 250 |
| 2/0 | 600 | 300 |
| 3/0 | 700 | 350 |
| 4/0 | 800 | 400 |

## Table 3 — Running overcurrent units

| Kind of motor | Supply system | Number and location of overcurrent units (such as trip coils, relays, or thermal cutouts) |
|---|---|---|
| 1-phase ac or dc | 2-wire, 1-phase ac or dc ungrounded | 1 in either conductor |
| 1-phase ac or dc | 2-wire, 1-phase ac or dc, one conductor grounded | 1 in ungrounded conductor |
| 1-phase ac or dc | 3-wire, 1-phase ac or dc, grounded-neutral | 1 in either ungrounded conductor |
| 3-phase ac | Any 3-phase | *3, one in each phase |

*Exception: Unless protected by other approved means.

NOTE: For 2-phase power supply systems see *National Electrical Code*®, Section 430-37.

## Table 4 — Control transformer overcurrent protection (primary voltage)

| Rated primary (amperes) | Maximum rating of primary overcurrent protective device as a percent of transformer rated primary current |
|---|---|
| Less than 2 | 300[1] |
| 2 or more | 250 |

[1]500 percent is permitted for a circuit of a control apparatus or system that carries the electric signals directing the performance of the motor controller, but does not carry the main power current.

### Table 5 — Control transformer overcurrent protection
### (120 volt secondary)

| Control transformer size, volt-amperes | Maximum rating, amperes |
|:---:|:---:|
| 50 | 0.5 |
| 100 | 1.0 |
| 150 | 1.6 |
| 200 | 2.0 |
| 250 | 2.5 |
| 300 | 3.2 |
| 500 | 5 |
| 750 | 8 |
| 1000 | 10 |
| 1250 | 12 |
| 1500 | 15 |
| 2000 | 20 |
| 3000 | 30 |
| 5000 | 50 |

NOTE: For transformers larger than 5000 volt-amperes, the protective device rating shall be based on 125 percent of the secondary current rating of the transformer.

**Table 6 — Ratings for three-phase, single-speed, full-voltage magnetic controllers for nonplugging and nonjogging duty**

| Size of controller | Continuous current rating[2] (amperes) | Horsepower [1,4] at | | | | Service-limit current rating[3] (amperes) |
| | | 60 Hertz | | 50 Hertz 380 volts | 60 Hertz 460 or 575 volts | |
| | | 200 volts | 230 volts | | | |
|---|---|---|---|---|---|---|
| 00 | 9 | $1\frac{1}{2}$ | $1\frac{1}{2}$ | $1\frac{1}{2}$ | 2 | 11 |
| 0 | 18 | 3 | 3 | 5 | 5 | 21 |
| 1 | 27 | $7\frac{1}{2}$ | $7\frac{1}{2}$ | 10 | 10 | 32 |
| 2 | 45 | 10 | 15 | 25 | 25 | 52 |
| 3 | 90 | 25 | 30 | 50 | 50 | 104 |
| 4 | 135 | 40 | 50 | 75 | 100 | 156 |
| 5 | 270 | 75 | 100 | 150 | 200 | 311 |
| 6 | 540 | 150 | 200 | 300 | 400 | 621 |
| 7 | 810 | — | 300 | — | 600 | 932 |
| 8 | 1215 | — | 450 | — | 900 | 1400 |
| 9 | 2250 | — | 800 | — | 1600 | 2590 |

Reference: ANSI/NEMA ICS-2, Table 2-321-1.

NOTE 1: These horsepower ratings are based on locked-rotor current ratings given in ANSI/NEMA ICS-2, Table 2-237-3. For motors having higher locked-rotor currents, a larger controller should be used so that its locked-rotor current rating is not exceeded. (Refer to ANSI/NEMA ICS-2 for horsepower ratings of single-phase, reduced voltage, or multispeed motor controller application.)

NOTE 2: The continuous-current ratings represent the maximum rms current, in amperes, that the controller may be expected to carry continuously without exceeding the temperature rises permitted by ANSI/NEMA ICS-1, Part 109 Part ICS1-109.

NOTE 3: The service-limit current ratings represent the maximum rms current, in amperes, the controller may be expected to carry for protracted periods in normal service. At service-limit current ratings, temperature rises may exceed those obtained by testing the controller at its continuous current rating. The current rating of overload relays or the trip current of other motor protective devices used shall not exceed the service-limit current rating of the controller.

NOTE 4: Refer to ANSI/NEMA ICS-2 for horsepower ratings of single-phase, reduced voltage, or multispeed motor controller applications.

**Table 7 — Ratings for three-phase, single-speed, full-voltage magnetic controllers for plug-stop, plug-reverse, or jogging duty**

| Size of controller | Continuous current rating[2] (amperes) | Horsepower [1,4] at | | | | Service-limit current rating[3] (amperes) |
|---|---|---|---|---|---|---|
| | | 60 Hertz | | 50 Hertz 380 volts | 60 Hertz 460 or 575 volts | |
| | | 200 volts | 230 volts | | | |
| 0 | 18 | 1½ | 1½ | 1½ | 2 | 21 |
| 1 | 27 | 3 | 3 | 5 | 5 | 32 |
| 2 | 45 | 7½ | 10 | 15 | 15 | 52 |
| 3 | 90 | 15 | 20 | 30 | 30 | 104 |
| 4 | 135 | 25 | 30 | 50 | 60 | 156 |
| 5 | 270 | 60 | 75 | 125 | 150 | 311 |
| 6 | 540 | 125 | 150 | 250 | 300 | 621 |

Reference: ANSI/NEMA ICS-2, Table 2-321-3.

NOTE I: These horsepower ratings are based on locked-rotor current ratings given in ANSI/NEMA ICS-2, Table 2-237-3. For motors having higher locked-rotor currents, a larger controller should be used so that its locked-rotor current rating is not exceeded. (Refer to ANSI/NEMA ICS-2 for horsepower ratings of single-phase, reduced voltage, or multispeed motor controller application.)

NOTE 2: The continuous-current ratings represent the maximum rms current, in amperes, the controller may be expected to carry continuously without exceeding the temperature rises permitted by ANSI/NEMA ICS-1, Part 109 Part ICS1-109.

NOTE 3: The service-limit current ratings represent the maximum rms current, in amperes, the controller may be expected to carry for protracted periods in normal service. At service-limit current ratings, temperature rises may exceed those obtained by testing the controller at its continuous current rating. The current rating of overload relays or the trip current of other motor protective devices used shall not exceed the service-limit current rating of the controller.

NOTE 4: Refer to ANSI/NEMA ICS-2 for horsepower ratings of single-phase, reduced voltage, or multispeed motor controller applications.

**Table 8 — Color coding for pushbuttons, indicator (pilot) lights, and illuminated pushbuttons**

| Color | Device Type | Typical Function | Examples |
|---|---|---|---|
| RED | Pushbutton | Emergency Stop, Stop, Off | Emergency Stop button, Master Stop button, Stop of one or more motors. |
| | Pilot Light | Danger or alarm, abnormal condition requiring immediate attention | Indication that a protective device has stopped the machine, e.g., overload. |
| | Illuminated Pushbutton | | Machine stalled because of overload, etc. (the color RED for the emergency stop actuator shall not depend on the illumination of its light). |
| YELLOW (AMBER) | Pushbutton | Return, Emergency Return, Intervention—suppress abnormal conditions | Return of machine elements to safe position, override other functions previously selected. Avoid unwanted changes. |
| | Pilot Light | Attention, caution/marginal condition. Change or impending change of conditions | Automatic cycle or motors running; some value (pressure, temperature) is approaching its permissible limit. Ground fault indication. Overload that is permitted for a limited time. |
| | Illuminated Pushbutton | Attention or caution/Start of an operation intended to avoid dangerous conditions | Some value (pressure, temperature) is approaching its permissible limit; pressing button to override other functions previously selected. |
| GREEN | Pushbutton | Start-on | General or machine start; start of cycle or partial sequence; start of one or more motors; start of auxiliary sequence; energize control circuits. |

| Color | Type | Meaning | Function |
|---|---|---|---|
| | Pilot Light | Machine Ready; Safety | Indication of safe condition or authorization to proceed. Machine ready for operation with all conditions normal or cycle complete and machine ready to be restarted. |
| | Illuminated Pushbutton | Machine or Unit ready for operation/Start or On | Start or On after authorization by light; start of one or more motors for auxiliary functions; start or energization of machine elements. |
| BLACK | Pushbutton | No specific function assigned | Shall be permitted to be used for any function except for buttons with the sole function of Stop or Off; inching or jogging. |
| WHITE or CLEAR | Pushbutton | Any function not covered by the above | Control of auxiliary functions not directly related to the working cycles. |
| | Pilot Light | Normal Condition Confirmation | Normal pressure, temperature. |
| | Illuminated Pushbutton | Confirmation that a circuit has been energized or function or movement of the machine has been started/Start-on, or any preselection of a function | Energizing of auxiliary function or circuit not related to the working cycle; start or preselection of direction of feed motion or speeds. |
| BLUE or GRAY | Pushbutton, Pilot Light, or Illuminated Pushbutton | Any function not covered by the above colors | |

For illuminated pushbuttons the function(s) of the light is separated from the function(s) of the button by a virgule (/).

## Table 9 — Single conductor characteristics

| Size (AWG/kcmil) | Cross-sectional area—nominal (CM/mm²) | DC resistance at 25°C (ohms/1000 ft) | Minimum number of strands | | |
|---|---|---|---|---|---|
| | | | Nonflexing (ASTM class) | Flexing (ASTM class) | Constant flex (ASTM class/ AWG size) |
| 22 | 640/0.324 | 17.2 | 7(') | 7(') | 19(M/34) |
| 20 | 1020/0.519 | 10.7 | 10(K) | 10(K) | 26(M/34) |
| 18 | 1620/0.823 | 6.77 | 16(K) | 16(K) | 41(M/34) |
| 16 | 2580/1.31 | 4.26 | 19(C) | 26(K) | 65(M/34) |
| 14 | 4110/2.08 | 2.68 | 19(C) | 41(K) | 41(K/30) |
| 12 | 6530/3.31 | 1.68 | 19(C) | 65(K) | 65(K/30) |
| 10 | 10380/5.261 | 1.060 | 19(C) | 104(K) | 104(K/30) |
| 8 | 16510/8.367 | 0.6663 | 19(C) | (\) | (\) |
| 6 | 26240/13.30 | 0.4192 | 19(C) | (\) | (\) |
| 4 | 41740/21.15 | 0.2636 | 19(C) | (\) | (\) |
| 3 | 52620/26.67 | 0.2091 | 19(C) | (\) | (\) |
| 2 | 66360/33.62 | 0.1659 | 19(C) | (\) | (\) |
| 1 | 83690/42.41 | 0.1315 | 19(B) | (\) | (\) |
| 1/0 | 105600/53.49 | 0.1042 | 19(B) | (\) | (\) |
| 2/0 | 133100/67.43 | 0.08267 | 19(B) | (\) | (\) |
| 3/0 | 167800/85.01 | 0.06658 | 19(B) | (\) | (\) |

| | | | |
|---|---|---|---|
| 4/0 | 211600/107.2 | 0.05200 | 19(B) |
| 250 kcmil | -/127 | 0.04401 | 37(B) |
| 300 | -/152 | 0.03667 | 37(B) |
| 350 | -/177 | 0.03144 | 37(B) |
| 400 | -/203 | 0.02751 | 37(B) |
| 450 | -/228 | 0.02445 | 37(B) |
| 500 | -/253 | 0.02200 | 37(B) |
| 550 | -/279 | 0.02000 | 61(B) |
| 600 | -/304 | 0.01834 | 61(B) |
| 650 | -/329 | 0.01692 | 61(B) |
| 700 | -/355 | 0.01572 | 61(B) |
| 750 | -/380 | 0.01467 | 61(B) |
| 800 | -/405 | 0.01375 | 61(B) |
| 900 | -/456 | 0.01222 | 61(B) |
| 1000 | -/507 | 0.01101 | 61(B) |

(B, C, K) ASTM Class designation B and C per ASTM B8-86, Class designation K per ASTM B 174-71 (R1980).

(') A class designation has not been assigned to this conductor but is designated as size 22-7 in ASTM B286-1989 (R1979) and is composed of strands 10 mils in diameter (no. 30 AWG).

(l) Nonflexing construction shall be permitted for flexing service * Per ASTM Class designation B 174-71 (R1980), Table 3.

(—) Constant flexing cables are not constructed in these sizes.

**Table 10 — Single conductor insulation thickness of
insulation in mils\* [average/minimum (jacket)]**

| Wire Size | A | B |
|---|---|---|
| 22 AWG | 30/27 | 15/13(4) |
| 20 | 30/27 | 15/13(4) |
| 18 | 30/27 | 15/13(4) |
| 16 | 30/27 | 15/13(4) |
| 14 | 30/27 | 15/13(4) |
| 12 | 30/27 | 15/13(4) |
| 10 | 30/27 | 20/18(4) |
| 8 | 45/40 | 30/27(5) |
| 6 | 60/54 | 30/27(5) |
| 4–2 | 60/54 | 40/36(6) |
| 1–4/0 | 80/72 | 50/45(7) |
| 250–500 MCM | 95/86 | 60/54(8) |
| 550–1000 | 110/99 | 70/63(9) |

(\*) UL 1063 Table 1.1 *NEC* Construction A-no outer covering,
B-Nylon covering

**Table 11 — Conductor ampacity based on copper conductors with 60°C and 75°C insulation in an ambient temperature of 30°C**

| Conductor size AWG | Ampacity in cable or raceway | | Control enclosure 60°C† |
|---|---|---|---|
| | 60°C | 75°C | |
| 30 | — | 0.5 | 0.5 |
| 28 | — | 0.8 | 0.8 |
| 26 | — | 1 | 1 |
| 24 | 2 | 2 | 2 |
| 22 | 3 | 3 | 3 |
| 20 | 5 | 5 | 5 |
| 18 | 7 | 7 | 7 |
| 16 | 10 | 10 | 10 |
| 14 | 15 | 15 | 20 |
| 12 | 20 | 20 | 25 |
| 10 | 30 | 30 | 40 |
| 8 | 40 | 50 | 60 |
| 6 | 55 | 65 | 80 |
| 4 | 70 | 85 | 105 |
| 3 | 85 | 100 | 120 |
| 2 | 95 | 115 | 140 |
| 1 | 110 | 130 | 165 |
| 0 | 125 | 150 | 195 |
| 2/0 | 145 | 175 | 225 |
| 3/0 | 165 | 200 | 260 |
| 4/0 | 195 | 230 | 300 |
| 250 | 215 | 255 | 340 |
| 350 | 260 | 310 | 420 |
| 400 | 280 | 335 | 455 |
| 500 | 320 | 380 | 515 |
| 600 | 355 | 420 | 575 |
| 700 | 385 | 460 | 630 |
| 750 | 400 | 475 | 655 |
| 800 | 410 | 490 | 680 |
| 900 | 435 | 520 | 730 |
| 1000 | 455 | 545 | 780 |

† Sizing of conductors in wiring harnesses or wiring channels shall be based on the ampacity for cables.

NOTE 1: Wire types listed in 15.1 shall be permitted to be used at the ampacities as listed in this table.

NOTE 2: For ambient temperatures other than 30°C, see *NEC* Table 310-16 correction factors.

NOTE 3: The sources for the ampacities in this table are Table 310-16 and 310-17 of the *NEC*.

**Table 12 — Maximum conductor size for given motor controller size***

| Motor controller size | Maximum conductor size AEG or MCM |
|:---------------------:|:---------------------------------:|
| 00 | 14 |
| 0 | 10 |
| 1 | 8 |
| 2 | 4 |
| 3 | 0 |
| 4 | 000 |
| 5 | 500 |

*See ANSI/NEMA ICS 2-1988 Table 2, 110-1.

## Table 13 — Minimum radius of conduit bends

| Size of conduit (in.) | Radius of bend done by hand (in.)[1] | Radius of bend done by machine (in.)[2] |
|---|---|---|
| $\frac{1}{2}$ | 4 | 4 |
| $\frac{3}{4}$ | 5 | $4\frac{1}{2}$ |
| 1 | 6 | $5\frac{1}{4}$ |
| $1\frac{1}{4}$ | 8 | $7\frac{1}{4}$ |
| $1\frac{1}{2}$ | 10 | $8\frac{1}{4}$ |
| 2 | 12 | $9\frac{1}{2}$ |
| $2\frac{1}{2}$ | 15 | $10\frac{1}{2}$ |
| 3 | 18 | 13 |
| $3\frac{1}{2}$ | 21 | 15 |
| 4 | 24 | 16 |
| $4\frac{1}{2}$ | 27 | 20 |
| 5 | 30 | 24 |
| 6 | 36 | 30 |

For SI units: (Radius) 1 in. = 25.4 mm.

NOTE 1: For field bends done by hand, the radius is measured to the inner edge of the bend.

NOTE 2: For a single-operation (one-shot) bending machine designed for the purpose, the radius is measured to the center line of the conduit.

Conduit shall be securely held in place and supported as follows:

| Conduit Size (In.) | Maximum Spacing Between Supports (Ft) |
|---|---|
| $\frac{1}{2}$–1 | 3 |
| $1\frac{1}{4}$–2 | 5 |
| $2\frac{1}{2}$–3 | 6 |
| $3\frac{1}{2}$–5 | 7 |
| 6 | 8 |

**Table 14 — Size of grounding conductors**

| Column "A," amperes | Copper conductor size, AWG |
|---|---|
| 10 | 16* or 18* |
| 15 | 14, 16*, or 18* |
| 20 | 12, 14*, 16*, or 18* |
| 30 | 10 |
| 40 | 10 |
| 60 | 10 |
| 100 | 8 |
| 200 | 6 |
| 300 | 4 |
| 400 | 3 |
| 500 | 2 |
| 600 | 1 |
| 800 | 0 |
| 1000 | 2/0 |
| 1200 | 3/0 |
| 1600 | 4/0 |

*Permitted only in multiconductor cable where connected to portable or pendant equipment.

# Appendix F
# Application of Electric Heat

## F.1 Calculating Heat Requirements

This appendix deals primarily with applications on and in the following:

- Relatively large metal objects, such as extruder barrels and platens for plastic molding machines
- Fluids used with hydraulic power and control systems

For the basic design of a heating element there are two important questions to answer:

1. How many watts (W) or kilowatts (kW; 1 kW = 1000 W) of electrical power will be required to bring the application to the required temperature in a given period of time?
2. How many watts or kilowatts of electrical power will be required to maintain the required temperature?

Under controlling conditions, the time required to bring the heated item from ambient temperature to the preset temperature, known as the *warm-up period,* varies depending on the amount of power applied. Unless the warm-up period of time is critical, it is generally more economical and practical to try to match the power applied to the load. The load requirements vary depending on the type of material to be heated, the percentage of time that the heat is required at the preset level, and heat losses. In practice, the power will generally be on 80% of the time to maintain the required temperature.

From a practical viewpoint, the required temperature is generally within a range. For example, in the plastic molding industry most materials are typically available within a 300°F to 700°F range. Therefore, if calculations are made for the top of the range (700°F), then the power will be on a lower percentage of the time and the warm-up period will be shorter for the bottom of the range (300°F).

It can be difficult to make the actual calculation of the watts or kilowatts of electrical power required to heat an object to a desired temperature in a definite time period. The main problem lies in determining heat losses, which include the following:

- Radiation from the heated object (affected by contour, condition of the surface exposed, and temperature difference between the heated item and the ambient)
- Material being processed (affected by type, amount, and rate of processing)
- Conduction to other parts of the machine (affected by temperature differential and type and placement of an insulating medium)

Tables and charts showing various types of heat losses are available from manufacturers of resistance heating elements. These items are a great help in approximating losses.

If the heating system can be assumed to be 100% efficient (no losses), and assuming that the object to be heated is steel, the information on kilowatt-hours (kW·h) required can be easily calculated:

$$kW \cdot h = \frac{\text{Weight of object in pounds} \times \text{specific heat of steel (0.12)} \times \text{(final temperature — initial temperature in degrees Fahrenheit)}}{3412 \times \text{warm-up period in hours}}$$

For example, given a 300-pound steel object, to raise its temperature from 70°F to 670°F in 1 hour the kilowatts required are:

$$\frac{300 \times 0.12 \times (670 - 70)}{3412 \times 1} = 6.3 \text{ kW}$$

If the warm-up period requirement is cut to one half-hour, then the kilowatt requirement doubles to 12.6 kW.

With cylindrical barrels such as are used on plastic injection molding machines, working in the 300°F to 700°F range, the designer will generally start with as many kilowatts for losses as were calculated for heating the barrel. The losses are based on the maximum operating temperature of 700°F. Operation is in an ambient temperature of 70°F.

## F.2 Selection and Application of Heating Elements

Basically, a *heating element* is resistance wire in a magnesium-oxide filler with a sheath around it. The sheath material may be steel, stainless steel, or copper. The material may be tubular or flat rectangular in cross-sectional shape.

After the total kilowatt requirements are determined, the quantity and wattage of the individual heating elements must be determined. Heating elements with the exact wattage calculated are not usually commercially available. The practical approach is to select the next larger heating element. For example, if requirements are for 975 W per unit, and this size is not available, select a 1000-W heating element.

The number of heating elements is generally dictated by the available mounting space. Coverage of the area available should be as near to 100% as is practical. The most important factor in selecting and installing heating elements is the transfer of heat, which is affected by mating surfaces and how securely the heating elements are held in place. Whether the application is a flat plate, groove, hole, or curved surface, it must be clean and smooth. Efficiency of any heating system is reduced as oxides or foreign materials enter between the heating element and the heated surfaces, thus acting as an insulator.

In the case of heating elements with a high watt density inserted in a groove or hole, the clearances should be held to 0.005 inch to 0.010 inch. (*Watt density* is the number of watts per unit area of surface, for example, watts per square inch [W/in$^2$]). Two types of heating elements used in industry are shown in Figures F-1 and F-2.

**Figure F-1** Cutaway section through a strip heating element. *(Courtesy of Chromalox.)*

**Figure F-2** Cutaway section through a cartridge heating element. *(Courtesy of Chromalox.)*

## F.3  Heating Element Control Circuits

Contactors for the transfer of electrical power are often used with heating elements. In some cases, the heating elements are thermostatically controlled. A circuit with this arrangement is shown in Figure F-3. If possible, the load should be arranged for a balanced, three-phase circuit. This arrangement minimizes the conductor size and may decrease sizes of the contactors and protective devices, as compared to single-phase connection. This topic is discussed further in Section F.3.

Referring to Figure F-3, the sequence of this heating element circuit is as follows:

1. Set 1TS at 70°F.
2. Set the control OFF/ON switch to ON.
3. Set the heat OFF/ON switch to ON.
4. Contactor 1CR is energized (with the temperature below 70°F).

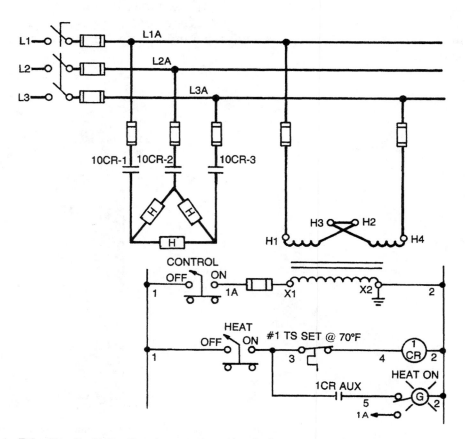

**Figure F-3**  Circuit with heating elements thermostatically controlled. Conductors carrying load current at line voltage are denoted by heavy lines.

  a. Power contacts 1CR-1, 1CR-2, and 1CR-3 close, energizing the heating elements.

  b. Auxiliary contact 1-CR Aux. closes, energizing the green pilot light.

5. When the temperature of the part or substance being heated exceeds 70°F, temperature switch 1TS opens.

6. The coil of contactor 1CR is deenergized.

7. Power contacts 1CR-1, 1CR-2, and 1CR-3 open, deenergizing the heating elements.

8. Auxiliary contact 1-CR Aux. opens, deenergizing the green pilot light.

From this description, it can be seen that as long as the temperature of the heated part or substance is below 70°F, the heating elements remain energized. When the temperature exceeds 70°F, the heating elements are deenergized.

There are applications in which it is necessary to combine temperature and motor control. For example, the temperature of bearings, operating fluid (oil), and material being processed must be monitored and controlled for the motor to be energized and remain energized during operation.

Figure F-4 shows such a combination of temperature and motor control. Temperature switches 1TS, 2TS, 3TS, and 4TS are in the machine oil reservoir. Temperature switches 5TS, 6TS, 7TS, and 8TS are located on four important separate bearings on the machine.

Referring to Figure F-4, the sequence of this circuit operation is as follows:

1. Set the control ON/OFF switch to ON.

2. Set the heat ON/OFF switch to ON.

With the oil temperature below 70°F, 1TS NC contact is closed, energizing contactor 10CR.

3. The power contacts on 10CR (10CR-1, 10CR-2, 10CR-3) close, energizing the heating elements.

  a. Auxiliary contact 10CR Aux. #1 closes, interlocking through 75°F TS #2 and around the 70°F TS #1.

  b. Auxiliary contact 10CR Aux. #2 closes, energizing the heat ON pilot light.

With the oil temperature below 125°F, normally closed #4TS is closed. With the four bearings in normal condition (temperature below 150°F), thermostat contacts #5TS, #6TS, #7TS, and #8TS are all closed.

4. Relay 2CR is energized.

  a. Normally open relay contact 2CR-1 closes.

  b. Normally closed relay contact 2CR-2 opens.

There are now two conditions under which the motor can be started:

  • The oil temperature is below 70°F. It will be necessary to wait until the heating elements raise the oil temperature to above 70°F, operating #1TS.

  • The oil temperature is above 70°F. The motor can be started immediately.

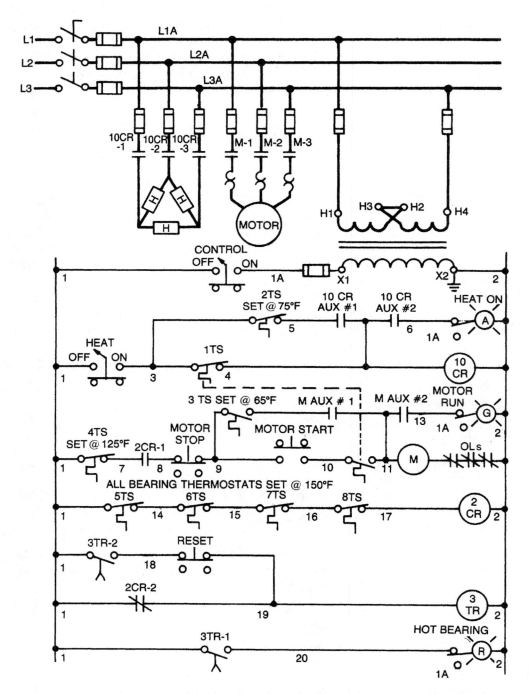

***Figure F-4*** **Circuit for combined temperature and motor control. Conductors carrying load current at line voltage are denoted by heavy lines.**

5. Operate the motor START push-button switch.
6. The motor starter contacts M1, M2, and M3 close, energizing the motor.
   a. Since the oil temperature is above 70°F, #3TS is closed.
   b. M Aux. #1 contact closes, interlocking through #3TS and around the motor START push-button switch and 1TS normally open contact.
   c. M Aux. #2 contact closes, energizing the motor run light.

7. The motor will remain energized until one or more of four events occur.
   a. The motor STOP push-button is operated.
   b. One or more of the four bearing temperatures exceeds 150°F. This condition deenergizes relay 2CR, opening relay contact 2CR-1.
   c. The oil temperature exceeds 125°F, opening #4TS.
   d. The oil temperature drops below 65°F, opening the interlock circuit around the motor START push-button switch.

8. Note that the heat control system keeps the oil temperature between 70°F and 75°F. Unless this system fails and the temperature drops below 65°F, thermostat #3TS will not open.

9. If one or more of the bearings exceed 150°F,
   a. Relay 2CR will be deenergized. Normally closed relay contact 2CR-2 closes, energizing time delay relay 3TR.
   b. Timing contact 3TR (on delay) 3TR-1 closes after a few seconds, energizing the hot bearing light.
   c. After the same few seconds delay, 3TR timing contact (on delay) 3TR-2 closes, interlocking around the normally closed relay contact 2CR-2.
   d. Even if the bearing cools down and the associated thermostat re-closes, 3TR will remain energized for inspection and troubleshooting.
   e. With the bearing problem corrected, the light can be deenergized by operating the reset push-button switch.

## F.4 Heating Element Connection Diagrams

The connection diagrams for electrical power heating circuits shown in this section are among the most useful found in industry.

In each case the heating elements are shown connected to a three-phase power line, represented by three parallel lines. Where the load is connected single phase, any two of the three lines are used. The connecting lines to the heating element load show the fuses used for protection and the contacts from a contactor. The contactor coil is in a heat control circuit (which is not shown).

These circuits are important in that they show the various arrangements in the connection of heating elements. For each connection, the current in each line feeding the load is calculated.

When either dc or ac is supplied to a resistance heating load, the resulting current (I) can be calculated from the applied voltage (V) and the rated wattage (W) of the heating element:

$$\text{Current in amperes} = \frac{\text{Watt rating}}{\text{Voltage applied}}$$

The first circuit, shown in Figure F-5, is connected single phase. The individual heating elements are rated at 240 V. Therefore, they are connected in par-

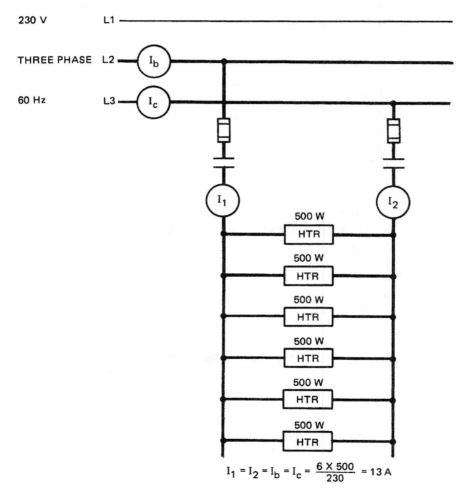

$$I_1 = I_2 = I_b = I_c = \frac{6 \times 500}{230} = 13 \text{ A}$$

*Figure F-5*  **Single-phase load using rated 240-V heating elements, connected on one phase of a three-phase circuit. Since only one single-phase load is connected to the three-phase line, line current $I_b$ = phase current and $I_1$ and line current $I_c$ = phase current $I_2$. Conductors carrying load current at line voltage are denoted by heavy lines.**

allel so that an applied voltage of 230 V is applied to each element. Note that the heating element industry rates the elements at 240 V even though they may be applied to a 230-V line.

In the circuit shown in Figure F-6, heating elements rated at 240 V are again used. However, they are connected single phase to a 460-V power line. Since the individual elements are only rated at 240 V, two units must be connected in series. It is important when making this connection that both elements be rated at 240 V and the same wattage to ensure that the line voltage is divided equally across the two elements. Note that since the resulting line current is equal to the wattage rating divided by the applied voltage, the resulting line current will be one half that of the parallel connection on 230 V.

Rather than having all six heating elements connected across one phase of a three-phase power line, they are now grouped in three groups of equal numbers. (There are two heating elements in each group.) These groups are connected across the three separate phases of a three-phase power line (L1-L2, L2-L3, L3-L1).

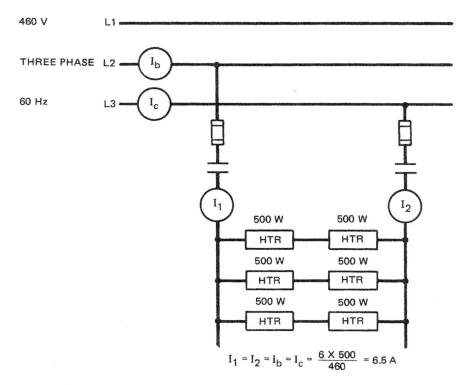

$$I_1 = I_2 = I_b = I_c = \frac{6 \times 500}{460} = 6.5 \text{ A}$$

*Figure F-6* Circuit with heating elements connected single phase to a 460-V three-phase power line. Since only one single-phase load is connected to the three-phase line, line current $I_b$ = phase current and $I_1$ and line current $I_c$ = phase current $I_2$. Conductors carrying load current at line voltage are denoted by heavy lines.

Where a three-phase power line is available, this type of connection has the advantage of reducing the current in the individual phase groups and in the three-phase line current. This condition often permits the use of smaller conductors, contactors, and fuses, resulting in cost savings. The line current in a three-phase delta connection is reduced by a factor of 1.73. Figure F-7 shows the relation-

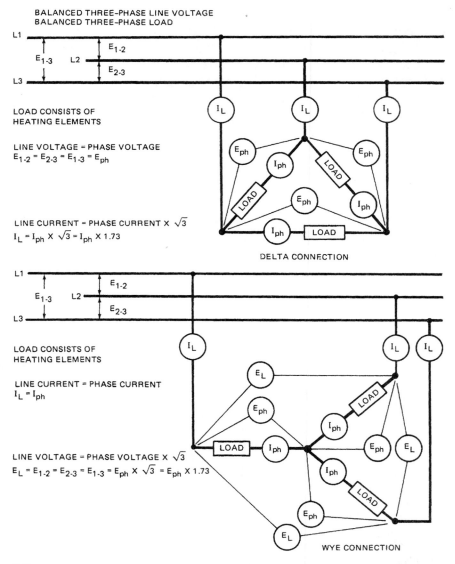

**Figure F-7** Relationship of current (I) and voltage (E) in three-phase circuits for delta (top) and wye (bottom) connection. Conductors carrying load current at line voltage are denoted by heavy lines.

ship of current (I) and voltage (E) in three-phase circuits for both delta and wye connections.

Figure F-8 shows the same heating elements connected to a 460-V, three-phase power line.

Figure F-9 shows a circuit using 240-V rated heating elements on a 230-V, three-phase line.

Sometimes it is desirable to reduce the heat to a "warm" level, such as on an installation for overnight or over a weekend.

Heat produced in a resistance heating element is proportional to the square of the voltage applied. Therefore, any reduction in the voltage applied will greatly reduce the heat. For example, if the voltage on a 240-V rated heating element is reduced from 230 V to 115 V, the heat reduction will be

$$\left(\frac{115}{230}\right)^2 = \left(\frac{1}{2}\right)^2 = \frac{1}{4}$$

If the heating elements can be arranged to first be applied to 230 V and then by switching be applied to 115 V, the resulting heat will be one-fourth the heat at 230 V.

Figure F-10 shows a single-phase circuit using either a selector switch or drum switch. The switching changes the connections from parallel to series.

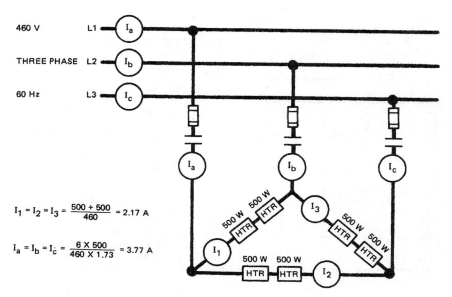

Figure F-8 Circuit for heating elements connected to a 460-V, three-phase power line. Conductors carrying load current at line voltage are denoted by heavy lines.

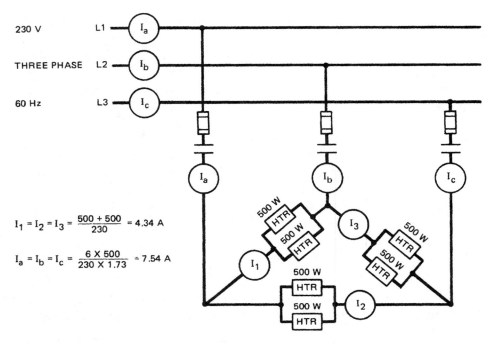

$$I_1 = I_2 = I_3 = \frac{500 + 500}{230} = 4.34 \text{ A}$$

$$I_a = I_b = I_c = \frac{6 \times 500}{230 \times 1.73} \approx 7.54 \text{ A}$$

***Figure F-9*** Circuit for 240-V rated heating elements on a 230-V, three-phase line. Conductors carrying load current at line voltage are denoted by heavy lines.

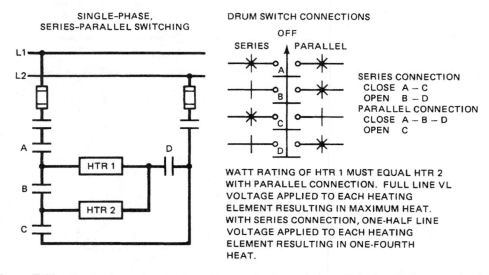

SINGLE-PHASE, SERIES-PARALLEL SWITCHING

DRUM SWITCH CONNECTIONS

OFF

SERIES    PARALLEL

SERIES CONNECTION
  CLOSE A – C
  OPEN   B – D
PARALLEL CONNECTION
  CLOSE A – B – D
  OPEN   C

WATT RATING OF HTR 1 MUST EQUAL HTR 2 WITH PARALLEL CONNECTION. FULL LINE VL VOLTAGE APPLIED TO EACH HEATING ELEMENT RESULTING IN MAXIMUM HEAT. WITH SERIES CONNECTION, ONE-HALF LINE VOLTAGE APPLIED TO EACH HEATING ELEMENT RESULTING IN ONE-FOURTH HEAT.

***Figure F-10*** Single-phase circuit using either a selector or drum switch. Conductors carrying load current at line voltage are denoted by heavy lines.

Figure F-11 shows the same type of series-parallel switching applied to a three-phase delta circuit.

Figure F-12 shows a three-phase connection where the full line voltage is applied to each heating element connected in a delta arrangement. By switching the heating elements into a wye connection, the voltage across each element is reduced. The heat is reduced to 33% of that when the elements were connected in delta.

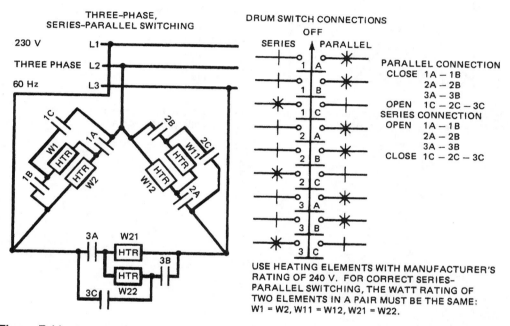

**Figure F-11**  Series-parallel switching applied to a three-phase delta circuit. Conductors carrying load current at line voltage are denoted by heavy lines.

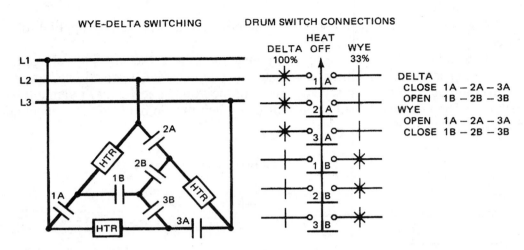

**Figure F-12**  Three-phase connection for full line voltage to each heating element connected in a delta arrangement. Conductors carrying load current at line voltage are denoted by heavy lines.

# Appendix G
# Power Factor Correction

## G.1  Apparent Power and Actual Power

Consideration of power factor is essential in manufacturing plants that use relatively large amounts of alternating-current power, such as in plants using a large number of three-phase, squirrel-cage induction motors that are lightly loaded. In this appendix, the problem of power factor is explained.

*Power factor* describes the ratio between *apparent power* and *actual* or *useful power*. These two power components will be present when squirrel-cage induction motors are used.

To understand apparent power, assume that a coil is wound on a magnetic core and connected to a source of direct current. A switch is provided to close and open the circuit (Figure G-1). When switch A is closed, a current flows through the coil, setting up a magnetic field around the coil (Figure G-2). The magnetic field remains as long as the switch is closed and current flows in the coil. If the switch is opened, the magnetic field collapses, causing the lines of force to cut through the winding. This action causes a current to flow through the coil in the direction opposite to the current that flowed into the coil.

Now, assume that in place of using a direct-current source of power, the coil is connected to a source of alternating-current power (Figure G-3). With alternating current, the current flows in one direction, dies to zero, and then flows in the opposite direction (Figure G-4). Each time the current goes to zero, the lines of force collapse and cut through the winding. The current that flows out of the coil builds to a maximum value just as the supply current drops to zero. The current that flows out of the coil is said to be *out of phase* with the supply current.

If there is no loss in the coil due to heating, the current flowing into the coil (supply current) and the current flowing out of the coil due to the collapsing field will have the same value. If an ammeter is placed in the circuit, it will indicate a current. The amount of current will depend on the design of the coil.

*483*

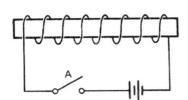

**Figure G-1**   Electromagnetic circuit with switch connected to dc power.

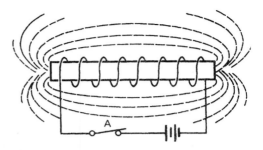

**Figure G-2**   With switch closed, current flows through the coil, producing magnetic field.

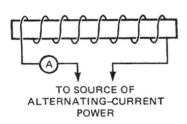

TO SOURCE OF
ALTERNATING–CURRENT
POWER

**Figure G-3**   Circuit connected to ac power.

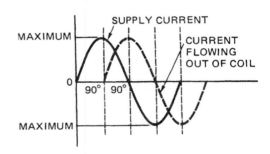

**Figure G-4**   Flow of ac current.

It would appear that the coil (electromagnet) is consuming power, but that is not the case. The current is simply flowing back and forth. A rough mechanical analogy can be made by considering a coil spring being compressed and released. It is from this action of the current flowing back and forth that the term *apparent power* is obtained.

Actual, or *useful, power* is that power used in a heating element such as an electric iron or light bulb. Actual or useful power is also present in the induction motor when the load is applied on the shaft.

When a squirrel-cage induction motor is connected to a source of alternating current, a current flows in the windings of the stator. This current produces a revolving magnetic field. The lines of force from this magnetic field cut through the squirrel-cage rotor and set up a second magnetic field. The second magnetic field reacts against the field of the stator, forcing the rotor to rotate.

With no load connected to the motor, all the actual or useful power required is that amount necessary to overcome the iron and copper losses. There are also friction and windage losses caused by the rotor turning. Therefore, most of the power used by a motor running idle is apparent power.

When a load is applied to the motor, such as connecting it to a pump, real power is supplied. This real power increases directly with the increase of load.

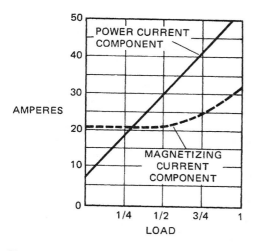

**Figure G-5** Graph comparing power current and magnetizing current at various loads.

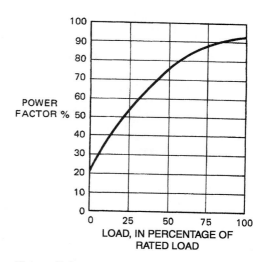

**Figure G-6** Typical power factor curve of a square-cage induction motor.

Thus, there are two components of current flowing in the motor: *magnetizing current* and *power current* (Figure G-5).

The ratio between useful power and apparent power can be expressed as

$$\text{Power factor} = \frac{\text{Useful power}}{\text{Apparent power}}$$

A typical curve of a squirrel-cage induction motor (Figure G-6) shows that there is a low power factor when the motor is running at light loads. Therefore, the motor requires a large amount of apparent power compared to the amount of useful power.

## G.2 Magnetizing Current and Power Current

Our explanation of apparent power began with describing current in a coil. The two current components, magnetizing and power, are present in the induction motor. However, they cannot be added directly because the magnetizing current is 90 degrees out of phase with the power current. The magnetizing current is said to *lag* the power current by 90 degrees.

Magnetizing current and power current can be added by using a triangle (Figure G-7). By knowing the amount of magnetizing current and power current, a triangle can be drawn to scale. By scaling the drawing, the total current can be determined.

When working with power, it is generally more convenient to work with the units *kilowatts* (kW), *kilovolt-amperes* (kVA), and *reactive kilovolt-amperes* (kVAR). Since the prefix *kilo* means 1000,

- 1 kilowatt-1000 watts (watts = volts × amperes)

- kilovolts-amperes (kVAC) $= \dfrac{\text{amperes} \times \text{volts}}{1000}$

- reactive kilovolts-amperes (kVAR) $= \dfrac{\text{reactive amperes} \times \text{volts}}{1000}$

The triangle can now be called a *power triangle,* as shown in Figure G-8.

For example, using values of current in amperes and power in kilowatts, assume the following: A three-phase, 480-V squirrel-cage induction motor is carrying a line current of 121 A in each phase. The kilowatt load, measured with a wattmeter, is 80 kW. The kVA of a three-phase circuit is

$$kVA = \frac{A \times V \times \sqrt{3}*}{1000}$$

Therefore,

$$kVA = \frac{121 \times 480 \times 1.73}{1000}$$

$$= 100$$

The horizontal line of the power triangle is 80 kW. The power triangle is drawn to scale in Figure G-9.

As shown in Figure G-9, the extreme points of the horizontal line are marked A and B, and a vertical line is drawn down from A. Using point B, an arc

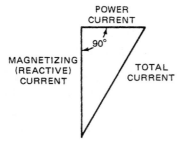

**Figure G-7**  Power triangle showing power current is 90° out of phase with magnetizing current.

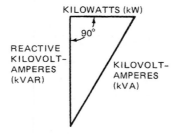

**Figure G-8**  Power triangle showing power (WATTS) is 90° out of phase with reacting power (kVAR).

*The factor of $\sqrt{3}$, or 1.73, is used because it is a three-phase circuit.

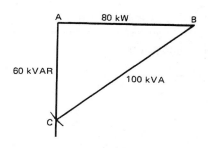

**Figure G-9** Power triangle without power factor correction.

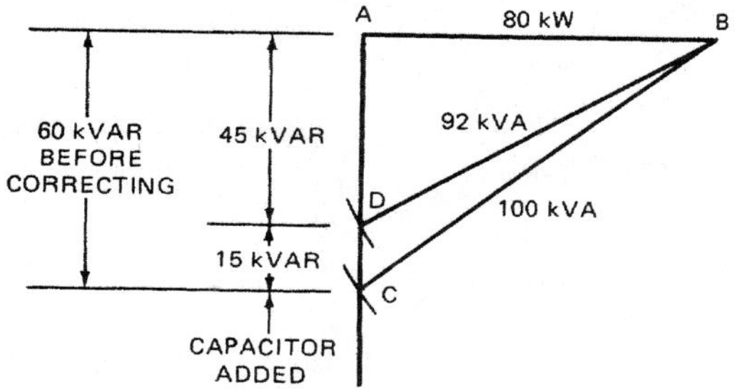

**Figure G-10** Power triangle with power factor correction.

is struck with a radius equal to the 100 kVA. The intersection of this arc with the vertical line is marked C. The reactive kVA can now be determined by measuring distance AC. The power factor may be expressed as the ratio of kilowatts, or working power, to the total kilovolt-amperes, or apparent power:

$$\text{Power factor} = \frac{AB}{BC} = \frac{kW}{kVA} = \frac{80}{100} = 80\%$$

The objective is to improve the power factor. Therefore, the magnetizing current, or kVAR, must be reduced in proportion to the power current, or kilowatts. To achieve this proportional reduction, a capacitor can be connected across the three lines feeding the motor. The capacitor will supply magnetizing current.

Using the example in Figure G-9, assume that a 15-kVAR capacitor is connected across the motor feed lines. As the capacitor supplies a leading power factor current, the kVAR of the capacitor is subtracted from the kVAR now being supplied to the motor (Figure G-10).

As represented by BD, the new kVA is 92. The power factor (PF) is now 87%, as follows:

$$PF = \frac{kW}{kVA} = \frac{80}{92} = 87\%$$

# G.3 Determining the Amount of Correction Required

A reasonable approximation of the capacitor size can be obtained from the following rule of thumb: For a 1200-rpm motor, use a figure of 1 kVAR for each 5 hp of induction motor load. This figure will vary slightly with motor speed and the amount of power factor correction required.

A more exact method is to use a table such as the one shown in Figure G-11, which gives the values of capacitors to correct power factor to approximately 95%.

| Horsepower | Motor synchronous Speed (rpm) | Capacitor rating correct power factor to approximately 95%* (kVAR) |
|---|---|---|
| 20 | 3600<br>1800<br>1200<br>900 | 3.0<br>4.5<br>5.0<br>8.0 |
| 30 | 3600<br>1800<br>1200<br>900 | 4.0<br>5.0<br>7.5<br>8.5 |
| 40 | 3600<br>1800<br>1200<br>900 | 5.0<br>5.5<br>9.0<br>14.0 |
| 50 | 3600<br>1800<br>1200<br>900 | 6.5<br>7.0<br>10.0<br>14.5 |
| 60 | 3600<br>1800<br>1200<br>900 | 7.5<br>10.0<br>12.0<br>17.5 |
| 75 | 3600<br>1800<br>1200<br>900 | 9.5<br>10.5<br>16.5<br>18.0 |
| 100 | 3600<br>1800<br>1200<br>900 | 13.0<br>15.0<br>17.5<br>22.0 |
| 150 | 3600<br>1800<br>1200<br>900 | 20.0<br>22.5<br>36.0<br>45.0 |

*When these exact sizes are not commercially available, use the next smaller size.

*Figure G-11*   Capacitor sizes for power factor correction.

There are many areas in an electrical system in which low power factor can present a problem. The concern can start with the alternator in the power company's station and continue through power distribution lines and transformers (both step up at the power station and step down at the user's plant). The distribution conductors in the user's plant are also involved.

Low power factor is of primary concern to the power company, as it must supply the magnetizing current that goes throughout the system. Therefore, it is to the power company's advantage if the power factor can be kept as high as practical (90–95%). As a result, most power companies include a penalty clause in their rate structure for low power factor and/or a discount for high power factor.

A high power factor also provides an advantage for the user. For any given distribution system in a user's plant, increased machine load can be added without increasing the size of the power plant feeders and associated equipment by improving the power factor.

In the example shown in Figure G-10, a capacitor is used to correct power factor. Using a capacitor is one of the most simple, direct, and relatively inexpensive methods for correcting power factor, and it can be made on the incoming lines or at individual motors. Remember, the correction from the capacitor always occurs toward the source of power.

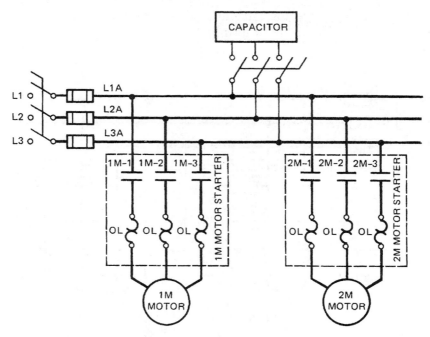

*Figure G-12*  Capacitors connected on the line bus.

If correction is desired only for the purpose of gaining rate advantage, one or more capacitors can be added on the incoming lines. A more satisfactory method for improving the user's load power factor is to connect individual capacitors on the load terminals of each motor starter. However, the optimum condition in installation economy for any given user's plant, may be the use of both methods. It is not unusual to find that savings in the user's power bill pay for the capacitors in one or two years.

Figures G-12 and G-13 show two installation circuits using capacitors.

Generally, it is not economical to apply capacitors directly to motors below 20 hp. If a plant has a predominance of small motors (below 20 hp), connecting capacitors on the line bus is practical (Figure G-12). However, this type of application can become difficult. It is suggested that the local power company be consulted on specific applications.

A check can be made to determine if too large a capacitor has been added on a motor. A voltmeter is attached across any two terminals of the motor (load side of starter). The load is disconnected by deenergizing the motor starter. The resultant rise in voltage as observed on the meter should not exceed 150% of the rated voltage on the motor nameplate.

The most practical location for the installation of a capacitor is on the outside of the motor starter enclosure. However, with long leads (more than 10 feet) from the starter to the motor, the capacitor should be mounted near the motor and the connections made in the motor terminal box.

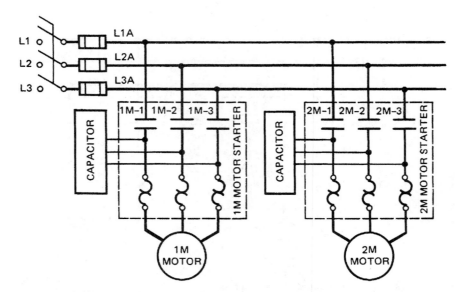

**Figure G-13** Capacitors connected across motors for power factor correction. Conductors carrying load current at line voltage are denoted by heavy lines.

**Figure G-14** Power factor correction capacitors mounted on a machine. *(Courtesy of HPM Corporation.)*

Figure G-14 illustrates the use of capacitors mounted on a machine. The connections are made in the motor terminal box.

## G.4 Typical Capacitor Design Features*

Figure G-15 shows a typical capacitor as used for power factor correction. The following design features are listed in reference to the number designations on the photograph.

1. *Case*—The capacitor case is made from heavy-gauge mild steel with all joints welded and reinforced at point of wear for protection of working element during handling, shipment, and service. Capacitor to be either single phase or three phase internally delta connected.
2. *Gland-Sealed Bushing*—The gland-sealed bushing assembly is constructed such that the silicon rubber, thick-walled gland is to be compressed between the porcelain insulator, copper stud, and the top cover of the capacitor. This compression seal will prevent leakage of the dielectric fluid even if the bushing assembly is tilted 15 degrees, which occurs during installation of capacitor elements.
3. *Working Element*—The working element consists of individually wound sections of special capacitor-grade paper, low-loss capacitor-grade synthetic

---

*Information provided by TECO-Westinghouse Motor Company.

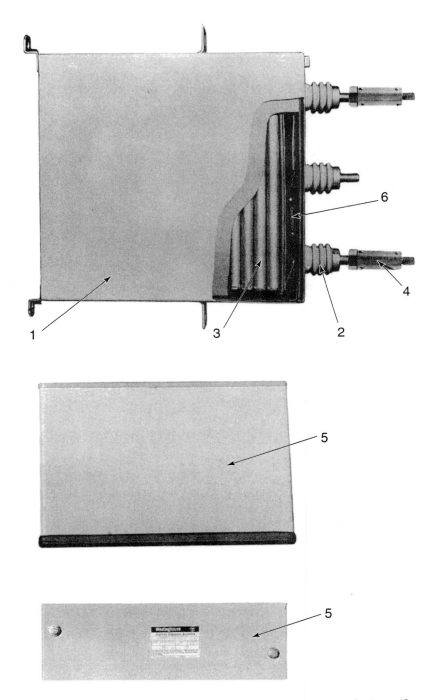

**Figure G-15**   Typical capacitor used for power factor correction. See text for key.   *(Courtesy of TECO-Westinghouse Motor Company.)*

film, and soft aluminum foil. Multiple sections to be arranged to yield the specified KVAC at rated voltage. The grouped sections are insulated from the steel case by means of high purity Kraft insulation material.

4. *Fuses*—Each three-phase capacitor unit is protected by full range current-limiting fuses that have an interrupting capacity of 200,000 amps.

5. *Dust Cover*—A dust cover is provided consisting of a bushing enclosure and cover. The bushing enclosure is formed from 12-gauge mild steel with a neoprene gasket positioned between the bottom of the enclosure and the top of the capacitor unit. A bolt-on steel cover compresses a vinyl gasket to assure a dustproof and weatherproof seal between the top cover and the bushing enclosure. The dust cover is bolted to the top of the capacitor unit by means of weld nuts and slotted head studs enclosing the bushings, fuses, and connections.

6. *Discharge Resistors*—Individual internal discharge resistors are connected across the terminals to reduce the residual voltage to 50 V or less within one minute after removal from the energized circuit.

# Appendix H
# Concepts Used in Programmable and Solid-State Controllers

## H.1  Number System

One familiar numbering system is known as the decimal system. A numbering system that uses ten digits (0 through 9) is called base 10. Take, for example, the decimal number 1111. The development of this number is shown in Figure H-1.

In programmable control it has been noted that it is necessary to deal with only two digits (1 or 0) because the signals are either ON or OFF (switches closed or open). A two-digit system is a *binary* system. A numbering system that uses only two digits (1 or 0) is called base 2.

| Thousands | Hundreds | Tens | Units |
|:---:|:---:|:---:|:---:|
| $Base^3 = 10^3$ | $Base^2 = 10^2$ | $Base^1 = 10^1$ | $Base^0 = 10^0$ |
| $10^3 = 10 \times 10 \times 10 = 1000$ | $10^2 = 10 \times 10 = 100$ | $10^1 = 10 \times 1$ | $10^0 = 1$ |
| $1000 \times 1 = 1000$ | $100 \times 1 = 100$ | $10 \times 1 = 10$ | $1 \times 1 = 1$ |

|  |  |  |
|:---|:---|---:|
| Adding the | Thousands | 1000 |
| | Hundreds | 100 |
| | Tens | 10 |
| | Units | 1 |
| | | 1111 |

*Figure H-1*  The decimal (base 10) numbering system.

In binary counting, when any column holding a 1 receives another count it goes back to 0 and develops a carry count to the next significant column to the left. Examples of counting in binary are shown in Figure H-2.

Note that in decimal notation, digits 0–9 require only one column. In binary notation they require four columns. If the decimal notation requires three columns, for example 173, the binary notation will require eight columns.

Using this larger decimal number $173_{10}$, the binary number is $10101101_2$. Note the decimal number is to the base 10 while the binary number is to the base 2.

The binary number can be converted to the decimal number as shown in Figure H-3.

| Decimal$_{10}$ | Binary$_2$ |
|:---:|:---:|
| 0 | 0000 |
| 1 | 0001 |
| 1 | 0001 |
| + 1 | + 0001 |
| = 2 | = 0010 |
| 2 | 0010 |
| + 1 | + 0001 |
| = 3 | = 0011 |
| 3 | 0011 |
| + 1 | + 0001 |
| = 4 | = 0100 |

Using the above development of binary equivalents of digits (0–4) all of the decimal digits of 0 through 9 are then expressed as follows:

| Decimal Notation | Binary Notation |
|:---:|:---:|
| 0 | 0000 |
| 1 | 0001 |
| 2 | 0010 |
| 3 | 0011 |
| 4 | 0100 |
| 5 | 0101 |
| 6 | 0110 |
| 7 | 0111 |
| 8 | 1000 |
| 9 | 1001 |

*Figure H-2*   **Examples of decimal and binary counting and notation.**

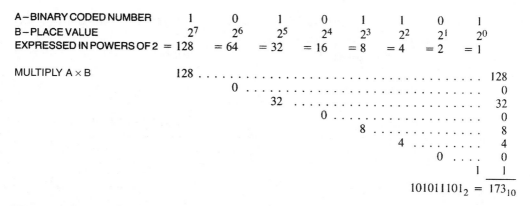

*Figure H-3* **Conversion of binary to decimal numbers.**

The decimal number 173 can now be converted back to the binary number by dividing by 2 (the base of the binary system). The remainder at each division (if any) then becomes the binary number in its proper position (Figure H-4).

Each digit of a binary number is called a *bit,* from BInary digiT. The binary number 10101101 as shown in Figure H-4 has 8 bits. A group of 8 bits is called a *byte.* A group of one or more bytes is called a *word.*

Figure H-5 shows a binary number composed of 16 bits, with the least significant bit and the most significant bit.

In storing large numbers, it is more practical to convert to what is known as the *binary coded decimal* (BCD). In this system, each decimal number is

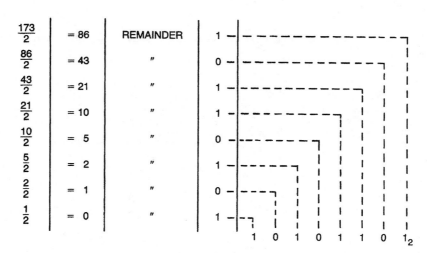

*Figure H-4* **Conversion of the decimal number 173 to the binary number 10101101.**

represented by four binary digits. This system is identified by adding the letters BCD at the right of the units place (Figure H-6).

The BCD number can be converted to its decimal equivalent as shown in Figure H-7.

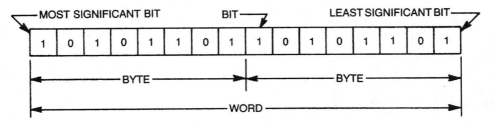

*Figure H-5*  **A binary number composed of 16 bits.**

| DECIMAL$_{10}$ | BINARY$_2$ |
|:---:|:---:|
| 1 | 0001 |
| 7 | 0111 |
| 3 | 0011 |

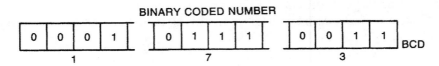

*Figure H-6*  **Using the decimal number 173 as an example of converting to the binary coded decimal system.**

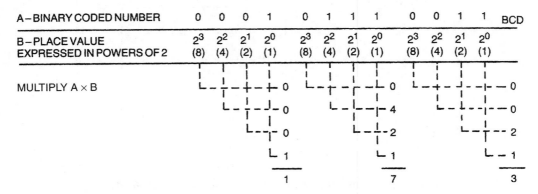

*Figure H-7*  **Converting a binary coded decimal number to its decimal equivalent.**

Another binary system used is the *octal* system (base 8). This system is made up of eight digits, 0 through 7. The first digit to the left of the octal point has a power of $8^0$. The next place to the left has a power of $8^1$ or 8. The next place to the left has a power of $8^2$ or 64. Any octal number will be designated by placing an 8 to the right of the units place, as shown in Figure H-8.

The *binary coded octal* (BCO) system uses three binary digits to represent one octal (base 8) digit. For example, the number $173_8$ is broken down into three groups: (3 bits), (3 bits), and (3 bits), as shown in Figure H-9.

The *hexadecimal* system (sometimes referred to as HEX) is another binary system that generally appears in manufacturer-supplied and after-market programming systems for programmable controllers.

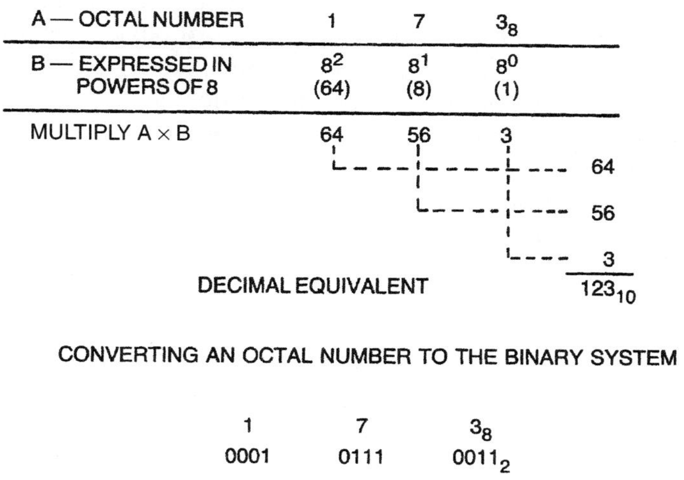

| A — OCTAL NUMBER | 1 | 7 | $3_8$ |
|---|---|---|---|
| B — EXPRESSED IN POWERS OF 8 | $8^2$ (64) | $8^1$ (8) | $8^0$ (1) |
| MULTIPLY A × B | 64 | 56 | 3 |

$$64$$
$$56$$
$$3$$
DECIMAL EQUIVALENT $\qquad 123_{10}$

CONVERTING AN OCTAL NUMBER TO THE BINARY SYSTEM

| 1 | 7 | $3_8$ |
|---|---|---|
| 0001 | 0111 | $0011_2$ |

*Figure H-8*   Converting an octal number to its decimal and binary equivalents.

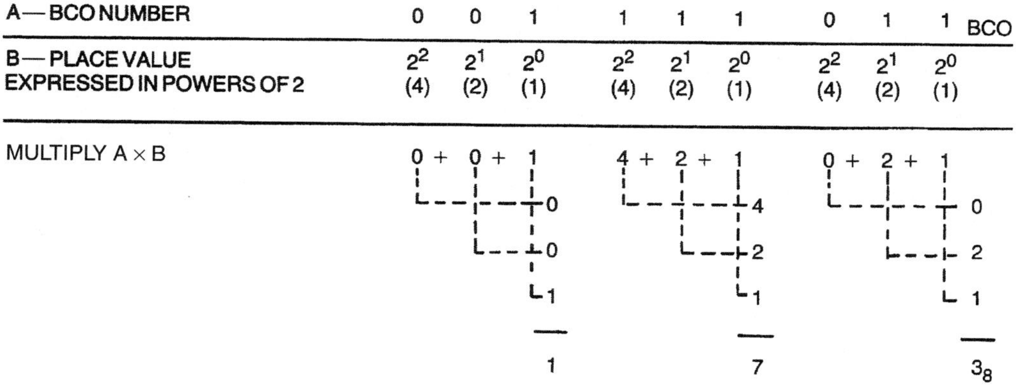

| A — BCO NUMBER | 0 | 0 | 1 | 1 | 1 | 1 | 0 | 1 | 1 | BCO |
|---|---|---|---|---|---|---|---|---|---|---|
| B — PLACE VALUE EXPRESSED IN POWERS OF 2 | $2^2$ (4) | $2^1$ (2) | $2^0$ (1) | $2^2$ (4) | $2^1$ (2) | $2^0$ (1) | $2^2$ (4) | $2^1$ (2) | $2^0$ (1) | |
| MULTIPLY A × B | 0 + | 0 + | 1 | 4 + | 2 + | 1 | 0 + | 2 + | 1 | |

|   | 1 | 7 | $3_8$ |
|---|---|---|---|

*Figure H-9*   Converting a binary coded octal number to its octal equivalent.

The HEX system has a base of 16. It consists of 16 digits: numbers 0 through 9 and letters A through F. These letters are substituted for the numbers 10 through 15.

In the table shown in Figure H-10, the decimal number and binary and hexadecimal equivalents are listed.

See Figure H-11 on converting a decimal number to a hexadecimal number; Figure H-12 on converting a hexadecimal number to a decimal number; and Figure H-13 on converting a hexadecimal number to a binary number.

In one additional numbering system, letters, numbers, and symbols are used. This is the American Standard Code for Information Interchange (ASCII). The entire alphabet of 26 letters (A through Z), numbers 0 through 9, and mathematical and punctuation symbols such as #, @, $, %, &, and others are used. This code can be made up of 6, 7, or 8 bits. The standard ASCII character sets use a 7-bit code. This provides all possible combinations of characters used when communicating with interfaces and peripheral equipment.

The manufacturers of programmable controllers vary in the numbering system they use to store information in the form of binary digits (bits) in the memory system of the CPU.

Regardless of the system used, it is important to remember that the information is always stored as 1s or 0s.

| Decimal | Binary | Hexadecimal |
|---------|--------|-------------|
| 0 | 0000 | 0 |
| 1 | 0001 | 1 |
| 2 | 0010 | 2 |
| 3 | 0011 | 3 |
| 4 | 0100 | 4 |
| 5 | 0101 | 5 |
| 6 | 0110 | 6 |
| 7 | 0111 | 7 |
| 8 | 1000 | 8 |
| 9 | 1001 | 9 |
| 10 | 1010 | A |
| 11 | 1011 | B |
| 12 | 1100 | C |
| 13 | 1101 | D |
| 14 | 1110 | E |
| 15 | 1111 | F |

*Figure H-10*   Decimal, binary, and hexadecimal equivalents.

$$\frac{4476}{16} = 279 \quad \text{REMAINDER} \quad 12$$

$$\frac{279}{16} = 17 \quad " \quad 7$$

$$\frac{17}{16} = 1 \quad " \quad 1$$

$$\frac{1}{16} = 0 \quad " \quad 1$$

|   |   |   |   |    |
|---|---|---|---|----|
| 1 | 1 | 7 | C | 16 |

*Figure H-11* Converting the decimal number $4476_{10}$ to a hexadecimal number.

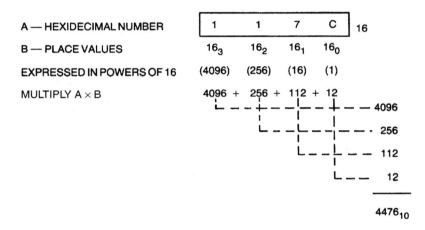

A — HEXIDECIMAL NUMBER

B — PLACE VALUES

EXPRESSED IN POWERS OF 16

MULTIPLY A × B

*Figure H-12* Converting a hexadecimal number to a decimal number.

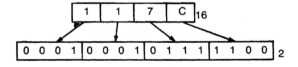

*Figure H-13* Converting a hexadecimal number to a binary number.

# H.2 Logic Gate Symbols and Circuits

Logic gate circuits are made using discrete combinations of diodes, transistors, and resistors.

The basic function of a *transistor* is that of a switch capable of accepting and combining multiple command signals. The resulting signal is then directed to produce an output or work function. The early use of solid-state elements involved mounting the individual units (such as diodes, transistors, resistors, and capacitors) on a small board. The individual units were known as *discrete components.* The board circuit was then wired or soldered according to a specific circuit design.

Almost all industrial electronic equipment contains one or more integrated circuits. An integrated circuit (IC) is a single monolithic chip of semiconductor material in which the electronic circuit elements are fabricated.

For example, a typical integrated circuit might contain 2 zener diodes, 5 diodes, 11 transistors, and 5 resistors combined on a single silicon chip. The size of the chip would measure only about 0.05 inch. Thus, the use of integrated circuits has given the electronics field a great space savings.

Regardless of whether discrete components or integrated circuits are used, several different functions can be developed. Each function can be represented by a symbol. Some of the basic symbols and their uses in designing a control circuit are discussed in the following section. These symbols are generally used in the design of solid-state control circuits.

## H.3 Symbols (IEEE STD. 91-1973—ANSI Y32, 14-1973; Distinctive Shape Used)

Contacts are either open or closed in electromechanical control. They have only two conditions. In solid-state control there also are only two conditions or states. These conditions deal with two voltage levels:

- HIGH or LOW

Other descriptions of such dual conditions include:

- TRUE or FALSE
- ON or OFF
- CLOSED or OPEN
- 1 or 0

In the nineteenth-century, a form of algebra was developed by an Englishman named George Boole. It was used at that time as a way of expressing statements that could be either true or false and is called Boolean algebra.

With the advent of solid-state control in which conditions called for a two-valued logic concept, i.e., TRUE-FALSE, ON-OFF, 1-0, etc., Boolean algebra fit in as an easy way to analyze and express logic statements.

In the material that follows, the system used is known as *positive logic.* In this system, two binary digits (1 or 0) are used to represent the two conditions. The digit 1 represents the higher voltage. The digit 0 represents the lower voltage.

Referring again to electromechanical control, a normally open contact on a relay closes when the relay coil is energized. The relay contact opens when the relay coil is deenergized. That is, the relay contacts have only two conditions: they are either open or closed. In a somewhat similar manner, when solid-state logic functions switch from 1 to 0, or from 0 to 1, they have only two conditions. Therefore, in this way they may be considered as operating in a similar manner to switch contacts. They are known as *logic gates* or *gates*. They also may be referred to as Boolean gates.

The basic operators of Boolean algebra as they are related to digital logic functions are AND, OR, and NOT. The AND gate is composed of two or more inputs and one output. Figure H-14 shows the AND gate logic symbol. (This figure also illustrates the equivalent electromechanical circuit.) The AB notation on the output of the AND symbol means A *and* B, not A times B. There must be a HIGH (1) signal at A *and* at B in order to obtain a HIGH (1) signal at the output AB.

A *truth table* displays the various conditions under which an output is one or the other of two values. The truth table for the AND function is shown in Figure H-14. In this truth table it can be seen that there is a 1 output only when there is a 1 input at A *and* a 1 input at B.

The OR gate logic symbol is shown in Figure H-15, which also shows the equivalent electromechanical circuit. The OR gate is composed of two or more inputs and a single output. The A + B notation on the output of the OR gate symbol means A *or* B, not A plus B. There must be a HIGH (1) signal at A *or* at B in order to obtain a HIGH (1) signal at the output A + B.

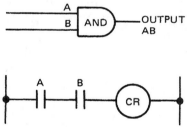

| Truth Table for Positive Logic | | |
|---|---|---|
| INPUT A | INPUT B | OUTPUT AB |
| 0 | 0 | 0 |
| 0 | 1 | 0 |
| 1 | 0 | 0 |
| 1 | 1 | 1 |

**Figure H-14** AND symbol with truth table.

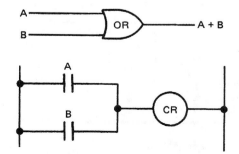

| Truth Table for Positive Logic | | |
|---|---|---|
| INPUT A | INPUT B | OUTPUT A + B |
| 0 | 0 | 0 |
| 1 | 0 | 1 |
| 0 | 1 | 1 |
| 1 | 1 | 1 |

**Figure H-15** OR symbol with truth table.

The truth table for a two-input OR gate is shown in Figure H-15. In this truth table it can be seen that there is a 1 output if there is a 1 input at A *or* a 1 input at B, *or* if there is a 1 input at both A and B.

A third gate is the NOT or INVERTER gate. It has a single input and a single output. The *bubble,* or circle, at the output is the standard symbol used to represent inversion. Figure H-16 shows the NOT or INVERTER gate logic symbol and its equivalent electromechanical circuit. The $\overline{A}$ (read *not A*) at the output indicates that the output is the opposite or complement of the input A. That is,

- If there is a HIGH (1) at the input A, there will be LOW (0) at the output $\overline{A}$.
- If there is a LOW (0) at the input A, there will be a HIGH (1) at the output $\overline{A}$.

The truth table for the NOT or INVERTER gate symbol is shown in Figure H-16.

The relationship between the logic symbol, logical statement, and Boolean equation is shown in Figure H-17.

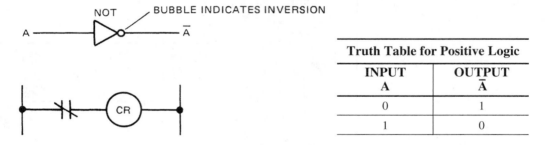

| Truth Table for Positive Logic | |
|---|---|
| INPUT A | OUTPUT $\overline{A}$ |
| 0 | 1 |
| 1 | 0 |

**Figure H-16**   NOT symbol with truth table.

| LOGIC SYMBOL | LOGICAL STATEMENT | BOOLEAN EQUATION |
|---|---|---|
| A, B → Y | Y IS 1 IF A AND B ARE 1. | Y = A•B OR Y = AB |
| C, D → Y | Y IS 1 IF C OR D IS 1. | Y = C + D |
| A → Y | Y IS 1 IF A IS 0. Y IS 0 IF A IS 1. | Y = $\overline{A}$ |

**Figure H-17**   Logic symbols, logic statement, and Boolean equation.

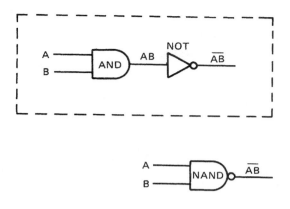

**Figure H-18** NAND symbol with truth table.

| INPUT A | INPUT B | OUTPUT $\overline{AB}$ |
|---------|---------|------------------------|
| 0 | 0 | 1 |
| 0 | 1 | 1 |
| 1 | 0 | 1 |
| 1 | 1 | 0 |

There are two important gate combinations that are coming into more frequent use: the NAND gate and the NOR gate. The NAND gate consists of an AND gate followed by a NOT or INVERTER gate. NAND gates may have more than two inputs, but have only a single output (Figure H-18). The symbol for the NAND gate is shown in Figure H-18. The symbol consists of an AND gate symbol followed by a small circle or bubble. Note that, as in the NOT or INVERTER symbol, the circle or bubble indicates inversion (the opposite). The term $\overline{AB}$ reads *not A and not B,* or *not AB.*

The truth table for the NAND gate is shown in Figure H-18. This table helps to explain the resulting outputs from various inputs. Notice that the output of the NAND gate is 1 until both inputs are 1.

One important feature of the NAND logic gate is that it can be used to produce any of the basic logic functions. Therefore, the NAND logic gate is generally referred to as a *universal gate.*

As shown in Figure H-19, the AND gate can be obtained by using two NAND logic gates and joining the inputs to the second NAND gate.

In Figure H-20, the OR gate is obtained by using three NAND gates. The inputs to the first two NAND gates are joined for a single input to each of the gates.

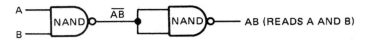

*Figure H-19*   **AND gate from two NAND gates.**

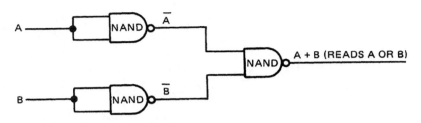

*Figure H-20*   **OR gate from three NAND gates.**

The NOT or INVERTER gate is obtained with a single NAND gate (Figure H-21). The inputs to the NAND gate are joined for a single input.

The second logic gate combination is the NOR gate. The NOR gate may have more than two inputs and a single output. The NOR gate consists of an OR gate followed by a NOT or INVERTER gate.

The combination of the OR gate and the NOT or INVERTER gate is shown in Figure H-22. The logic symbol for the NOR gate is shown in Figure H-22. The symbol consists of an OR gate symbol followed by a small circle or bubble. As in the NOT or INVERTER symbol, the circle or bubble indicates inversion (the opposite). The term $\overline{A + B}$ on the output reads *not A or B.*

*Figure H-21*   **NOT gate from single NAND gate.**

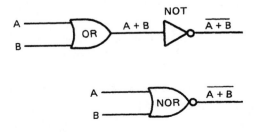

| INPUT A | INPUT B | OUTPUT $\overline{A + B}$ |
|---|---|---|
| 0 | 0 | 1 |
| 0 | 1 | 0 |
| 1 | 0 | 0 |
| 1 | 1 | 0 |

*Figure H-22*   **NOR gate and truth table.**

The truth table for the NOR gate is shown in Figure H-22. Note that the output of the NOR gate is 1 only when there is a 0 input at both A and B. If there is a 1 at input A or a 1 at input B, or a 1 at both inputs A and B, the output is 0.

An important feature of the NOR logic gate is that it can be used to produce any of the basic logic functions. Therefore, like the NAND gate, the NOR gate is generally referred to as a *universal gate.*

Figure H-23 shows that the function of an AND gate can be obtained by using three NOR gates. Two inputs are joined together so there is only a single input provided on each of two NOR gates (A-B). The outputs from these two NOR gates are connected to the input of a third NOR gate. The output (AB) is an AND function.

It can be seen in Figure H-24 that the function of an OR gate can be obtained by using two NOR gates. The inputs to the second NOR gate are joined together so that only a single input, A + B, is provided. The output from the second NOR gate is an OR function ($\overline{A + B}$).

In Figure H-25, the NOT or INVERTER gate is obtained with a single NOR gate by joining the inputs so that only a single input, A, is provided. The output is a NOT function ($\overline{A}$).

In our explanation of the OR logic gate, it is noted that when there is a HIGH or 1 at A or B, there is an output A + B. It is also true that when there is

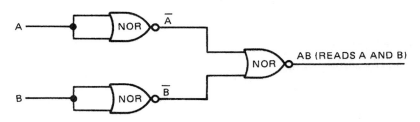

**Figure H-23**   AND gate from three NOR gates.

**Figure H-24**   OR gate from two NOR gates.

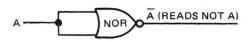

**Figure H-25**   NOT gate from single NOR gate.

a HIGH or 1 at A and a HIGH or 1 at B, there is an output A + B. Contrast OR logic gate to the following explanation of the EXCLUSIVE-OR gate.

The EXCLUSIVE-OR gate produces an output when input A is HIGH or 1 or when input B is HIGH or 1, but not when both input A and input B are HIGH or 1.

The development of the EXCLUSIVE-OR logic gate is shown in Figure H-26. It is made up of two NOT or INVERTER gates, two AND gates, and one OR gate. Note that the output is written in a manner similar to that of the OR output, except that a circle is drawn around the + sign ⊕. This is usually read *not AB*, or *A not B*.

Figure H-26 shows the symbol for the EXCLUSIVE-OR logic gate. It is similar to the OR symbol, except that an additional arc precedes the OR symbol.

The truth table for the EXCLUSIVE-OR logic gate functions is shown in Figure H-26. This table demonstrates that an output exists when there is an input at A or at B, but not at both A and B.

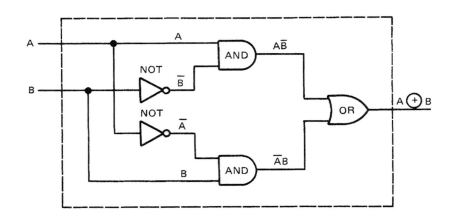

| INPUT A | INPUT B | OUTPUT A ⊕ B |
|---------|---------|--------------|
| 0 | 0 | 0 |
| 0 | 1 | 1 |
| 1 | 0 | 1 |
| 1 | 1 | 0 |

*Figure H-26*  EXCLUSIVE-OR gate and truth table.

## H.4 Circuit Applications

The control circuits using solid-state symbols in this section include the solid-state control circuit, a word description of the operation required, and a ladder-type diagram using electromechanical components.

It is generally better to design the solid-state control circuit from a word description of the operation. Working directly from a ladder-type electromechanical control circuit may reveal redundancies in the relay circuit, but it is also important to understand the relay ladder-type control circuit. The knowledge and use of the ladder-type circuit is of even greater importance when dealing with programmable control.

Changes may be of concern with solid-state control as compared to the ladder-type electromechanical circuit in the area of input and output signals.

Problems are sometimes encountered when standard industrial components are used to provide input information to a logic system operating at a low dc voltage level. These components, such as push-button switches, limit switches, and pressure switches are generally designed to operate at 120 V ac. When used at a low dc voltage level, such factors as noise, dry switching, and contact bounce may enter into the control system to create problems.

Where these problems exist, 120 V ac can be used in the input or information section. An interface consisting of a signal converter module can be used to convert the input signal into signals compatible with the low voltage dc.

Figure H-27 shows a typical relay equivalent circuit for a signal converter module.

In industrial plants, a voltage of 120 V ac is generally available on machines from an isolated secondary control transformer. Therefore, a power supply module is used to reduce and rectify the 120 V ac to a low-level dc required in the logic section.

A circuit showing a typical power supply arrangement is illustrated in Figure H-28.

In the output or work section of the solid-state control system, sufficient power must be available to operate components such as solenoids, contactors, and motor starters. Unless the power requirements are low, it is advisable to energize these components at 120 V ac.

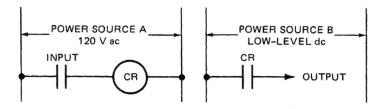

**Figure H-27**   Relay equivalent circuit.

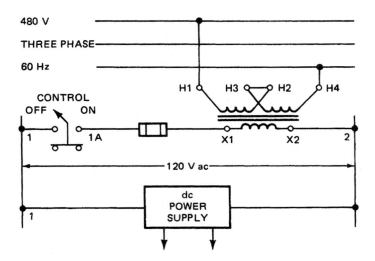

**Figure H-28**    Control circuit showing dc power supply.

The output interface (connection) to the power component may be a reed relay or an ac static switching unit. Many output components are inductive, so care must be taken that the interface device can handle a relatively high inrush current. In some cases, this current may reach 50 A.

Figure H-29 shows the symbol for an ac static switch. This module provides a solid-state means of switching ac power. It is made up of a bidirectional triode ac switch.

In the following circuits, a voltage of 120 V ac is available from the isolated secondary of a step-down control transformer. Low-voltage dc is available from a dc power supply, as shown.

A cylinder and piston assembly is shown in Figure H-30. The piston is to move to the right until it engages and operates limit switch 1LS. At that point, the

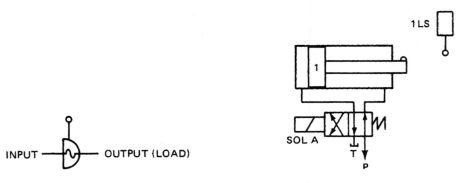

**Figure H-29**    Symbol for ac static switch.          **Figure H-30**    Cylinder and piston assembly.

piston is to return to its START position. Provision must be made to return the piston from any point in its forward travel. A single-solenoid, spring-return operating valve is used to supply fluid power to the cylinder. Figure H-31 shows the electromechanical circuit used to operate this system.

Figure H-32 shows the solid-state circuit used to operate the system. The inputs in this circuit operate at a low-level dc. The outputs operate at 120 V ac. An interface module is shown in the output section of the circuit.

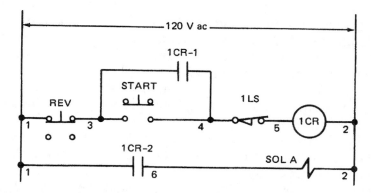

**Figure H-31**    Electromechanical circuit for Figure H-30.

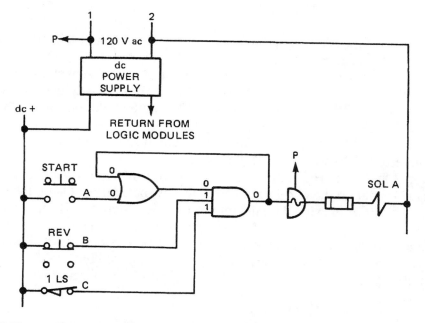

**Figure H-32**    Solid-state circuit for Figure H-31 showing the deenergized or start condition.

Up to this point in explaining the various logic module symbols, the names have been added to the symbols to help associate each symbol with its name. In the following circuits, the names are omitted; the symbol is identified only by its individual shape or form.

Note that when a low-level dc is given, the voltages generally range from 4 V to 12 V for most manufacturers of solid-state equipment. The input signals are obtained through an information component such as a push-button switch, limit switch, pressure switch, or temperature switch. The source is one side of the low-level dc voltage power supply. The return side of the dc power supply is not drawn in, but is indicated on the circuit.

The operation of this solid-state circuit is explained in the steps below to help visualize the changes that take place. The digit 1 is used to indicate a HIGH, and the digit 0 is used to indicate a LOW.

A two-input OR gate and a three-input AND gate are used in the logic circuit. An ac static switch is used in the output so that 120 V ac can be applied to the solenoid.

**Step 1.** In the explanation of the logic module symbols, it was shown that to obtain a HIGH or 1 output from a two-input OR gate, there must be a HIGH or 1 input at one of the inputs.

In Figure H-32, showing the deenergized or start condition, the START push-button switch has not been operated. There is thus a LOW (0) at one of the inputs. There is also a LOW (0) at the other input. (The reason for this condition will be clear later.) With a LOW (0) at both inputs of the OR gate, there is a LOW (0) output.

In a three-input AND gate, to obtain a HIGH (1) from the AND gate there must be HIGH (1) at each of the three inputs.

While there is a HIGH (1) at B (REVERSE push-button) as well as a HIGH (1) at C (limit switch 1LS), there is a LOW (0) input from the OR gate. This condition results in a LOW (0) output from the AND gate. Note that this LOW (0) is connected back as one of the OR gate inputs.

With a LOW (0) output from the AND gate, the ac static switch is not operated. Therefore, the solenoid is deenergized.

**Step 2.** In Figure H-33, the START push-button switch is operated, providing a HIGH (1) input on the OR gate. There will now be a HIGH (1) on the OR gate output. This HIGH (1) signal on the input to the three-input AND gate now completes the conditions for a HIGH (1) output from the AND gate. This signal is connected into the static switch, providing 120 V ac to energize solenoid A. Note that the HIGH (1) output from the AND gate is connected back to one of the OR gate inputs. This circuit is known as a *latching circuit;* its purpose will be evident in Step 3.

**Step 3.** In Figure H-34, the START push-button has been released, resulting in a LOW (0) signal from A into the two-input OR gate. However, since the OR

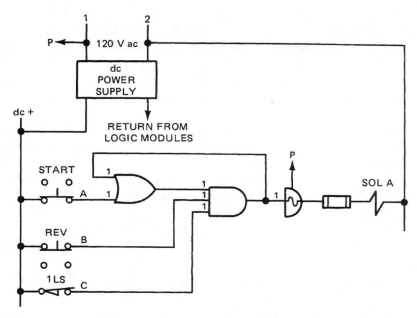

**Figure H-33**  Solid-state control circuit for Figure H-31 showing START switch operated.

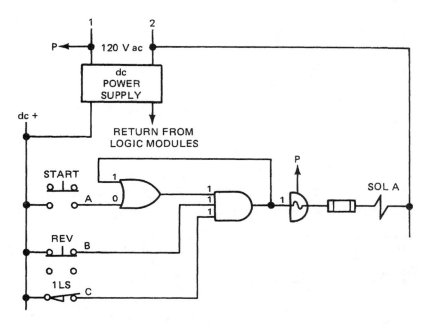

**Figure H-34**  Solid-state control circuit for Figure H-31 showing START switch released.

gate input was latched from the AND gate output (Step 2), the conditions are satisfied to continue a HIGH (1) output from the OR gate. Note that latching provides a function similar to that of the interlock circuit in the electromechanical circuit. That is, it allows the output to remain even though the initial energizing path has been opened. In this case, the path was through the START push-button switch. The remaining conditions are the same as in Step 2, with solenoid A energized.

**Step 4.** In Figure H-35, the piston has reached and operated the normally closed limit switch contact 1LS. This action opens the NC limit switch contact. The signal C now goes to LOW (0). Since this signal is one of the three inputs to the AND gate, the output of the AND gate goes to LOW (0). The static switch now opens, deenergizing solenoid A. Note that the OR gate is unlatched when the output of the AND gate goes to LOW (0).

**Step 5.** The circuit requirements state that at any point in the forward travel of the piston, before operating limit switch 1LS, it must be possible to reverse the direction of the piston travel. Figure H-36 shows that circuit. This circuit operation is the same as that explained in Step 4, except that here the LOW (0) signal comes from the open contact on the REVERSE push-button switch.

Assume that the circuit shown in Figure H-37 is now changed to add the requirement that the temperature of the operating fluid must be at a given level for

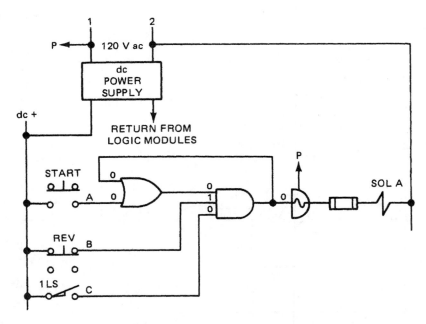

***Figure H-35*** **Solid-state control circuit for Figure H-31 showing the open contact at the NC limit switch.**

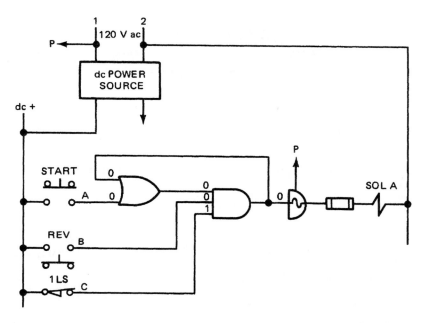

**Figure H-36** Solid-state control circuit for Figure H-31 showing the open contact at the REVERSE push-button switch.

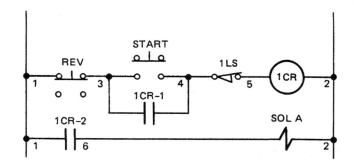

**Figure H-37** Typical electromechanical circuit used in this section to illustrate solid-state equivalent circuits.

start conditions. This means that the connecting wiring must be changed and a component (temperature switch) must be added. The temperature switch 1TS is added in series with the START push-button switch (Figure H-38).

The same concern of physically reconnecting elements when a circuit is changed is true when using solid-state modules. For example, the electro-mechanical circuit shown in Figure H-37 is now shown in Figure H-39, using solid-state modules.

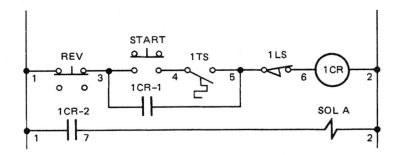

**Figure H-38**   Electromechanical circuit with a temperature switch in series with the START push-button switch.

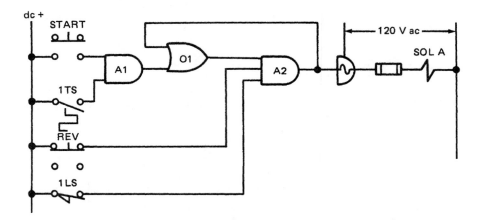

**Figure H-39**   Solid-state version of the electromechanical circuit shown as Figure H-38.

This section illustrates that physical changes to electromechanical or solid-state circuits must be made in order to alter the control circuit. These changes may involve only wiring, but in many cases components will have to be added, changed, or omitted. Costs increase as changes are made since labor is now required to "fit" the changes in.

# Appendix I
# Selecting a Transformer

The first item to consider in selecting a transformer for a specific control circuit application is to determine the primary and secondary voltages and frequency needed.

Then, consideration should be given to the continuous or sealed and inrush current characteristics of the coils on relays, contactors, motor starters, solenoids, etc. This information is available from the manufacturer of the component.

Using this information, calculate the total maximum continuous or sealed current by adding the continuous or sealed current drawn by all the coils that will be in an energized condition at the same time and multiply this figure by $\frac{5}{4}$. Then calculate the total maximum inrush current by adding the inrush current of all the coils that will be energized together at any one time and multiply this figure by $\frac{1}{4}$.

Using the larger of the two figures you have just calculated, multiply this current by the control voltage. This product is the volt-amperes (VA) required by the transformer load. If the resulting figure drops below a commercially available transformer size, it is generally advisable to use the next larger size. For example, you may have calculated that the maximum VA required is 698. Then use a 750-VA transformer, which is commercially available.

Another method is used if specific regulation requirements must be met. *Regulation* is the rise in secondary voltage when the full load is removed and is expressed as a percent (%) of the full-load voltage. A problem can arise with the drop in secondary voltage when a momentary heavy load is applied. If the secondary voltage drops too low, many of the component coils that are in an energized condition are deenergized. This action is called "dropping out." The point of dropping out may vary with the size of component. More on this action is given in Chapter 4.

Regulation curves for transformers are available from the manufacturers and are generally shown for 100% PF (power factor) loads and 20% PF loads.

Since a coil presents a low power factor load, it is generally advisable to use the 20% PF curve. (More about power factor in Appendix G.) Figure I-1 shows the regulation curves for both 100% and 20% power factor loads. Note that these curves give the relationship between secondary current and secondary voltage.

Now calculate the continuous or sealed current of all the coils that will be in an energized condition at the same time. Multiply this figure by the secondary voltage to obtain the continuous or sealed VA. Next, calculate the inrush current of all the coils that will be energized together at any one time. Multiply this figure by the secondary voltage for the inrush VA.

The transformer VA required, then, is:

$$\sqrt{(\text{VA continuous or sealed})^2 + (\text{VA inrush})^2}$$

Go now to the 20% power factor regulation curve for the transformer size you have just determined. The location where the power factor curve intersects at the 85% voltage output point is the maximum secondary amperes that can be allowed for the condition you have calculated. The 85% figure is used since NEMA standards require most magnetic devices to operate at 85% of rated voltage. The maximum amperes is shown in Figure I-1 for 85% secondary voltage. The curves shown are for a 1500-VA transformer.

Another important factor in selecting a transformer is temperature rise. Temperature rise is caused by losses in both the copper winding and the iron core. The ability to dissipate heat from the surface of the coil and core is important. The transformer is designed to carry full load continuously if the heat is properly dissipated. The continuous load current is thus an important factor in determining the transformer's volt–ampere rating to use for a given job.

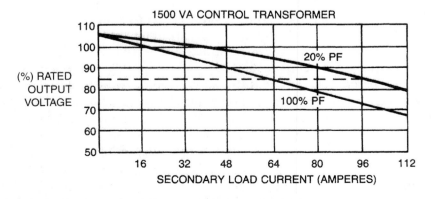

*Figure I-1*   Regulation curves for 100% and 20% power factor loads.

# Glossary

**AC Input Interface (Module)**—An input circuit that conditions various ac signals from connected devices to logic levels required by the processor.

**AC Output Interface (Module)**—An output circuit that switches the user-supplied control voltage required to control connected ac devices.

**Accessible (as applied to equipment)**—Admitting close approach because not guarded by locked doors, elevation, or other effective means.

**Across-the-Line Starter**—A motor starter that connects the motor to full voltage supply.

**Actuator**—A cam, arm, or similar mechanical or magnetic device used to trip limit switches.

**Address**—A reference number assigned to a unique memory location. Each memory location has an address, and each address has a memory location.

**Alphanumeric**—A character set that contains both numeric and alphabetic characters.

**Alternating Current (ac)**—Electric current that periodically changes direction and magnitude.

**Alternation**—One half-cycle in alternating current, either positive or negative half.

**Ambient Conditions**—The condition of the atmosphere adjacent to the electrical apparatus; the specific reference may apply to temperature, contamination, humidity, etc.

**Ampacity**—The current-carrying capacity, expressed in amperes.

**Ampere (A)**—The unit of current.

**Analog Device**—Apparatus that measures continuous information (e.g., current voltage). The measured analog signal has an infinite number of possible values. The only limitation on resolution is the accuracy of the measuring device.

**Analog Input Interface**—An input circuit that employs an analog-to-digital converter to convert an analog value, measured by an analog measuring device, to a digital value that can be used by the processor.

**Analog Output Interface**—An output circuit that employs a digital-to-analog converter to convert a digital value sent from the processor to an analog value that will control a connected analog device.

**Analog Signal**—One having the characteristic of being continuous and changing smoothly over a given range, rather than switching suddenly between certain levels as with discrete signals.

**Analog-to-Digital Conversion**—Hardware and/or software process that converts a scaled analog signal or quantity into a scaled digital signal or quantity.

**AND**—An operation that yields a logic "1" output if all the inputs are "1" and a logic "0" if any of the inputs are "0."

**Apparatus**—The set of control devices used to accomplish the intended control functions.

**Arithmetic Capability**—The ability to perform such math functions as addition, subtraction, multiplication, division, square roots, etc. A given controller may have some or all of these functions.

**Auxiliary Contacts**—Contacts in addition to the main circuit contacts in a switching device. They function with the movement of the main circuit contacts.

**Auxiliary Device**—Any electrical device other than motors and motor starters necessary to fully operate the machine or equipment.

**Available Short-Circuit Current**—The maximum short-circuit current that can flow in an unprotected circuit.

**Binary Coded Decimal**—A binary number system in which each decimal digit from 0 to 9 is represented by four binary digits (bits). The four positions have weighted values of 1, 2, 4, and 8, respectively, starting with the least significant bit. A thumbwheel switch is a BCD device; when connected to a programmable controller, each decade requires four wires. For example: decimal 9 = 1001 BCD.

**Binary Number System**—A number system that uses two numerals (binary digits), "0" and "1." Each digit position for a binary number has a place value of 1, 2, 4, 8, 16, 32, 64, 128, and so on, beginning with the least significant (right-hand) digit. Also called base 2.

**Binary Word**—A related grouping of 1s and 0s having coded meaning assigned by position or as a group; has some numerical value. 10101101 is an 8-bit binary word, in which each bit could have a coded significance, or as a group represents the number 173 in decimal.

**Bit**—One binary digit. The smallest unit of binary information. Bit is an abbreviation for BInary digiT. A bit can have a value of 1 or 0.

**Block Diagram**—A diagram that shows the relationship of separate subunits (blocks) in the control system.

**Bonding**—The permanent joining of metallic parts to form an electrical conductive path that will assure electrical continuity and the capacity to conduct safely any current likely to be imposed (NFPA 79-1984, *National Electrical Code®*).

**Boolean Algebra**—A mathematical shorthand notation that expresses logic functions, including AND, OR, EXCLUSIVE-OR, NAND, NOR, and NOT.

**Branch Circuit**—That portion of a wiring system extending beyond the final overcurrent device protecting the circuit. (A device not approved for branch circuit protection, such as a thermal cutout or motor overload protective device, is not considered an overcurrent device protecting the circuit).

**Breakdown Voltage**—The voltage at which a disruptive discharge takes place, either through or over the surface of insulation.

**Breaker**—An abbreviated name for the circuit breaker.

**Bus**—Power distribution conductors.

**Buss**—A bank of storage capacitors whose function is to store energy and supply dc power to the transistors in the output bridge as it is required.

**Byte**—A group of adjacent bits usually operated on as a unit, such as when moving data to and from the memory. One byte consists of 8 bits.

**Capacitor**—Two conductors separated by an insulator.

**Capacitor Sensor**—Used in limit switches, these sensors contain an oscillator. When the object to be detected moves into the sensor's range, it creates enough capacitance to set the oscillator in motion.

**Cassette Recorder**—A peripheral device for transferring information between the PC memory and magnetic tape. In the record mode it is used to make a permanent record of a program existing in the processor memory. In the playback mode it is used to enter a previously recorded program into the processor memory.

**Cassette Tape**—A magnetic recording tape permanently enclosed in a protective housing.

**Cathode Ray Tube (CRT)**—A vacuum tube with a viewing screen as an integral part of its envelope.

**Central Processing Unit (CPU)**—That part of the programmable controller that governs systems activities, including interpretation and execution of programmed instructions. In general, the CPU consists of a logic unit, timing and control circuitry, program counter, address stack, and an instruction register.

**Character**—One symbol of a set of elementary symbols such as a letter of the alphabet or a decimal number. Characters can be expressed in many binary codes.

**Chassis**—A sheet metal box, frame, or simple plate on which electronic components and their associated circuitry can be mounted.

**Chip**—A very small piece of semiconductor material on which electronic components are formed. Chips are generally made of silicon and are generally less than one-quarter-inch square and one-thousandth-inch thick.

**Circuit Breaker**—A device that opens and closes a circuit by nonautomatic means or that opens the circuit automatically on a predetermined overload of current, without damage to itself when properly applied within its rating.

**Circuit Interrupter**—A nonautomatic, manually operated device designed to open, under abnormal conditions, a current-carrying circuit without damage to itself.

**CMOS**—An abbreviation for complimentary metal oxide semiconductor. A family of very low power, high-speed integrated circuits.

**Coaxial Cable**—A cable that is constructed with an outer conductor forming a cylinder around a central conductor. An insulating dielectric separates the inner and outer conductors. The complete assembly is enclosed in a protective outer sheath.

**Color CRT**—Television that is specially designed to be used with computer-based devices. Information and data are displayed graphically in color.

**Combination Starter**—A magnetic motor starter with a manually operated disconnecting device built into the same enclosure with the motor starter. Protection is always added. A control transformer may be added to provide 120 V for control. START-STOP pushbuttons and a pilot light may be added in the enclosure door.

**Command**—In data communication, an instruction represented in the control field or a frame and transmitted by the primary device. It causes the addressed secondary to execute a specific data link control function.

**Communication Network**—A communication system that consists of adapter modules and cabling. It is used to transfer system status and data between network PLCs and computers.

**Compartment**—A space within the base, frame, or column of the equipment.

**Compatibility**—The ability of various specified units to replace one another with little or no reduction in capability.

**Complement**—A logical operation that inverts a signal or bit. The complement of 1 is 0. The complement of 0 is 1.

**Component**—*See* **Device.**

**Computer Interface**—An interface that consists of an electronic circuit designed to communicate data and instructions between a PLC and a computer. It also can be used to communicate with another computer-based intelligent device.

**Conductor**—A substance that easily carries an electrical current.

**Conduit, Flexible Metal**—A flexible raceway of circular cross section specially constructed for the purpose of pulling in or withdrawing wires or cables after the conduit and its fittings are in place.

**Conduit, Flexible Nonmetallic**—A flexible raceway of circular cross section specially constructed of nonmetallic material for the purpose of pulling in or withdrawing wires or cables after the conduit and its fittings are in place.

**Conduit, Rigid Metal**—A raceway specially constructed for the purpose of pulling in or withdrawing wires or cables after the conduit and its fittings are in place. It is made of metal pipes of standard weight and thickness, permitting the cutting of standard threads.

**Contact**—One of the conducting parts of a relay, switch, or connector that is engaged or disengaged to open or close an electrical circuit. When considering software, it is the junction point that provides a complete path when closed.

**Contact Bounce**—The continuing making and breaking of a contact after the initial engaging or disengaging of the contact.

**Contact Symbology**—A set of symbols used to express the control program using conventional relay symbols.

**Contactor**—A device for repeatedly establishing and interrupting an electrical power circuit.

**Continuity**—A complete conductive path for an electrical current from one point to another in an electrical circuit.

**Continuous Rating**—A rating that defines the substantially constant load that can be carried for an indefinite period of time.

**Control Circuit**—The circuit of a control apparatus or system that carries the electrical signals directing the performance of the controller but that does not carry the main power circuit.

**Control Circuit Transformer**—A voltage transformer used to supply a voltage suitable for the operation of control devices.

**Control Circuit Voltage**—The voltage provided for the operation of shunt coil magnetic devices.

**Control Compartment**—A space within a base, frame, or column of the machine used for mounting the control panel.

**Control Logic**—The program. Control plan for a given system.

**Control Panel**—*See* **Panel.**

**Control Station**—*See* **Push-Button Station.**

**Controller, Electric**—A device (or group of devices) that serves to govern, in some predetermined manner, the electric power delivered to the apparatus to which the device is connected.

**CRT**—Abbreviation for cathode ray tube, an electronic display tube similar to the familiar TV picture tube.

**CRT Programmer**—A programming device containing a cathode ray tube (CRT). This programming device is primarily used to create and monitor the control program. It can also be used to display data.

**CRT Terminal**—An enclosure that contains a CRT (cathode ray tube), a special-purpose keyboard, and a microprocessor; used to program a PLC.

**Current-Carrying Capacity**—The maximum amount of current that a conductor can carry without heating beyond a predetermined safe limit.

**Current-Limiting Fuse**—A fuse that limits both the magnitude and duration of current flow under short-circuit conditions.

**Cycle**—An alternating current waveform that begins from a zero reference point. It reaches a maximum value and then returns to zero. It then reaches a negative maximum value and returns to the zero reference point.

**Data**—Information encoded in digital form. It is stored in an assigned address of data memory for later use by the processor.

**Data Bus**—A system of wires (or conductors on a PC board) through which data is transmitted from one part of a computer to another.

**Data Highway**—A network of connections, which may be made up of any combination of twisted pair cables, coaxial cables, and fiber optic cables. It carries data allowing PLCs and other microcomputer-based devices to communicate.

**Data Terminal**—A peripheral device that can load, monitor, program, or save the contents of a PLC's memory.

**Delta Connection**—Connection of a three-phase system so that the individual phase elements are connected across pairs of the three-phase power leads A–B, B–C, C–A.

**Derate**—To reduce the current, voltage, or power rating of a device to improve its reliability or to permit operation at high ambient temperatures.

**Device (Component)**—An individual apparatus used to execute a control function.

**Dielectric**—The insulating material between metallic elements of any electrical or electronic component.

**Digital**—The representation of numerical quantities by means of discrete numbers. It is possible to express in binary form all information stored, transferred, or processed by dual state conditions; for example, ON-OFF, CLOSED-OPEN.

**Disconnect Switch (Motor Circuit Switch)**—A switch intended for use in a motor branch circuit. It is rated in horsepower and is capable of interrupting the maximum operating overload current of a motor of the same rating at the same rated voltage.

**Disconnecting Means**—A device that allows the current-carrying conductors of a circuit to be disconnected from their source of supply.

**Documentation**—An orderly collection of recorded hardware and software information covering the control system. These records provide valuable reference data for installation, debugging, and maintenance of the programmable controller.

**Downtime**—The time in which a system is not available for production due to required maintenance, either scheduled or unscheduled.

**Drop-Out**—The current, voltage, or power value that will cause energized relay contacts to return to their normal deenergized condition.

**Dual-Element Fuse**—Often confused with time delay, dual element is a manufacturer's term describing fuse element construction.

**EEPROM**—Electrically erasable programmable read-only memory (provides permanent storage of the program and can be easily changed with the use of a manual programming unit or standard CRT).

**Electrical Equipment**—The electromagnetic, electronic, and static apparatus as well as the more common electrical devices.

**Electrical Optical Isolator**—A device that couples input to output using a light source and detector in the same package. It is used to provide electrical isolation between input circuitry and output circuitry.

**Electrical System**—An organized arrangement of all electrical and electromechanical components and devices in a way that will properly control the machine or industrial equipment.

**Electromechanical**—A term applied to any device in which electrical energy is used to magnetically cause mechanical movement.

**Electronic Control**—Electronic, static, precision, and associated electrical control equipment.

**Elementary (Schematic) Diagram**—A diagram in which symbols and a plan of connections are used to illustrate in simple form the control scheme.

**Enclosure**—A case, box, or structure surrounding the electrical equipment to protect it from contamination; the degree of tightness is usually specified (such as NEMA 12).

**Encoder**—An electronic, mechanical, or optical device used to monitor the circular motion of a device.

**Energize**—To apply electrical power.

**EPROM**—Erasable programmable read-only memory (PROM that can be erased with ultraviolet light and then reprogrammed).

**Ethernet**—A common system of switching devices and cables connected in a data highway.

**Examine On**—Refers to a normally open contact instruction in a relay ladder diagram.

**Examine Off**—Refers to a normally closed contact instruction in a relay ladder diagram.

**Exposed (as applied to electrically live parts)**—Capable of being inadvertently touched or approached nearer than a safe distance by a person. It is applied to parts not suitably guarded, isolated, or insulated (NFPA 70-1984, *National Electric Code®*).

**External Control Devices**—Any control device mounted external to the control panel.

**Fail-Safe Operation**—An electrical system designed so that the failure of any component in the system will prevent unsafe operation of the controlled equipment.

**Fault**—An accidental condition in which a current path that bypasses the connected load becomes available.

**Feeder**—The circuit conductors between the service equipment or the generator switchboard of an isolated plant and the branch circuit overcurrent device.

**Ferroresonance (transformer)**—Phenomenon usually characterized by overvoltages and very irregular wave shapes and associated with the excitation of one or more saturable inductors through capacitance in series with the inductor.

**Fiber Optics**—Transparent strands of glass or plastic used to conduct light energy.

**Filter (electrical)**—A network consisting of a resistor, capacitor, and inductor used to suppress electrical noise.

**Frequency**—The number of recurrences of an event within a specific time period, expressed in cycles per second or hertz.

**Fuse**—An overcurrent protective device containing a calibrated current-carrying member that melts and opens a circuit under specified overcurrent conditions.

**Fuse Element**—A calibrated conductor that melts when subjected to excessive current. The element is enclosed by the fuse body and may be surrounded by an arc-quenching medium such as silica sand. The element is sometimes referred to as a *link*.

**Gate**—A circuit having two or more input terminals and one output terminal, and in which an output is present when, and only when, the prescribed inputs are present.

**Grounded**—Connected to earth or to a conducting body that serves in place of the earth.

**Grounded Circuit**—A circuit in which one conductor or point (usually the neutral or neutral point of a transformer or generator windings) is intentionally grounded (earthed), either solidly or through a grounding device.

**Grounded Conductor**—A conductor that carries no current under normal conditions. It serves to connect exposed metal surfaces to an earth ground to prevent hazards in case of breakdown between current-carrying parts and exposed surfaces. If insulated, the conductor is colored green, with or without a yellow stripe.

**Guarded**—Covered, shielded, fenced, enclosed, or otherwise protected by suitable covers or easings, barriers, rails, screens, mats, or platforms to prevent contact or approach of persons or objects to a point of danger.

**Hard Copy**—A printed document of what is stored in the memory.

**Hardware**—Includes all the physical components of the programmable controller, including the peripheral.

**Hardwired Logic**—Logic control functions that are determined by the way devices are interconnected, as contrasted to programmable control in which logic control functions are programmable and easily changed.

**Heat Rise**—An increase in the temperature of a solenoid caused by resistance to current flow, eddy currents, and hysteresis.

**Hermetic Seal**—A mechanical or physical closure that is impervious to moisture or gas, including air.

**Hertz (Hz)**—Cycles per second; the unit measuring the frequency of alternating current.

**Hexadecimal Numbering System (HEX)**—A number system that uses the numerals 0 through 9 and the letters A through F. The system uses base 16.

**Hysteresis (Magnetic)**—Refers to a certain magnetic property of iron, which causes a power loss when iron is magnetized and demagnetized due to the change in magnetic flux in the iron lagging behind the change that causes it.

**Image Table**—An area in the PC memory dedicated to the I/O data. ON and off are represented by 1s and 0s, conditions, respectively. During every I/O scan, each input controls a bit in the input image table. Each output is controlled by a bit in the output image table.

**Impedance**—Measure of opposition to flow of current, particularly alternating current. It is a vector quantity and is the vector sum of resistance and reactance. The unit of measurement is the ohm.

**Inching**—*See* **Jogging.**

**Input**—Information sent to the processor from connected devices through the input interface.

**Input Device**—Any connected equipment that will supply information to the central processing unit, such as switches, push-buttons, sensors, or peripheral devices. Each type of input device has a unique interface to the processor.

**Inrush Current**—In a solenoid or coil, the steady-state current taken from the line with the armature blocked in the rated maximum open position.

**Instruction**—A command or order that will cause a PC to perform one prescribed operation.

**Interconnecting Wire**—A term referring to connections among subassemblies, panels, chassis, and remotely mounted devices; does not necessarily apply to internal connections of these units.

**Interconnection Diagram**—A drawing showing all terminal blocks in the system, with each terminal identified.

**Interface**—A circuit that permits communication between the central processing unit and a field input or output device.

**Interlock**—A device actuated by the operation of another device with which it is directly associated. The interlock governs succeeding operations of the same or allied devices and may be either electrical or mechanical.

**Interrupting Capacity**—The highest current at rated voltage that a device can interrupt.

**I/O**—Abbreviation for input/output.

**I/O Address**—A unique number assigned to each input and output. The address number is used when programming, monitoring, or modifying a specific input or output.

**I/O Module**—A plug-in type assembly that contains more than one input or output circuit. A module usually contains two or more identical circuits, for example, 2, 4, 8, or 16 circuits.

**I/O Update**—The continuous process of revising each and every bit in the I/O tables, based on the latest results from reading the inputs and processing the outputs according to the control program.

**Isolated I/O**—Input and output circuits that are electrically separated from any and all other circuits of a module. They are designed to allow for connecting field devices that are powered from different sources to one module.

**Isolation Transformer**—A transformer used to separate one circuit from another.

**Jogging (Inching)**—A quickly repeated closure of the circuit to start a motor from rest to accomplish small movements of the driven machine.

**Joint**—A connection between two or more conductors.

**Kilohertz (kHz)**—1000 Hertz.

**Kilowatts (kW)**—1000 watts.

**Ladder Diagram**—An industry standard for representing relay-logic control systems.

**Ladder Element**—Any one of the device that can be used in a ladder diagram, including relays, switches, timers, counters, etc.

**Ladder Program**—A type of control program that uses relay-equivalent contact symbols as instructions.

**Language**—A set of symbols and rules for representing and communicating information among people or between people and machines. The method used to instruct a programmable device to perform various operations.

**Latch**—A ladder program output instruction that retains its state even though the conditions that caused it to latch on may go off. A latched output must be unlatched. A latched output will retain its last state (on or off) if power is removed.

**Latching Relay**—A relay that can be mechanically latched in a given position manually, or when operated by one element, and released manually or by the operation of a second element.

**LED**—Abbreviation for light-emitting diode. A semiconductor diode, the junction of which emits light when passing a current in the forward direction.

**Legend Plate**—A plate that identifies the function of operating controls, indicating lights, etc.

**Limit Switch**—A switch operated by some part or motion of a power-driven machine or equipment to alter the associated electric circuit.

**Line Printer**—A hard copy device that prints one line of information at a time.

**Location**—In reference to memory, a storage position or register identified by a unique address.

**Locked Rotor Current**—The steady-state current taken from the line with the rotor locked and with rated voltage (and rated frequency in the case of alternating current motors) applied to the motor.

**Logic**—A process of solving complex problems through the repeated use of simple functions that can be either true or false (on or off).

**Logic Control Panel Layout**—The physical position or arrangement of the devices on a chassis or panel.

**Logic Diagram**—A drawing that shows the relationship of standard logic elements in a control system; it is not necessary to show internal detail of the logic elements.

**Logic Level**—The voltage magnitude associated with signal pulses that represent 1s and 0s in digital systems.

**Machine Language**—A program written in binary form.

**Magnetic Device**—A device operated by electromagnetic means.

**Magnetostrictive Material**—The phenomenon of magnetostriction relates to the stresses and changes in dimensions produced in a material by magnetization and the inverse effect of changes in the magnetic properties produced by mechanical stresses. Nickel, alloys of nickel and iron, invar, nichrome, and various other alloys of iron exhibit pronounced magnetostrictive effects.

**Main Memory**—The block of data storage location connected directly to the central processing unit.

**Master Terminal Box**—The main enclosure on the equipment containing terminal blocks for the purpose of terminating conductors from the control enclosure. Normally associated with equipment requiring a separately mounted control enclosure.

**Mean**—A statistical value useful in analyzing quality, the mean is the true average of the sampled data measurements.

**Median**—The median is the middle value of all the sampled data collected. It is a statistic useful in quality analysis.

**Memory**—That part of the programmable controller in which data and instructions are stored either temporarily or semipermanently. The control program is stored in the memory.

**Microprocessor**—A digital electronics logic package capable of performing the program control, data processing functions, and execution of the central processing unit.

**Microsecond**—One millionth of a second.

**Millisecond**—One thousandth of a second.

**Mnemonic**—Aiding or designed to aid the memory.

**Mode**—Useful in quality analysis, the mode is the value that occurs most frequently in the sampled data.

**Modem**—Acronym for MODulator/DEmodulator. It is a device that employs frequency-shift keying for the transmission of data over a telephone line. It converts a two-level binary signal to a two-frequency audio signal and vice versa.

**MOS**—Metal-oxide semiconductor.

**Motor Circuit Switch**—*See* **Disconnect Switch.**

**Motor Junction (Conduit) Box**—An enclosure on a motor for the purpose of terminating a conduit run and joining the motor to power conductors.

**NAND**—A logic operation that yields a logic "1" output if any input is "0" and a logic "0" if all inputs are "1."

**Network**—An organization of devices that are interconnected for the purpose of intercommunication.

**Noise**—Random, unwanted electrical signals normally caused by radio waves or electrical or magnetic fields generated by one conductor and picked up by another.

**Nominal Voltage**—The utilization voltage (see the appropriate NEMA standard for device voltage ratings).

**Nonvolatile Memory**—A type of memory whose contents are not lost or disturbed if operating power is lost.

**NOR**—A logic operation that yields a logic "1" output if all inputs are "0" and a logic "0" output if any input is "1."

**Normally Closed, Normally Open**—When applied to a magnetically operating switch device such as a contactor or relay or to the contacts thereof, these terms signify the position taken when the operating magnet is deenergized. The terms apply only to nonlatching types of devices.

**NOT**—A logic operation that yields a logic "1" at the output if a logic "0" is entered at the input and a logic "0" at the output if a logic "1" is entered at the input.

**Octal Numbering System**—A number system that uses eight numeral digits (0 through 7); base 8.

**Ohm ($\Omega$)**—Unit of electrical resistance.

**Operating Floor**—A floor or platform used by the operator under normal operating conditions.

**Operating Overload**—The overcurrent to which electrical apparatus is subjected under normal operating conditions; such overloads are currents that may persist for a very short time only, usually a matter of seconds.

**Operator's Control Station**—*See* **Push-Button Station.**

**Optical Coupler**—A device that couples signals from one circuit to another by means of electromagnetic radiation, usually visible or infrared.

**Optical Isolation**—Electrical separation of two circuits with the use of an optical coupler.

**OR**—A logic operation that yields a logic "1" output if one or any number of inputs is "1" and a logic "0" if all inputs are "0."

**Oscilloscope**—An instrument used to visually show voltage or current waveforms or other electrical phenomena, either repetitive or transient.

**Outline Drawing**—A diagram that shows approximate overall shape with no detail.

**Output (PC)**—Information sent from the processor to a connected device through an interface.

**Output Device (PC)**—Any connected equipment that will receive information or instructions from the central processing unit. It may consist of solenoids, motors, lights, etc.

**Overcurrent**—Current in an electrical circuit that causes excessive or dangerous temperature in the conductor or conductor insulation.

**Overcurrent Protective Device**—A device that operates on excessive current and causes and maintains the interruption of power in the circuit.

**Overlapping Contacts**—Combinations of two sets of contacts actuated by a common means; each set closes in one of two positions and is arranged so that its contacts open after the contacts of the other set have been closed (NEMA IC-1).

**Overload**—Operation of equipment in excess of normal full-load rating or a conductor in excess of rated ampacity, which, if it were to persist for a sufficient length of time would cause damage or overheating.

**Overload Relay**—A device that provides overload protection for the electrical equipment.

**Panel**—A subplate on which the control devices are mounted.

**Panel Layout**—The physical position or arrangement of the components on a panel or chassis.

**Parallel Circuit**—A circuit in which two or more of the connected components or contact symbols in a ladder diagram are connected to the same set of terminals, so that current may flow through all the branches. This arrangement is in contrast to a series circuit where the parts are connected end to end so that current flow has only one path.

**Pendant (Station)**—A push-button station suspended from overhead, connected by means of flexible cord or conduit but supported by a separate cable.

**Plugging**—A control function that provides braking by reversing the motor line voltage polarity or phase sequence so that the motor develops a counter torque that exerts a retarding force (NEMA IC-1).

**Plug-In Device**—A plug arranged so that it may be inserted in its receptacle only in a predetermined position.

**Potting**—A method of securing a component or a group of components by encapsulation.

**Power Factor**—The ratio between apparent power (volt-amperes) and actual or true power (watts). The value is always unity or less than unity.

**Power Supply**—The unit that supplies the necessary voltage and current to the system circuitry.

**Precision Device**—A device that operates within prescribed limits and consistently repeats operations within those limits.

**Pressure Connector**—A conductor terminal applied with pressure to make the connection mechanically and electrically secure.

**Processor**—*See* **Central Processing Unit.**

**Program**—A planned set of instructions stored in memory and executed in an orderly fashion by the central processing unit.

**Programming Device**—A device for inserting the control program into the memory. The programming device is also used to make changes to the stored program.

**PROM**—Programmable read-only memory. It can be programmed once and cannot be altered after that.

**Protocol**—A structured format used to organize binary data. Protocols are necessary for clear communication between devices connected into a data highway because they provide a means of communication independent of the individual system languages used by these devices.

**Push-Button Station**—A unit assembly of one or more externally operable push-button switches. It sometimes includes other pilot devices, such as indicating lights and selector switches in a suitable enclosure.

**Raceway**—Any channel designed expressly for and used solely for the purpose of holding wires, cables, or bus bars.

**RAM**—Random access memory. Referred to as "read/write" as it can be written into as well as read from.

**Read/Write Memory**—A type of memory that can be read from or written to. Can be altered quickly and easily by writing over the part to be changed or inserting a new part to be added.

**Readily Accessible**—Capable of being reached quickly for operation, renewal, or inspection without a worker having to climb over or to remove obstacles or use a portable ladder, etc.

**Rejection Fuse**—A current-limiting fuse with high interrupting rating and with unique dimension or mounting provisions.

**Relay**—A device that is operative by a variation in the conditions of one electric circuit to affect the operation of other devices in the same or another electric circuit.

**ROM**—Read only memory. A type of memory that permanently stores information.

**Rung**—A ladder-program term that refers to the programmed instructions that drive one output.

**Scan Time**—The time required to read all the inputs, execute the control program, and update local and remote I/O.

**Schematic Circuit**—*See* **Elementary (Schematic) Diagram.**

**SCR**—Silicon-controlled rectifier. A semiconductor device that functions as an electrically controlled switch for dc loads.

**Semiconductor**—A device that can function either as a conductor or nonconductor, depending on the polarity of the applied voltage, such as a rectifier or transistor that has a variable conductance depending on the control signal applied.

**Semiconductor Fuse**—An extremely fast-acting fuse intended for the protection of power semiconductors. Sometimes referred to as a *rectifier fuse.*

**Sequence of Operations**—A detailed written description of the order in which electrical devices and other parts of the equipment should function.

**Series Circuit**—A circuit in which all the components or contact symbols are connected end to end. All must be closed to permit current flow.

**Shielded Cable**—A single- or multiple-conductor cable surrounded by a separate conductor (shield) in order to minimize the effects of other electrical circuits.

**Short-Time Rating**—A rating that defines the load that can be carried for a short, definitely specified time with the machine, apparatus, or device being at approximately room temperature at the time the load is applied.

**Single Phasing**—An occurrence in a three-phase system in which one phase is lost.

**Software**—Any written documents associated with the system hardware, such as the stored program or instructions.

**Solenoid**—An electromagnet with an energized coil, approximately cylindrical in form, and an armature whose motion is reciprocating within and along the axis of the coil.

**Solid State**—Circuitry designed using only integrated circuits, transistors, diodes, etc.

**Standard Deviation**—A statistic used to determine the spread between all the data sampled. It is value useful to analyze quality.

**Starter**—An electric controller that accelerates a motor from rest to normal speed. (A device for starting a motor in either direction of rotation includes the additional function of reversing and should be designated a controller.)

**Static Device**—As associated with electronic and other control or information-handling circuits, a device with switching functions that has no moving parts.

**Status**—The condition or state of a device; for example, on or off.

**Stepping Relay (Switch)**—A multiposition relay in which wiper contacts mate with successive sets of fixed contacts in a series of steps, moving from one step to the next in successive operations of the relay.

**Subassembly**—An assembly of electrical or electronic components, mounted on a panel or chassis, that forms a functional unit by itself.

**Subplate**—A rigid metal panel on which control devices can be mounted and wired.

**Swingout Panel**—A panel that is hinge mounted in such a way that the back of the panel may be made accessible from the front of the machine.

**Symbol**—A widely accepted sign, mark, or drawing that represents an electrical device or component thereof.

**Temperature Controller**—A control device responsive to temperature.

**Terminal**—A point of connection in an electrical circuit.

**Terminal Block**—An insulating base or slab equipped with one or more terminal connectors for the purpose of making electrical connections.

**Termination**—The load connected to the output end of a transmission line and provision for ending a transmission line and connecting to a bus bar or other terminating device.

**Three Phase**—Three different alternating currents or voltages, 120 degrees out of phase with each other.

**Tie Point**—A distribution point in circuit wiring other than a terminal connection where junction of leads are made.

**Time-Delay Fuse**—A fuse that will carry an overcurrent of a specified magnitude for a minimum specified time without opening. The current/time requirements are defined in the UL 198 fuse standards.

**Tolerance**—A measurement that defines the acceptable range of values in meeting the specifications of quality.

**Torque**—The turning effect about an axis; it is measured in foot-pounds or inch-ounces and is equal to the product of the length of the arm and the force available at the end of the arm.

**Transient (Transient Phenomena)**—Rapidly changing action occurring in a circuit during the interval between closing of a switch and settling to steady-state condition or any other temporary actions occurring after some change in a circuit or its constants.

**Triac**—A semiconductor device that functions as an electrically controlled switch for ac loads.

**Truth Table**—A table that shows the state of a given output as a function of all possible input combinations.

**TTL**—Abbreviation for transistor-transistor logic. A semiconductor logic family in which the basic logic element is a multiple-emitter transistor. This family of devices is characterized by high speed and medium power dissipation.

**Undervoltage Protection**—The effect of a device operative upon the reduction or failure of voltage that causes and maintains the interruption of power to the main circuit.

**Undervoltage Release**—The effect of a device operative upon the reduction or failure of voltage that causes the interruption of power to the main circuit; voltage will return to the device when nominal voltage is reestablished.

**User Memory**—The memory where the application control program is stored.

**Ventilated**—Provided with a means to permit circulation of air sufficient to remove excess heat, fumes, or vapors (NFPA 70-1984, *National Electrical Code*®).

**Viscosity**—The property of a body that, when flow occurs inside, forces a rise in such a direction as to oppose the flow.

**Volatile Memory**—A memory whose contents are irretrievable when operating power is lost.

**Volt (V)**—The unit of electrical pressure or potential.

**Watt (W)**—The unit of electrical power.

**Wheatstone Bridge**—A circuit employing four arms in which the resistance of one unknown arm may be determined as a function of the remaining three arms that have known values.

**Wireway**—A sheet-metal trough with a hinged cover for housing and protecting electrical conductors and cable, in which conductors are laid in place after the wireway has been installed as a complete system.

**Wobble Stick**—A rod extended from a pendant station that operates the STOP contacts; it functions when pushed in any direction.

**Word**—The unit number of binary digits (bits) operated on at any time by the central processing unit when it is performing an instruction or operating on data.

**Write**—The process of putting information into a storage location.

**Wye Connections**—A connection in a three-phase system in which one side of each of the three phases is connected to a common point or ground; the other side of each of the three phases is connected to the three-phase power line.

# Index

absolute encoder, 125
accelerating contactor, 297
across-the-line starters, 278–280
   circuits, 283–293
active matrix liquid crystal display
   (LCD), 52–53
alternating current (ac), 1
   motors, operational theory of,
     243–245
   solenoids for, 77
ambient compensated overload relay,
   278
ambient light receiver, 142
ambient temperature, 277
American National Safety Institute
   (ANSI), 374
American Society of Safety
   Engineers (ASSE), 374
American Society of Testing and
   Materials, 374
ampere squared seconds ($I^2t$), 18
analog set point, defined, 181
annunciators, 45
asynchronous data transmission, 352
attenuation range, 119, *120*
automatic reset, defined, 181
autotransformer motor starters,
   294–297

banana-plug test probe, *394*
band width, defined, 181
bar charts, circuit analysis and,
   218–219
basic twisted nematic liquid crystal
   display (LCD), 50–52
bearings, overheating motors,
   troubleshooting, 401
Bourdon-tube pressure switch,
   155–156
brackets, used in polyphase motors,
   246
brushless dc motors, 263–272
buss, 263

capacitive sensors, 114
capacitor start motors, 250–252
cassette recorder, 327
circuit breakers
   nonautomatic, 13
   types of, 22–24
circuits
   control
     bar charts and, 218–219
     components, placement of,
       216–219
     examples of, 219–240
     troubleshooting, 401–407

circuits (*cont.*)
    diagram of, ladder-type, used in
       PLC, *321*
    motion control and, 109–113
closed-loop control, 95–97
closed-transition,
    in autotransformer motor starters,
       297
    in wye-delta motor starters, 297
coils
    dual, 3
       electromechanical devices and,
          56–58, 61–66
    latch, 66–67
    relays and, 56–66
    troubleshooting, 401
    unlatch, 66
combination starter, 280
compensator motor starters (*see*
    autotransformer motor starters)
connections, loose, troubleshooting,
    382–383
constant voltage regulators (CVR), 5–6
    circuit cross-section, *7*
contact bounce, troubleshooting, 398
contactors, 70–73
contacts
    described, 58–61
    faulty, troubleshooting, 383
    wiring the wrong, troubleshooting,
       389
control
    chart, 368
    circuits
       bar charts and, 218–219
       components, placement of,
          216–219
       troubleshooting, 401–407
    closed-loop, 95–97
    count
       circuit applications and,
          210–212

       preset electrical impulses and,
          209–210
       solid-state counters and,
          212–214
    divisions of, *217*
       in programmable controllers,
          *311*
motion
    circuits, applications and,
       109–113
    importance of, 104
    limit switches and, 104–107
    stepping motors and, 128–137
    transducers and, 123–124
    vane switches and, 122
open-loop, 94
pressure
    circuit applications, 160–165
    importance of, 153–154
    switches, types of, 155–160
programmable
    data highway and, 330–331
    disk drives and, 327
    EXAMINE ON/EXAMINE
       OFF, 326–327
    industrial applications of,
       340–342
    input/output (I/O) and, 314–317
    logic controllers and, 312–314
    memory and, 318–320
    modems and, 328
    PC data ports and, 328–330
    power supplies and, 320–321
    printers and, 327
    processor and, 317
    programming and, 321–326
    solid-state concepts and, 311
    tape drives and, 327
    troubleshooting, 407–408
proportional, 97–100
relay
    circuits, *60, 61*

described, 56–58
design conditions, 58–59
sections of, *313*
servo positioning, 137–140
systems
    assembling, 422–432
    components
        accessibility of, 429–431
        numbering and identification,
            431
        plug-in, 432
    considerations in, 412–414
    diagrams and layouts, 414–421
    indicator lights and, 431–432
    installing, 422–432
    locating, 422–432
    quality, 363–370
        control chart, used in, 368
        data acquisitions systems,
            used in, 369
        electrical and electronic
            circuits, used in, 364
        errors, typical, 370
        machine and process
            monitoring, use of,
            364–366
        objective of, 364
        sensors, use in, 95–97
controllers
    millivoltmeter, 172–173
    potentiometric, 173
    programmable logic, 312–313
    proportional (*see* stepless
        controllers)
    stepless, 174
    temperature control and, 168–180
    three axis, 137–138
    zero-crossover-fired, 175
control technology, 348
count control
    circuit applications and, 210–212
    counters, solid-state, 212–214

electrical impulses, preset and,
    209–210
counters, solid-state, 212–214
current
    imbalance, locating, 396, *397*
    let-thru, 18–21
    measurements, 396–398
    motor starting, measuring, 398,
        *398*
    solenoids and, 75–76
    *See also* voltage
current-limiting fuses, *15*
    benefits of, 18
CVR (*see* constant voltage
    regulators)
cycle control, troubleshooting,
    406–407


data
    communications, defined, 348
        (*see also* industrial data
        communications)
    transfer rate, 358
    transmission, 351–352
data highway, 330–331
    ports, 328–330
delay, defined, 26
diaphragm pressure switch, 156
differential travel, defined, 106
digital set point, 181
direct current (dc), 1
    motors, 257–272
        brushless, 263
        classifications, 257
        compound wound, 257
        constant horsepower, 257, 262
        constant torque, 257, 259, 261,
            262
        load characteristics, 261–263
        output torque of, 258
        power in, 264
        series wound, 257

direct current (dc) (*cont.*)
  motors (*cont.*)
    speed
      controlling, 257–261
        armature voltage, use of,
          258–259, 259–260
        shunt field, use of, 259,
          259–260
        variation in rated speed,
          260–261
    stepping circuit, *133*
disconnect switches, fusible, 13
disk drive, 327
displacement
  angular position, transducers and,
    124–128
  linear position, transducers and,
    123–124
displays, liquid crystal, 49–53
double-solenoid valve, 84–85

electricity, protection factors, 12–13
Electric Power Research Institute,
  388
electronic troubleshooting hints,
  408–409
encoders
  motion control and, 125–128
  wiring circuit for, *127*
end bells, used in polyphase motors,
  246
Ethernet, 330, 351, 357
EXAMINE ON/EXAMINE OFF,
  326–327

faults, momentary, troubleshooting,
  389, 391
ferroresonance, 6
fiber optics, used in sensors, 143
flow sensors, 147–149
foot-operated switches, 38, *39*
force motor, 90

frequency, surges, 21–22
frequency distribution, 368
full-voltage starters, 278–280
fuses
  construction and operation of, 14
  cross section of, *14*
  current-limiting, *15*, 18
  melting time-current, *17*
  one-time, *15*, 16
  standard voltage, 15
  time-delay, *15*, 16
  troubleshooting, 382
  types of, 14–18
fusible disconnect switch, *13*

grounds
  troubleshooting, 384–388
    detection means, 385–387

Hall effect, 113
handshaking, 352
heat control, troubleshooting, 406
heavy-duty switches, 38
holding current, solenoids and, 76
hysteresis, 118–119

incremental encoder, 125
indicator
  lights, 38–40
    control systems and, 431–432
induction motors
  part winding starters and, 301–304
  squirrel cage, 245–247
inductive load, 55
inductive sensors, 114
industrial data communications
  data
    transfer rate, 358
    transmission, 351–352
      asynchronous, 352
      parallel, 352
      serial, 352

synchronous, 352
Ethernet and information
highway, 357
industrial data highway, 352–353
information technology, 348
architecture, 348–350
interference, 358
network
concepts, 350–351
industrial, 355
layers, 359
topologies, 353
typical systems, 359
open vs. proprietary systems,
358–359
transmission media, 357–358
information technology (IT), 348
architecture, 348–350
inrush current, 57
solenoids and, 76
Institute of Electrical and Electronics
Engineers (IEEE), 374
interference, 358
interlock circuit, 60, 233

L-frame circuit breaker, 23
ladder-type circuit diagram, *321*
laminations, motors and, 245
latch coil, 66–67
latching relays, 66–69
circuits, *68*
symbols for, *67*
LCDs (*see* liquid crystal displays)
LEDs (*see* light-emitting diode)
let-thru current, 18–21
light-emitting diode (LED), 45, 140
indicators, 118
lightning
arresters, 22
surges by, 21–22
lights, indicator, 38–40
control systems and, 431–432

limit switches
motion control and, 104–107,
109–113
operating movement of, 106
problems of, 426–429
proximity, 113–117
symbols of, 107–109, *108*
vane-operated, 122
line contactor (LC), 297
linear potentiometer, 123
lines, common, troubleshooting, 389
liquid crystal display (LCD), 49
active matrix, 52–53
basic twisted nematic, 50–52
supertwist, 52
locked rotor current, 293
low-voltage
protection, 283
release, 283
troubleshooting, 384

magnet, proximity limit switch, 120
magnetic
motor starters, 276
tape drives, 327
trip circuit breaker, 22
manual motor starters, 275
manual-set timers, 201–202
mean, definition, 368
median, definition, 368
membrane switch, 47–49
memory
programmable control and,
318–320
relay, 66
metering, electrical, 25–26
meters, safety checklist for, 399–400
millivoltmeter controller, 172–173
mode
in statistics, defined, 368
in temperature control, defined, 181
modems, 328

modulated LED sensor, 141
molded-case circuit breaker, 23
motion control
    circuits, applications and,
        109–113
    flow sensors, 147–149
    importance of, 104
    limit switches and, 104–107,
        109–113
        proximity, 113–117
    motors, stepping, 128–137
    output, solid-state, 118
    sensors
        ambient light receiver, 142
        attenuation range in, 119
        capacitive sensors, 114
        fiber optic, 143
        hysteresis in, 118–119
        inductive, 114
        modes of
            opposed, 144, *144*
            proximity, 145–147
                convergent beam, 145, *146*
                diffuse, 145, *145*
                divergent, 145, *146*
                fixed-field, 146, *147*
                retroreflective, 144, *144*
        modulated light-emitting diode
            (LED), 141
        photoelectric, 140, 142–143
        phototransistor, used in, 141
        remote, 142
        self-contained, 142–143
        sensing range of, 118
        ultrasonic, 142, 143
    switch
        light-emitting diode (LED)
            indicators, used in, 118
        proximity limit, mercury, 121
        response time of, 120

    transducers
        angular position displacement
            and, 124–128
        linear position displacement
            and, 123–124
        vane switches and, 122
motor starters
    across-the-line, 278–280
    circuits, 283–293
    horsepower chart, 276
    low-voltage, part-winding,
        301–304
    multispeed, 282–283
        jogging operations and, 287
        normal operations and, 287
    overload relays and, 275–278
    reduced-voltage, 293–304
        autotransformer, 294–297
        compensator (*see* motor starters,
            reduced-voltage,
            autotransformer)
        part winding, 301–304
        primary resistor type, 297–298
        wye-delta, 299–301
    reversing, 280–281
    solid-state, 304–306
    starting sequence of, 306–307
motors
    alternating current (ac), operational
        theory of, 243–245
    capacitor start, 250–252
    control circuits and,
        troubleshooting, 401–407
    direct current (dc), 257–272
        brushless, 263–272
        classifications, 257
        compound wound, 257
        constant horsepower, 257, 262
        constant torque, 257, 259, 261,
            262

load characteristics, 261–263
output torque of, 258
power in, 264
series wound, 257
speed
    controlling, 257–261
        armature voltage, use of,
            258–259, 259–260
        shunt field, use of, 259,
            259–260
        variation in rated speed,
            260–261
    stalled current capability,
        261–262
force, 90
polyphase, troubleshooting, 401
protection of, programmable,
    24–25
shaded-pole, 254–256
single-phase, 248
split-capacitor, permanent,
    252–254
split-phase, resistance, 249–250
squirrel-cage
    part-winding starters and, 304
    polyphase induction, 245–247
stepping, motion control and,
    128–137
troubleshooting, 401
multimeters, 392–394
multiple-interval timer, 197, *200*
multispeed motor starters, 282–283
    jogging operations and, 287
    normal operations and, 287

*National Electrical Code*® (NEC), 26,
    374, 395
National Electrical Manufacturers
    Association (NEMA), 57, 312
*National Electrical Safety Code*®, 374

National Fire Protection Association
    (NFPA), 26
network
    buses
    concepts, 350–351
    industrial, 355
    layers, 359
    peer-to-peer, 357
    topologies, 353
    typical systems, 359
node, 355
noise, motors and, troubleshooting,
    401
nonautomatic circuit breakers, 13, 22
noninductive load, 58
normal distribution, 367
no-voltage
    protection, 283
    release, 283

Occupational Safety and Health Ad-
    ministration (OSHA), 372, 374
oil-tight switches, 41
    miniature, 40–41
open-loop control, 94
open system, 358
open transition, 297
operating force, defined, 106
optoisolator, 315
overload relays, starters and, 275–278
overtravel, defined, 106
overvoltage, defined, 26

panels, mounting space, relays and,
    59
parallel data transmission, 352
part winding starters, 301–304
PC data ports, programmable control
    and, 328–330
peak let-thru current ($I_p$), 18

personal computer software, 370
phase
    loss, defined, 26
    reversal, defined, 26
    unbalance, defined, 26
photoelectric sensor, 140, 142–143
    phototransistor, used in, 141
    remote, 142
    self-contained, 142–143
phototransistor, 141
plug-in
    components, use of, 432
    relays, 69
polyphase motors, troubleshooting,
    401
position indication, importance of,
    104
potentiometric controller, 173
power, in dc motors, 264
power line, three phase, 2
power sources, uninterruptible, 8, 9
power supplies, programmable
    control and, 320–321
precision limit switch, 104
pressure
    control
        circuit applications, 160–165
        importance of, 153–154
        pressure switches and, 155–160
    sensors, 158–160
    switches
        Bourdon-tube, 155–156
        circuit applications of,
            160–165
        described, 153
        diaphragm, 156
        sealed piston, 155
        solid-state, 156–157
        symbols for, *154*
    transducers, 158
pretravel, defined, 106

primary resistor motor starter, 297–298
printers, programmable control and,
    327
programmable control
    data highway and, 330–331
    disk drives and, 327
    EXAMINE ON/EXAMINE OFF,
        326–327
    industrial applications of, 340–342
    input/output (I/O) and, 314–317
    logic controllers and, 312–314
    memory and, 318–320
    modems and, 328
    PC data ports and, 328–330
    power supplies and, 320–321
    printers and, 327
    processor and, 317
    programming and, 321–326
    solid-state, concepts of, 311
    tape drives and, 327
    troubleshooting, 407–408
programmable logic controller (PLC),
    312–313
    relay logic, conversion from,
        332–339
    safety circuits, used in, 377–378
    timers, used in, 332, 336
programmable motor protection,
    24–25
programming, programmable control
    and, 321–326
proportional
    control, 97–100
    controller, (*see* stepless controller)
    valves, solenoids and, 88–89
proprietary system, 358
protection
    devices
        selecting, 26–29
        symbols of, *28*
    voltage, 25–26

protocol, definition, 330, 351
proximity limit switch, 113–117, 429
    mercury, 121
push-button switch, 33–36
push-to-test indicating light, 39

quality, definition of, 363
quality control systems, 363–370
    control chart, used in, 368
    data acquisitions systems, used in,
        369
    electrical and electronic circuits,
        used in, 364
    errors, typical, 370
    machine and process monitoring,
        use of, 364–366
    objective of, 364

rate, defined, 181
RC (*see* run contactor)
reduced-voltage starters, 293–304
    autotransformer, 294–297
    part winding, 301–304
    primary resistor type, 297–298
    wye-delta, 299–301
regulation
    parallel, 5
    temperature rise in, 5
regulators, constant voltage (CVR),
    5–7
relay
    contacts, 56–59
    control
        circuits, *60, 61*
        described, 56–58
        design considerations, 58–59
    latching, 66–69
        circuits, *68*
        symbols for, *67*
    logic, converting to PLC, 332–339
    memory, 66

mounting space for, 59
overload, 275–278
    ambient compensated, 278
plug-in, 69
time-delay, described, 194
timing, 61–66
    circuits, *65*
    illustrated, *63, 64*
    symbols for, *62*
uses of, 56–61
release force, defined, 106
remote photoelectric sensors, 142
repeat-cycle timers, 200–201
reset timers, 196–200
resistance, 58
    split-phase motors, 249–250
response time, 120
reversing motor starters, 280–281
rotary encoder, 125–128
rotating-cam limit switch, 105
rotor motion, described, 243–245
RTD unit, described, 171
run contactor (RC), 295, 299

safety  372–378
    agencies, involved in, 374
    circuits
        devices, used in, 373
        electrical controls, methods for
            use, 373
    control techniques, types, 373
    detectors, used in, 375
    diagnostic systems, used in, 375
    machine, 374–375
    programmable controllers, used in,
        377–378
    troubleshooting and, 380–381
    worker, 372–374
SC (*see* start contactor)
SCR (*see* silicon-controlled
    rectifiers)

sealed current, 57
   solenoids and, 75–76
sealed piston pressure switch, 155
selector switches, 36–38
   symbols for, *37–38*
self-contained photoelectric sensors,
   142–143
sensing theory, 140–147
sensor
   control systems, use in, 95–97
   definition, 95
   motion control, used in ambient
      light receiver, 142
      attenuation range in, 119
      capacitive sensors, 114
      fiber optic, 143
      hysteresis in, 118–119
      inductive, 114
      modes of
         opposed, 144, *144*
         proximity, 145–147
            convergent beam, 145, *146*
            diffuse, 145, *145*
            divergent, 145, *146*
            fixed-field, 146, *147*
         retroreflective, 144, *144*
      modulated light-emitting diode
        (LED), 141
      photoelectric, 140, 142–143
        phototransistor, used in, 141
        remote, 142
        self-contained, 142–143
        sensing range of, 118
      ultrasonic, 142, 143
   pressure, 158–160
   structure of, 97
   temperature, 185–186
sequence bar chart, 218, *219*
serial data transmission, 352
server, 357
servo positioning control, 137–140

controller, three axis, used in,
   137–139
servo valves, 90
shaded-pole motors, 254–256
silicon-controlled rectifiers (SCR),
   174–176
single-phase motors, 248
   troubleshooting, 401
slip, motors and, 247
slots, motors and, 245
solenoid
   50-cycle, compared to 60-cycle,
      81–82
   actions of, 75–78
   alternating current (ac), on direct
      current (dc), 79–80
   application problems of, 78
   circuits of, *87, 88*
      applications of, 87–88
   cross section of, air-gap, *76*
   direct current (dc), on alternating
      current (ac), 80
   double valve, 84–85
   force and voltage, 78
   heat rise in, 82–86
   low voltage in, 78–79
   motion control and, 109–113
   overvoltage in, 79
   proportional valves and, 88–89
   variable, 88
solid-state
   counters, 212–214
   motor starters, 304–306
   output, 118
   pressure switch, 156–157
   programmable control, 311
   timers, 195, 202–204
speed
   motors and, troubleshooting, 401
   regulation, in dc motors, 257–261
   variation, in dc motors, 260–261

split-capacitor motors, permanent, 252–254

split-phase motors, resistance, 249–250

squirrel-cage motors
    part winding starters and, 304
    polyphase induction, 245–247

standard deviation, definition, 368

start contactor (SC), 294–295

starters (*see* motor starters)

starting sequence, main drive motors, 306–307

steady-state error, 99

step-down transformer, 2

step-up transformer, 2

stepless controller, 174

stepping circuit, direct current (dc), *133*

stepping motors
    motor control and, 128–137
        load characteristics, 130–132
        sample calculations for, 135–137
        shaft speeds and, 130
    stepping circuit, *133*

supertwist liquid crystal display (LCD), 52

surges, voltage/frequency, 21–22

switches
    circuits for, 42–45
    color designations of, 34–35
    contact block, 33–34
    described, 33
    disconnect, fusible, 13
    foot-operated, 38, *39*
    heavy-duty, 38
    indicating lights and, 38–40
    limit
        motion control and, 104–107, 109–113
        operating movement of, *106*
        problems of, 426–429

symbols, 107–109, *108*
vane-operated, 122
membrane, 47–49
motion control, used in
    LED indicators, used in, 118
    proximity limit mercury, 121
    response time of, 120
oil-tight, 32, 41
    miniature, 40
panel mounting, 33
push-button, 33–36
selector, 36–38
    symbols for, *37–38*
temperature, 182–183
    circuit applications, 187–191
vane, problems of, 429

synchronous data transmission, 352

synchronous motor-driven timers, 196–202
    manual-set, 201–202
    repeat-cycle, 200–201
    reset timers, 196–200

synchronous motors
    motion control and, 128–137
        load characteristics, 130–133
        sample calculations for, 135–137
        shaft speeds and, 130
    stepping circuit, *133*

tape drives, programmable control and, 327

temperature control
    circuit applications and, 187–191
    controller outputs, 181
    controllers of, 168–180
        electronic, 169–180
        selection of, 168–169
    importance of, 167–168
    sensors, 185–186
    thermostats and, 182–184

temperature measurements, 399, *400*
temperature rise, 5, 82
temperature sensors, 185
test probe
    banana-plug, 394
    industrial, 394
thermal
    element circuit breaker, 22
    trip circuit breaker, 22–23
thermistor, described, 170
thermocouple, described, 169
thermometer, digital, 394
thermostats, temperature control and,
        182–184
time control
    circuit applications and, 204–207
    operations described, 193–195
    timers
        solid-state, 202–204
        synchronous motor-driven,
            196–202
        types of, 195
time-delay relays, described, 194
time-proportional burst firing, 175
time sequence chart, *197*
timers
    clutch, symbol for, *196*
    described, 194
    manual-set, 201–202
    motor, symbol for, *196*
    repeat-cycle, 200–201
    reset, 196–200
    solid-state, 195, 202–204
    synchronous motor-driven,
        196–202
    types of, 195
timing, symbols, *196*
timing relays, 61–66
    circuits, *65*
    illustrated, *63, 64*
    symbols for, *62*
tolerance, definition, 367

total travel, defined, 106
transducers
    displacement
        angular position and, 124–128
        linear position and, 123–124
    pressure, 158
transformers
    control, 1–4
        symbol for, *3*
    voltage regulation and, 4–5
transmission control protocol and
        Internet protocol (TCP/IP), 330
trip travel, defined, 106
troubleshooting
    combination problems, 383–384
    common line, opens in, 389
    connections, loose, 382–383
    contacts
        faulty, 383
        wiring the wrong, 389
    control circuit, 401–407
    equipment for, 391–400
        contact bounce, and, 398
        current measurements and,
            396–398
        multimeters, 392–394
        temperature measurements and,
            399, *400*
        voltage measurements and,
            395–396
    faults, momentary, 389, 391
    fuses, 382
    grounds, 384–388
        detection means, 385–387
    hints, 408–409
    housekeeping and, 388
    meters and safety checklist for,
        399–400
    motors, 401
    problem analysis and, 381–382
    programmable logic controller,
        407–408

safety and, 380–381
trouble patterns and, 388–389
voltage, low, 384
wire markers, 383
turns ratio, 2–3

ultrasonic sensors, 142, 143
undervoltage, defined, 26
uninterruptible power source (UPS), 8
diagram of, *9*
unlatch coil, 66
UPS (*see* uninterruptible power
source)

valves
proportional, solenoids and, 88–89
servo, 90
vane switches
operated limit, 122
problems of, 429
variable solenoid, 88

voltage
drop, measuring, *396*
imbalance, 396, *397*
low, troubleshooting, 384
measuring, 395–396
over, defined, 26
protection, 25–26
factors, 12–13
reduction ratio, 2–3
regulation, 4–5
surges, 21–22
under, defined, 26
*See also* current

wire markers, troubleshooting, 383
wiring diagram, 285
worker safety, 372–374
wye-delta starters, 299–301
zero-crossing controller (*see* zero-
crossover-fired controllers)
zero-crossover-fired controllers, 175